# ng-book 2

The Complete Guide to
Angular 2

Felipe Coury
Ari Lerner
Nate Murray
Carlos Taborda

Technical Editor:
Frode Fikke

ISBN: 978-0991344611

Questions? Email us at: us@fullstack.io

Sample code download available at ng-book.com/code

(code) means there is code for this
see "ng-book-code 2" folder

# Contents

# Book Revision

Revision 44p for paper print - Covers up to Angular 2 (2.2.0-rc.0, 2016-11-08)

# Sample Code Download

This book comes with example code - see the next page for details on how to download it.

# Bug Reports

If you'd like to report any bugs, typos, or suggestions just email us at: us@fullstack.io[1].

# Chat With The Community!

We're experimenting with a community chat room for this book using Gitter. If you'd like to hang out with other people learning Angular 2, come join us on Gitter[2]!

# Be notified of updates via Twitter

If you'd like to be notified of updates to the book on Twitter, follow @fullstackio[3]

# We'd love to hear from you!

Did you like the book? Did you find it helpful? We'd love to add your face to our list of testimonials on the website! Email us at: us@fullstack.io[4].

---

[1]mailto:us@fullstack.io?Subject=ng-book%202%20feedback

[2]https://gitter.im/ng-book/ng-book

[3]https://twitter.com/fullstackio

[4]mailto:us@fullstack.io?Subject=ng-book%202%20testimonial

# Download the Example Code

This book contains several example apps and code samples. Because you purchased the paperback version you can **download this code for free** at our website.

Note: because of page restrictions, the *Typescript* chapter has been removed and is **available as a free PDF** when you download the example code:

**"Thank you for the awesome book, it's the bible for Angular."**
- *Vijay G.*

| | |
|---|---|
| URL | https://ng-book.com/code/ |
| BOOK SERIAL CODE | AMZ-82CA |

To download the code, visit the URL above and enter your email and serial code and we'll email you the code download.

Learn more at: ng-book.com/code/

DOWNLOAD

# Writing your First Angular 2 Web Application

## Simple Reddit Clone

In this chapter we're going to build an application that allows the user to **post an article** (with a title and a URL) and then **vote on the posts**.

You can think of this app as the beginnings of a site like Reddit[6] or Product Hunt[7].

In this simple app we're going to cover most of the essentials of Angular 2 including:

- Building custom components
- Accepting user input from forms
- Rendering lists of objects into views
- Intercepting user clicks and acting on them

By the time you're finished with this chapter you'll have a good grasp on how to build basic Angular 2 applications.

Here's a screenshot of what our app will look like when it's done:

[6]http://reddit.com
[7]http://producthunt.com

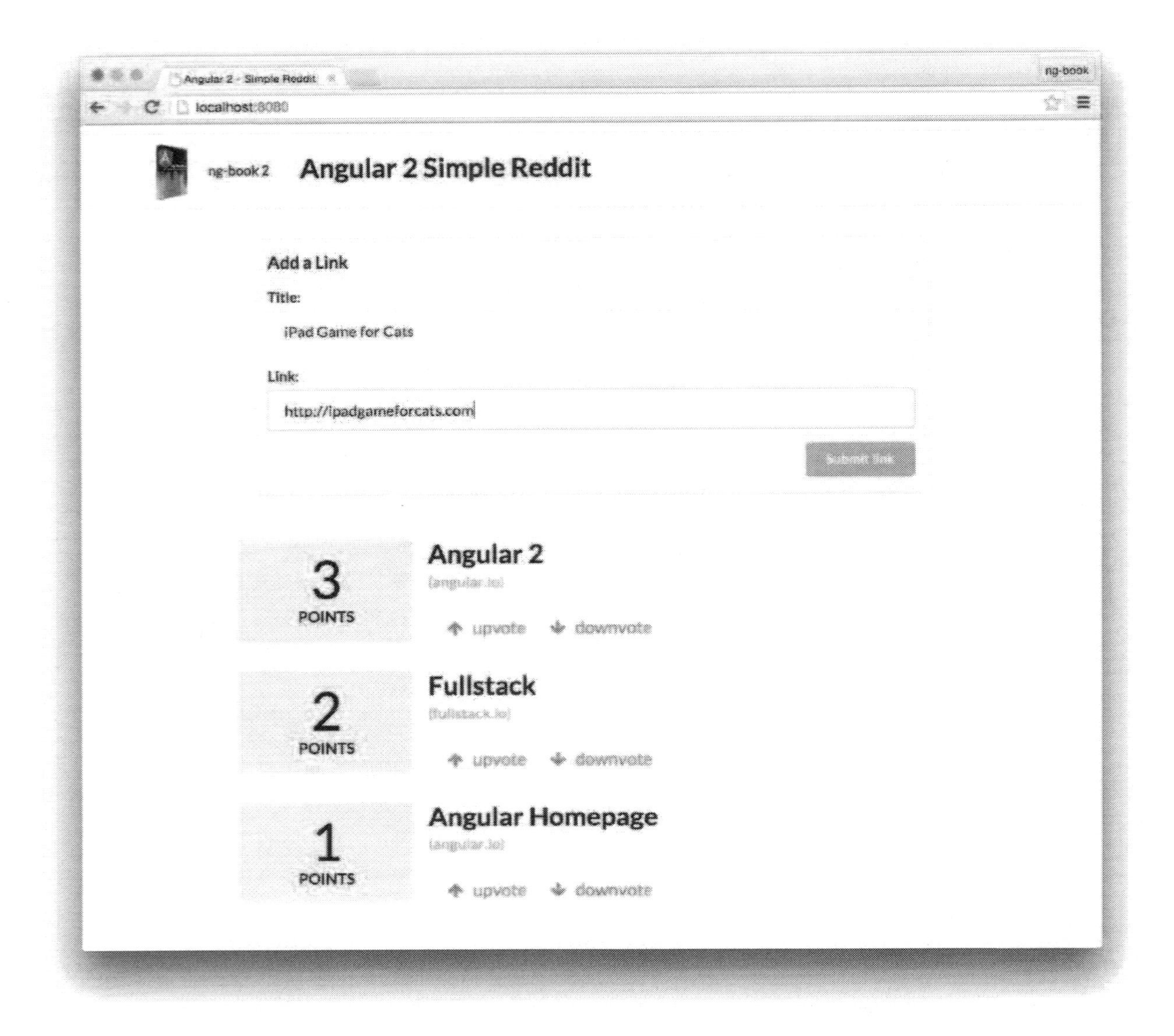


**Completed application**

First, a user will submit a new link and after submitting the users will be able to upvote or downvote each article. Each link will have a score and we can vote on which links we find useful.

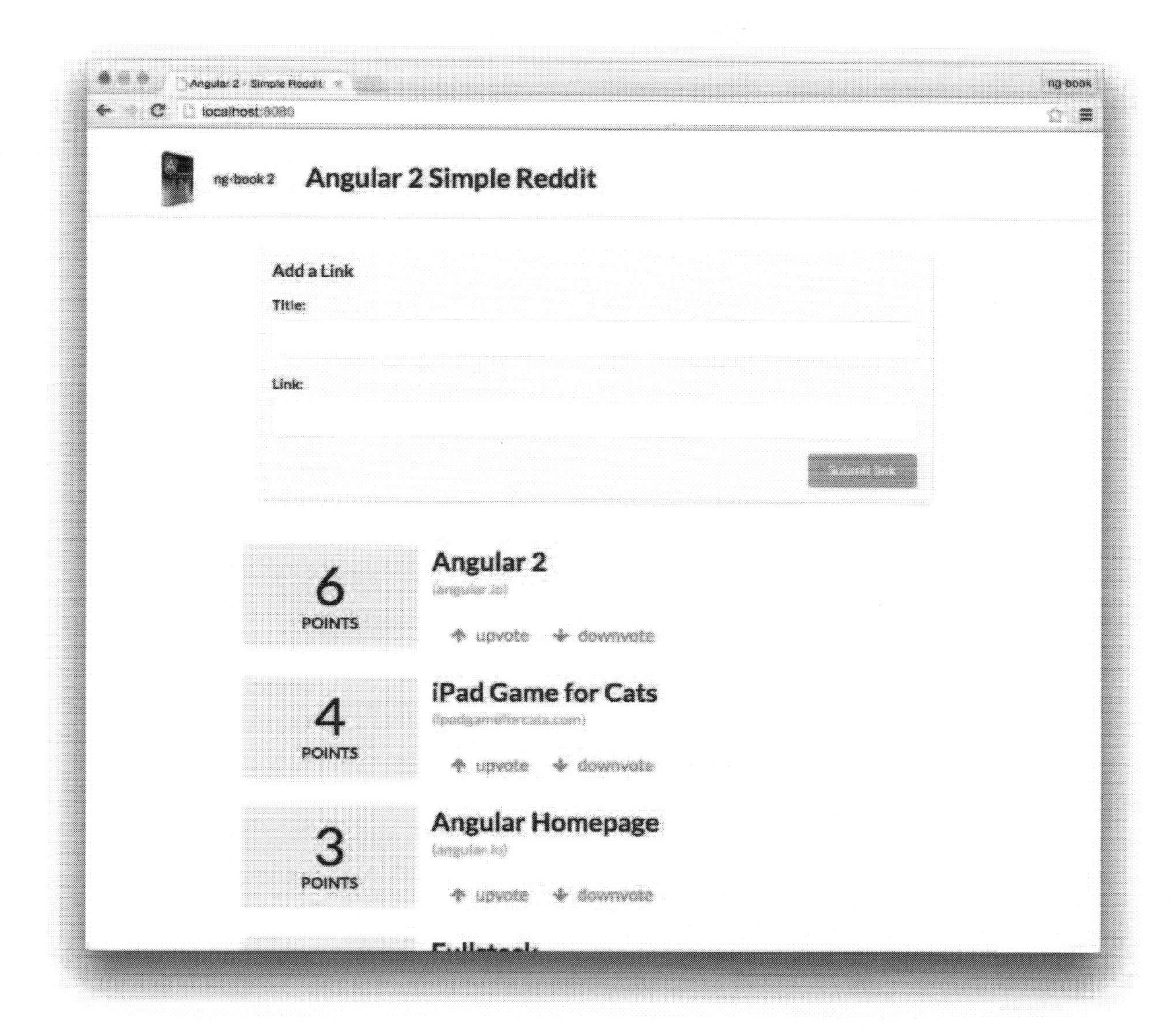


**App with new article**

In this project, and throughout the book, we're going to use TypeScript. TypeScript is a superset of JavaScript ES6 that adds types. We're not going to talk about TypeScript in depth in this chapter, but if you're familiar with ES5 ("normal" javascript) / ES6 (ES2015) you should be able to follow along without any problems.

**We'll go over TypeScript more in depth in the next chapter**. So don't worry if you're having trouble with some of the new syntax.

# Getting started

## TypeScript

To get started with TypeScript, you'll need to have Node.js installed. There are a couple of different ways you can install Node.js, so please refer to the Node.js website[8] for detailed information.

**Do I have to use TypeScript?** No, you don't *have* to use TypeScript to use Angular 2, but you probably should. ng2 does have an ES5 API, but Angular 2 is written in TypeScript and generally that's what everyone is using. We're going to use TypeScript in this book because it's great and it makes working with Angular 2 easier. That said, it isn't strictly required.

Once you have Node.js setup, the next step is to install TypeScript. Make sure you install at least version 1.7 or greater. To install it, run the following `npm` command:

```
1 $ npm install -g typescript
```

`npm` is installed as part of Node.js. If you don't have `npm` on your system, make sure you used a Node.js installer that includes it.

Windows Users: We'll be using Linux/Mac-style commands on the command line throughout this book. We'd highly recommend you install Cygwin[9] as it will let you run commands just as we have them written out in this book.

## angular-cli

Angular provides a utility to allow users to create and manage projects from the command line. It automates tasks like creating projects, adding new controllers, etc. It's generally a good idea to use `angular-cli` as it will help create and maintain common patterns across our application.

To install `angular-cli`, just run the following command:

```
1 $ npm install -g angular-cli@1.0.0-beta.18
```

npm install -g @angular/cli
(most recent version)

Once it's installed you'll be able to run it from the command line using the `ng` command. When you do, you'll see a lot of output, but if you scroll back, you should be able to see the following:

https://www.npmjs.com/package/@angular/cli

[8] https://nodejs.org/download/

[9] https://www.cygwin.com/

```
1 $ ng
2 Could not start watchman; falling back to NodeWatcher for file system events.
3 Visit http://ember-cli.com/user-guide/#watchman for more info.
4 Usage: ng <command (Default: help)>
```

The reason we got that huge output is because when we run `ng` with no arguments, it runs the default `help` command, which explains how to use the tool.

If you're running OSX or Linux, you probably received this line among the output:

```
1 Could not start watchman; falling back to NodeWatcher for file system events.
```

This means that we don't have a tool called **watchman** installed. This tool helps `angular-cli` when it needs to monitor files in your filesystem for changes. If you're running OSX, it's recommended to install it using Homebrew with the following command:

```
1 $ brew install watchman
```

If you're on OSX and got an error when running brew, it means that you probably don't have Homebrew installed. Please refer to the page http://brew.sh/ to learn how to install it and try again.

If you're on Linux, you may refer to the page https://ember-cli.com/user-guide/#watchman for more information about how to install watchman.

If you're on Windows instead, you don't need to install anything and `angular-cli` will use the native Node.js watcher.

And with that we have `angular-cli` and its dependencies installed. Throughout this chapter we're going to use this tool to create our first application.

## Example Project

Now that you have your environment ready, let's start writing our first Angular application!

Let's open up the terminal and run the `ng new` command to create a new project from scratch:

```
1 $ ng new angular2_hello_world
```

Once you run it, you'll see the following output:

```
installing ng2
  create .editorconfig
  create README.md
  create src/app/app.component.css
  create src/app/app.component.html
  create src/app/app.component.spec.ts
  create src/app/app.component.ts
  create src/app/app.module.ts
  create src/app/index.ts
  create src/app/shared/index.ts
  create src/assets/.gitkeep
  create src/assets/.npmignore
  create src/environments/environment.dev.ts
  create src/environments/environment.prod.ts
  create src/environments/environment.ts
  create src/favicon.ico
  create src/index.html
  create src/main.ts
  create src/polyfills.ts
  create src/styles.css
  create src/test.ts
  create src/tsconfig.json
  create src/typings.d.ts
  create angular-cli.json
  create e2e/app.e2e-spec.ts
  create e2e/app.po.ts
  create e2e/tsconfig.json
  create .gitignore
  create karma.conf.js
  create package.json
  create protractor.conf.js
  create tslint.json
Successfully initialized git.
� Installing packages for tooling via npm
```

This will run for a while while it's installing npm dependencies. Once it finishes we'll see a success message:

```
Installed packages for tooling via npm.
```

There are a lot of files generated! Don't worry too much about all of them yet. We'll walk through what each one means and is used for throughout the book. For now, let's focus on getting started with Angular code.

Let's go inside the angular2_hello_world directory, which the ng command created for us and see what has been created:

```
1  $ cd angular2_hello_world
2  $ tree -F -L 1
3  .
4  ├── README.md                   // an useful README
5  ├── angular-cli.json            // angular-cli configuration file
6  ├── e2e/                        // end to end tests
7  ├── karma.conf.js               // unit test configuration
8  ├── node_modules/               // installed dependencies
9  ├── package.json                // npm configuration
10 ├── protractor.conf.js          // e2e test configuration
11 ├── src/                        // application source
12 └── tslint.json                 // linter config file
13
14 3 directories, 6 files
```

For now, the folder we're interested in is src, where our application lives. Let's take a look at what was created there:

```
1  $ cd src
2  $ tree -F
3  .
4  |-- app/
5  |   |-- app.component.css
6  |   |-- app.component.html
7  |   |-- app.component.spec.ts
8  |   |-- app.component.ts
9  |   |-- app.module.ts
10 |   |-- index.ts
11 |   `-- shared/
12 |       `-- index.ts
13 |-- assets/
14 |-- environments/
15 |   |-- environment.dev.ts
16 |   |-- environment.prod.ts
17 |   `-- environment.ts
18 |-- favicon.ico
19 |-- index.html
20 |-- main.ts
21 |-- polyfills.ts
```

```
22 |-- styles.css
23 |-- test.ts
24 |-- tsconfig.json
25 `-- typings.d.ts
26
27 4 directories, 18 files
```

Using your favorite text editor, let's open `index.html`. You should see this code:

**code/first_app/angular2_hello_world/src/index.html**

```
1  <!doctype html>
2  <html>
3  <head>
4    <meta charset="utf-8">
5    <title>Angular2HelloWorld</title>
6    <base href="/">
7
8    <meta name="viewport" content="width=device-width, initial-scale=1">
9    <link rel="icon" type="image/x-icon" href="favicon.ico">
10 </head>
11 <body>
12   <app-root>Loading...</app-root>
13 </body>
14 </html>
```

Let's break it down a bit:

**code/first_app/angular2_hello_world/src/index.html**

```
1  <!doctype html>
2  <html>
3  <head>
4    <meta charset="utf-8">
5    <title>Angular2HelloWorld</title>
6    <base href="/">
```

If you're familiar with writing HTML file, this first part should be trivial, we're declaring the page charset, title and base href.

**code/first_app/angular2_hello_world/src/index.html**

```
8     <meta name="viewport" content="width=device-width, initial-scale=1">
```

If we continue to the template body, we see the following:

**code/first_app/angular2_hello_world/src/index.html**

```
12    <app-root>Loading...</app-root>
13  </body>
14  </html>
```

The `app-root` tag is where our application will be rendered. We'll see this later when we inspect other parts of the source code. The text **Loading...** is a placeholder that will be displayed before our app code loads. We can use this technique to inform the user the application is still loading by using either a message like we're doing here, or a spinner or other kind of progress notification we see fit.

### Writing Application Code

## Running the application

Before making any changes, let's load our app from the generated application into the browser. `angular-cli` has a built in HTTP server that we can use to start our app. Back in the terminal, at the root of our application (for the previously generated application, this will be in the directory the generated created `./angular2_hello_world`) and run:

```
1  $ ng serve
2  ** NG Live Development Server is running on http://localhost:4200. **
3  // a bunch of debug messages
4
5  Build successful - 1342ms.
```

Our application is now running on localhost port 4200. Let's open the browser and visit:

http://localhost:4200[10]

Note that if for some reason port `4200` is taken it may start on another port number. Be sure to read the messages on your machine to find your exact development URL

[10]http://localhost:4200

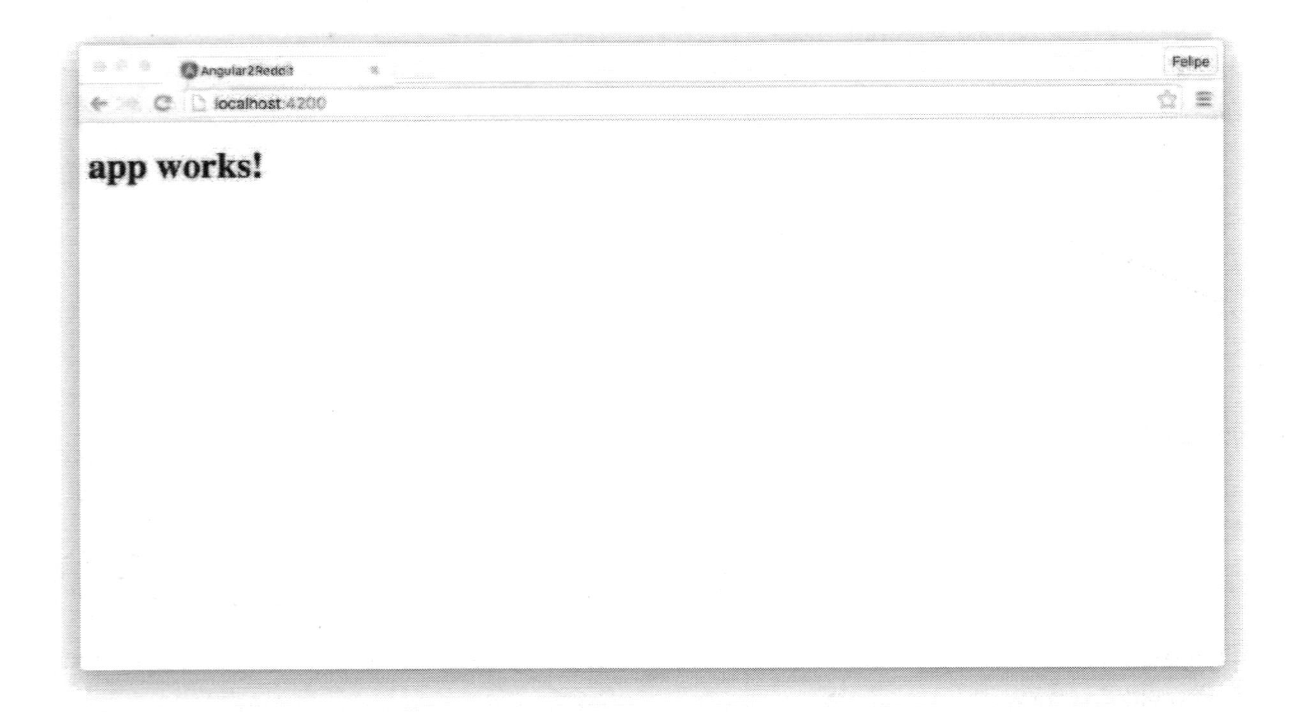


**Running application**

Alright, now that we the setup for the application in place, and we know how to run it, it's time to start writing some code.

## Making a `Component`

One of the big ideas behind Angular is the idea of *components*.

In our Angular apps, we write HTML markup that becomes our interactive application, but the browser understands only so many markup tags; Built-ins like `<select>` or `<form>` or `<video>` all have functionality defined by our browser creator.

What if we want to teach the browser new tags? What if we wanted to have a `<weather>` tag that shows the weather? Or what if we wanted to have a `<login>` tag that creates a login panel?

This is the fundamental idea behind components: we will teach the browser new tags that have custom functionality.

If you have a background in Angular 1, **Components are the new version of directives**.

Let's create our very first component. When we have this component written, we will be able to use it in our HTML document like so:

```
<app-hello-world></app-hello-world>
```

To create a new component using `angular-cli`, we'll use the **generate** command.

To generate the **hello-world** component, we need to run the following command:

```
$ ng generate component hello-world
installing component
  create src/app/hello-world/hello-world.component.css
  create src/app/hello-world/hello-world.component.html
  create src/app/hello-world/hello-world.component.spec.ts
  create src/app/hello-world/hello-world.component.ts
```

So how do we actually define a new Component? A basic Component has two parts:

1. A `Component` annotation
2. A component definition class

Let's look at the component code and then take these one at a time. Open up our first TypeScript file: `src/app/hello-world/hello-world.component.ts`.

**code/first_app/angular2_hello_world/src/app/hello-world/hello-world.component.ts**

```
import { Component, OnInit } from '@angular/core';

@Component({
  selector: 'app-hello-world',
  templateUrl: './hello-world.component.html',
  styleUrls: ['./hello-world.component.css']
})
export class HelloWorldComponent implements OnInit {

  constructor() { }

  ngOnInit() {
  }

}
```

Notice that we suffix our TypeScript file with `.ts` instead of `.js` The problem is our browser doesn't know how to interpret TypeScript files. To solve this gap, the `ng serve` command live-compiles our `.ts` to a `.js` file automatically.

This snippet may seem scary at first, but don't worry. We're going to walk through it step by step.

## Importing Dependencies

The `import` statement defines the modules we want to use to write our code. Here we're importing two things: `Component`, and `OnInit`.

We `import Component` from the module `"@angular/core"`. The `"@angular/core"` portion tells our program **where to find the dependencies** that we're looking for. In this case, we're telling the compiler that `"@angular/core"` defines and exports two JavaScript/TypeScript objects called `Component` and `OnInit`.

Similarly, we `import OnInit` from the same module. As we'll learn later, `OnInit` helps us to run code when we initialize the component. For now, let's not worry about it.

Notice that the structure of this `import` is of the format `import { things } from wherever`. In the `{ things }` part what we are doing is called *destructuring*. Destructuring is a feature provided by ES6 and TypeScript. We will talk more about it in the next chapter.

The idea with the `import` is a lot like `import` in Java or `require` in Ruby: we're pulling in these dependencies from another module and making these dependencies available for use in this file.

## Component Annotations

After importing our dependencies, we are declaring the component:

**code/first_app/angular2_hello_world/src/app/hello-world/hello-world.component.ts**

```
3 @Component({
4   selector: 'app-hello-world',
5   templateUrl: './hello-world.component.html',
6   styleUrls: ['./hello-world.component.css']
7 })
```

If you've been programming in JavaScript for a while then this next statement might seem a little weird:

```
1 @Component({
2   // ...
3 })
```

What is going on here? If you have a Java background it may look familiar to you. These are annotations.

> Angular 1's dependency injection used the annotation concept behind the scenes. Even if you're not familiar with them, annotations are a way to add functionality to code using the compiler.

We can think of annotations as **metadata added to your code**. When we use `@Component` on the `HelloWorld` class, we are “decorating” the `HelloWorld` as a `Component`.

We want to be able to use this component in our markup by using a `<app-hello-world>` tag. To do that we configure the `@Component` and specify the `selector` as `app-hello-world`.

```
@Component({
  selector: 'app-hello-world'
  // ... more here
})
```

Similar to CSS selectors, XPath, or JQuery selectors, there are lots of ways to configure a selector. Angular Components adds their own special sauce to the selector mix, and we’ll cover that later on. For now, keep in mind that we’re **defining a new HTML markup tag**.

The `selector` property here indicates which DOM element this component is going to use. This way any `<app-hello-world></app-hello-world>` tags that appear within a template will be compiled using this `Component` class and all of it’s definitions within it.

## Adding a template with `templateUrl`

In our component we are specifying a `templateUrl` of `./hello-world.component.html`. This means that we will load our template from the file `hello-world.component.html` in the same directory as our component. Let’s take a look at that file:

**code/first_app/angular2_hello_world/src/app/hello-world/hello-world.component.html**

```
<p>
  hello-world works!
</p>
```

Here we’re defining a `p` tag with some basic text in the middle. When Angular loads this component it will also read from this file and use it as the template for our component.

## Adding a `template`

We can define templates two ways, either by using the `template` key in our `@Component` object or by specifying a `templateUrl`.

We could add a template to our `@Component` by passing the `template` option:

```
@Component({
  selector: 'app-hello-world',
  template: `
    <p>
      hello-world works inline!
    </p>
  `
})
```

Notice that we're defining our `template` string between backticks (` ... `). This is a new (and fantastic) feature of ES6 that allows us to do **multiline strings**. Using backticks for multiline strings makes it easy to put templates inside your code files.

**Should you really be putting templates in your code files?** The answer is: it depends. For a long time the commonly held belief was that you should keep your code and templates separate. While this might be easier for some teams, for some projects it adds overhead because you have switch between a lot of files.

Personally, if our templates are shorter than a page, we much prefer to have the templates alongside the code (that is, within the `.ts` file). When we see both the logic and the view together, it's easy to understand how they interact with one another.

The biggest drawback to mixing views and our code is that many editors don't support syntax highlighting of the internal strings (yet). Hopefully, we'll see more editors supporting syntax highlighting HTML within template strings soon.

## Adding CSS Styles with `styleUrls`

Notice the key `styleUrls`:

```
  styleUrls: ['./hello-world.component.css']
```

This code says that we want to use the CSS in the file `hello-world.component.css` as the styles for this component. Angular 2 uses a concept called "style-encapsulation" which means that styles specified for a particular component *only apply to that component.* We talk more about this in-depth later on in the book in the Styling section of Advanced Components.

For now, we're not going to use any component-local styles, so you can leave this as-is (or delete the key entirely).

You may have noticed that this key is different from `template` in that it accepts *an array* as it's argument. This is because we can load multiple stylesheets for a single component.

## Loading Our Component

Now that we have our first component code filled out, how do we load it in our page?

If we visit our application again in the browser, we'll see that nothing changed. That's because we only **created** the component, but we're not **using** it yet.

In order to change that, we need to add our component tag to a template that is already being rendered. Open up the file: `first_app/angular2_hello_world/src/app/app.component.html`

Remember that because we configured our `HelloWorldComponent` with the `app-hello-world` selector, we need to use the `<app-hello-world></app-hello-world>` in a template. Let's add the `<app-hello-world>` tag to `app.component.html`:

**code/first_app/angular2_hello_world/src/app/app.component.html**

```
<h1>
  {{title}}

  <app-hello-world></app-hello-world>
</h1>
```

Now refresh the page and take a look:

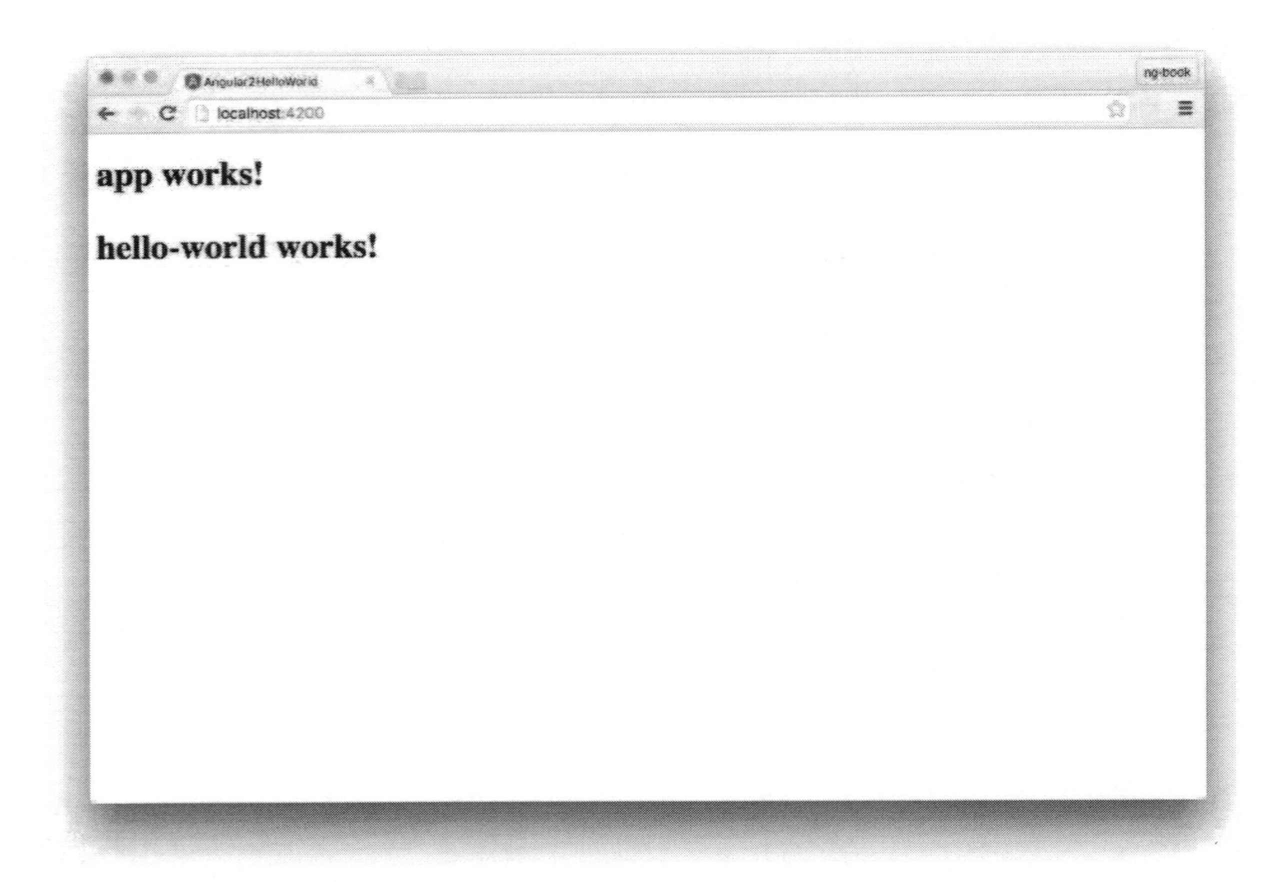


**Hello world works**

It works!

# Adding Data to the Component

Right now our component renders a static template, which means our component isn't very interesting.

Let's imagine that we have an app which will show a list of users and we want to show their names. Before we render the whole list, we first need to render an individual user. So let's create a new component that will show a user's name.

To do this, we will use the `ng generate` command again: first ctrl C – stop server

```
1 ng generate component user-item
```

Remember that in order to see a component we've created, we need to add it to a template.

Let's add our `app-user-item` tag to `app.component.html` so that we can see our changes as we make them. Modify `app.component.html` to look like this:

**code/first_app/angular2_hello_world/src/app/app.component.html**

```
<h1>
  {{title}}

  <app-hello-world></app-hello-world>

  <app-user-item></app-user-item>
</h1>
```

Then refresh the page and confirm that you see the `user-item works!` text on the page.

We want our `UserItemComponent` to show the name of a particular user .

Let's introduce `name` as a new *property* of our component. By having a `name` property, we will be able to reuse this component for different users (but keep the same markup, logic, and styles).

In order to add a name, we'll introduce a property on the `UserItemComponent` class to declare it has a local variable named `name`.

**code/first_app/angular2_hello_world/src/app/user-item/user-item.component.ts**

```
export class UserItemComponent implements OnInit {
  name: string; // <-- added name property

  constructor() {
    this.name = 'Felipe'; // set the name
  }

  ngOnInit() {
  }

}
```

Notice that we've changed two things:

### 1. `name` Property

On the `UserItemComponent` class we added a *property*. Notice that the syntax is new relative to ES5 Javascript. When we write `name: string;` it means `name` is the name of the attribute we want to set and `string` is the *type*.

The typing of the name is a feature of using TypeScript and gives some assurances of the value that it will be a `string`. This sets up a `name` property on *instances* of our `UserItemComponent` class and the compiler ensures that `name` is a `string`.

### 2. A Constructor

On the `UserItemComponent` class we defined a *constructor*, i.e. function that is called when we create new instances of this class.

In our constructor we can assign our `name` property by using `this.name`

When we write:

**code/first_app/angular2_hello_world/src/app/user-item/user-item.component.ts**

```
11    constructor() {
12      this.name = 'Felipe'; // set the name
13    }
```

We're saying that whenever a new `UserItemComponent` is created, set the name to `'Felipe'`.

#### Rendering The Template

With the value filled out, we can use the templating syntax (which is two squiggly brackets `{{ }}`) to display the value of the variable in our template. For instance:

**code/first_app/angular2_hello_world/src/app/user-item/user-item.component.html**

```
1  <p>
2    Hello {{ name }}
3  </p>
```

On the `template` notice that we added a new syntax: `{{ name }}`. The brackets are called "template-tags" (or "mustache tags"). Whatever is between the template tags will be expanded as an *expression*. Here, because the `template` is *bound* to our Component, the `name` will expand to the value of `this.name` i.e. `'Felipe'`.

#### Try it out

After making these changes reload the page and the page should display `Hello Felipe`

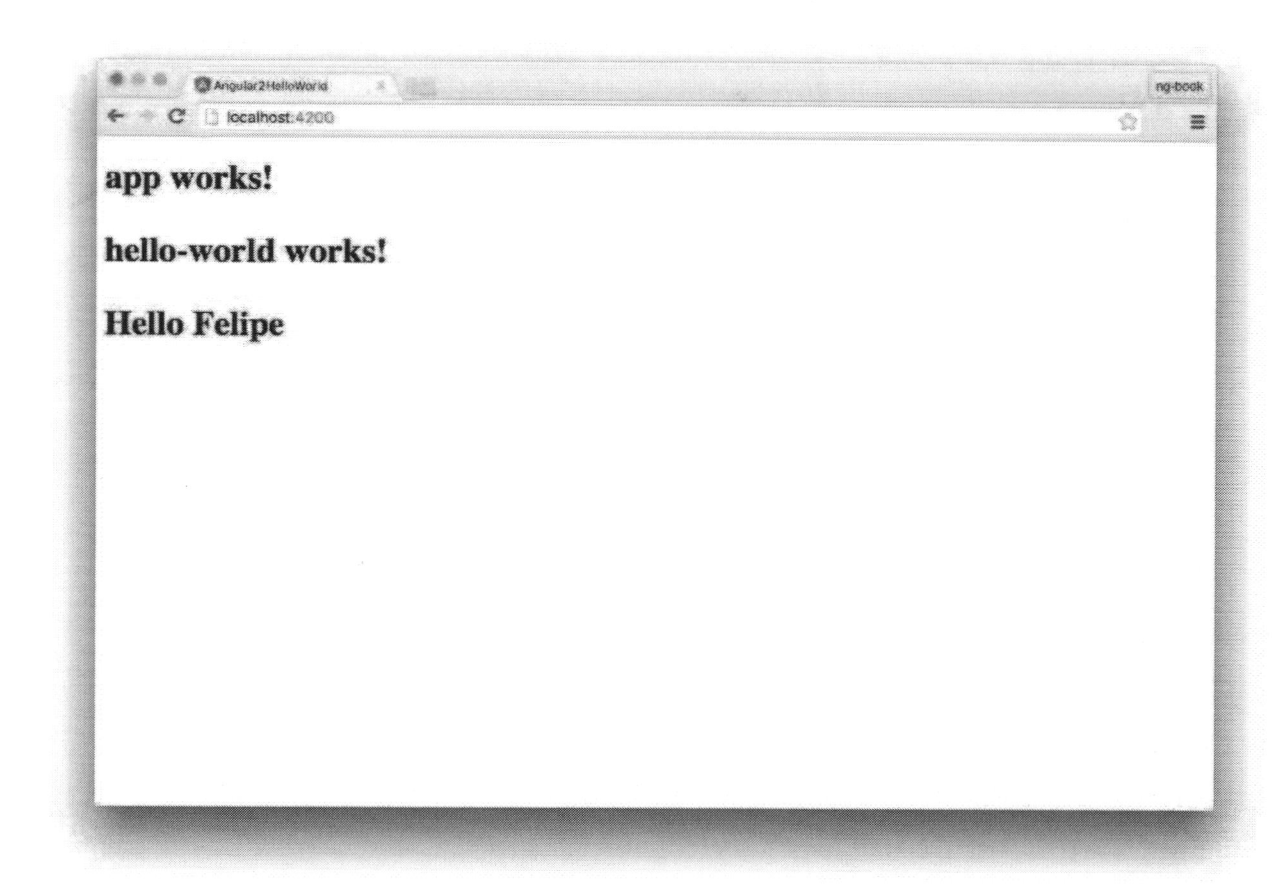


Application with Data

## Working With Arrays

Now we are able to say "Hello" to a single name, but what if we want to say "Hello" to a collection of names?

If you've worked with Angular 1 before, you've probably used the `ng-repeat` directive. In Angular 2, the analogous directive is called `NgFor` (we use it in the markup as `*ngFor`, which we'll talk about soon). Its syntax is slightly different but they have the same purpose: **repeat the same markup for a collection of objects**.

Let's create a new component that will render a *list* of users. We start by generating a new component:

```
1  ng generate component user-list
```

And let's replace our `<app-user-item>` tag with `<app-user-list>` in our `app.component.html` file:

**code/first_app/angular2_hello_world/src/app/app.component.html**

```
<h1>
  {{title}}

  <app-hello-world></app-hello-world>

  <app-user-list></app-user-list>
</h1>
```

In the same way that we added a `name` property to our `UserItemComponent`, let's add a `names` property to this `UserListComponent`.

However, instead of storing only a single string, let's set the type of this property to *an array of strings*. An array is notated by the `[]` after the type, and we can it like this:

**code/first_app/angular2_hello_world/src/app/user-list/user-list.component.ts**

```
export class UserListComponent implements OnInit {
  names: string[];

  constructor() {
    this.names = ['Ari', 'Carlos', 'Felipe', 'Nate'];
  }

  ngOnInit() {
  }

}
```

The first change to point out is the new `string[]` property on our `UserListComponent` class. This syntax means that `names` is typed as an `Array` of `string`s. Another way to write this would be Array<string>.

We changed our constructor to set the value of `this.names` to `['Ari', 'Carlos', 'Felipe', 'Nate']`.

Now we can update our template to render this list of names. To do this, we will use `*ngFor`, which will iterate over a list of items and generate a new tag for each one. Here's what our new template will look like:

**code/first_app/angular2_hello_world/src/app/user-list/user-list.component.html**

```
1 <ul>
2   <li *ngFor="let name of names">Hello {{ name }}</li>
3 </ul>
```

We updated the template with one `ul` and one `li` with a new `*ngFor="let name of names"` attribute. The `*` character and `let` syntax can be a little overwhelming at first, so let's break it down:

The `*ngFor` syntax says we want to use the `NgFor` directive on this attribute. You can think of `NgFor` akin to a `for` loop; the idea is that we're creating a new DOM element for every item in a collection.

The value states: `"let name of names"`. `names` is our array of names as specified on the `HelloWorld` object. `let name` is called a *reference*. When we say `"let name of names"` we're saying loop over each element in `names` and assign each one to a *local* variable called `name`.

The `NgFor` directive will render one `li` tag for each entry found on the `names` array and declare a local variable `name` to hold the current item being iterated. This new variable will then be replaced inside the `Hello {{ name }}` snippet.

We didn't have to call the reference variable `name`. We could just as well have written:

```
1 <li *ngFor="let foobar of names">Hello {{ foobar }}</li>
```

But what about the reverse? Quiz question: what would have happened if we wrote:

```
1 <li *ngFor="let name of foobar">Hello {{ name }}</li>
```

We'd get an error because `foobar` isn't a property on the component.

`NgFor` repeats the element that the `ngFor` is called. That is, we put it on the `li` tag and **not** the `ul` tag because we want to repeat the list element (`li`) and not the list itself (`ul`).

If you're feeling adventurous you can learn a lot about how the Angular core team writes Components by reading the source directly. For instance, you can find the source of the NgFor directive here[11]

When we reload the page now, we'll see that we now have have one `li` for each string on the array:

[11]https://github.com/angular/angular/blob/master/modules/%40angular/common/src/directives/ng_for.ts

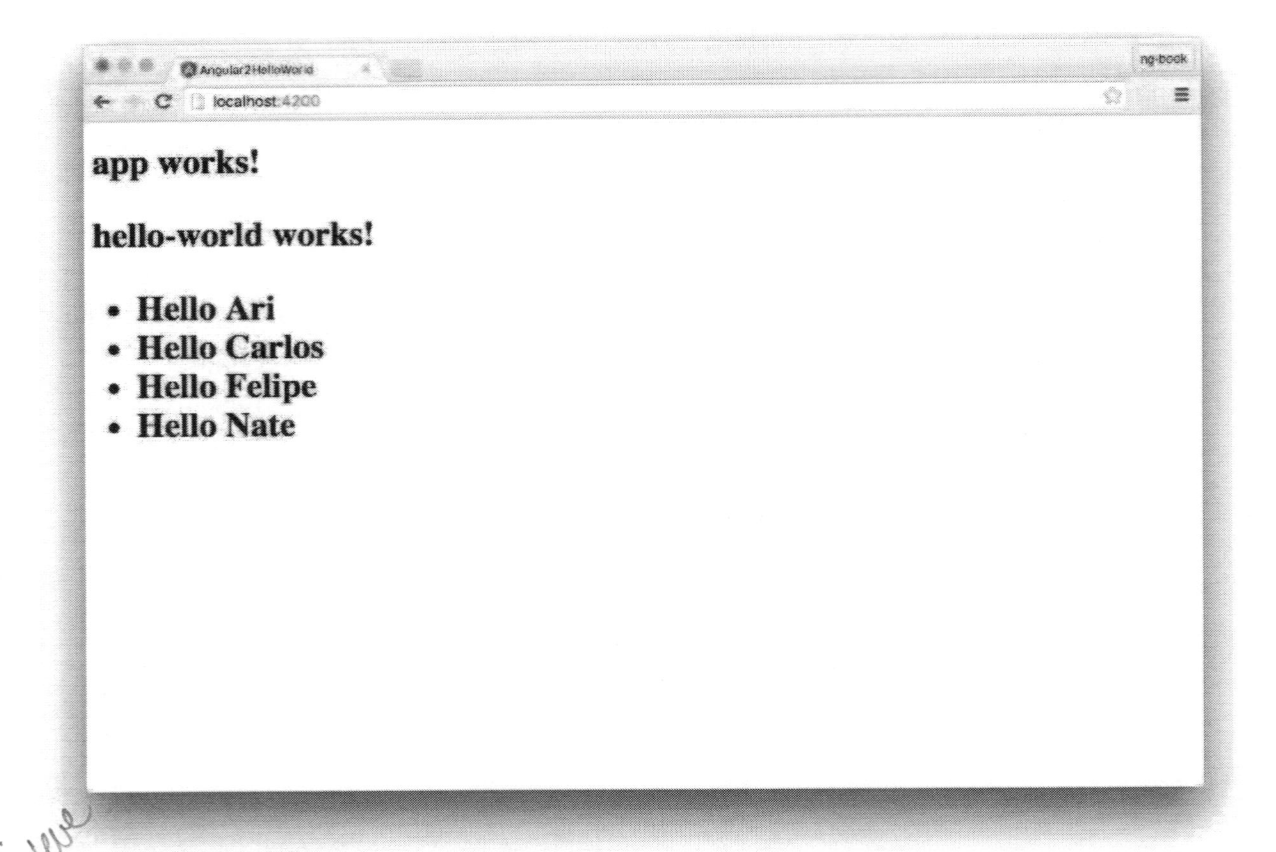


Application with Data

## Using the User Item Component

Remember that earlier we created a `UserItemComponent`? Instead of rendering each name within the `UserListComponent`, we ought to use `UserItemComponent` as a *child component* - that is, instead of repeating over `li` tags directly, we should let our `UserItemComponent` specify the template (and functionality) of each item in the list.

To do this, we need to do three things:

1. Configure the `UserListComponent` to render to `UserItemComponent` (in the template)
2. Configure the `UserItemComponent` to accept the `name` variable as an *input* and
3. Configure the `UserListComponent` template to **pass the name** to the `UserItemComponent`.

Let's perform these steps one-by-one.

## Rendering the `UserItemComponent`

Our `UserItemComponent` specifies the selector `app-user-item` - let's add that tag to our template. What we're going to do is replace the `<li>` tag with the `app-user-item` tag:

**code/first_app/angular2_hello_world/src/app/user-list/user-list.component.html**

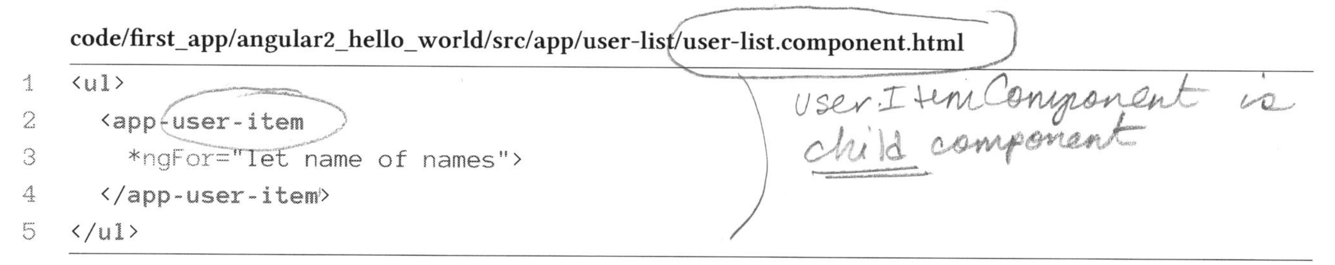

```
1 <ul>
2   <app-user-item
3     *ngFor="let name of names">
4   </app-user-item>
5 </ul>
```

Notice that while we swapped out the `li` tag for `app-user-item` we left in the `ngFor` attribute because we *still want to loop* over the list of names.

Notice that we also removed the inner content of this template because *the component* has it's own template. If we reload our browser, this is what we will see:

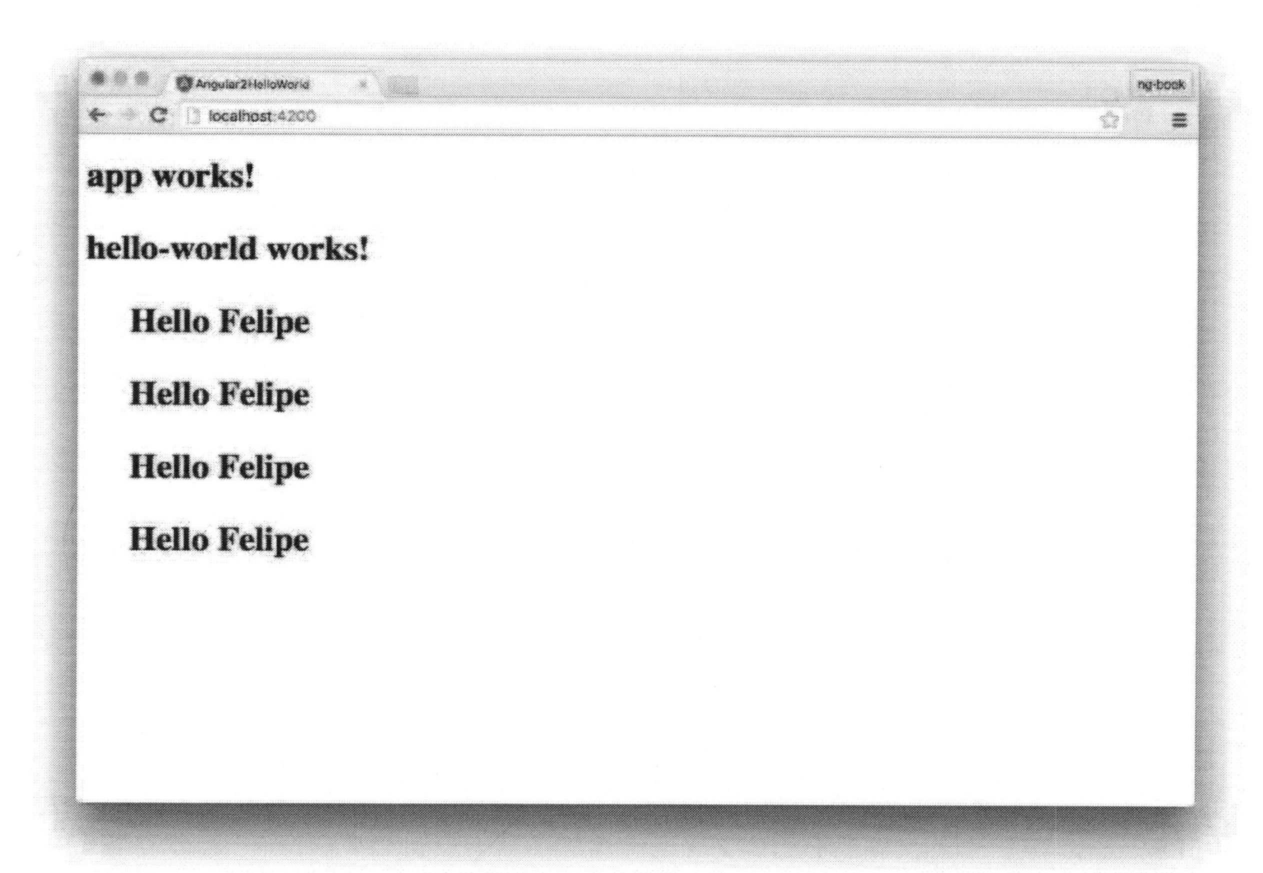


Application with Data

It repeats, but something is wrong here - every name says “Felipe”! We need a way to *pass data into the child component.*

Thankfully, Angular provides a way to do this: the `@Input` annotation.

## Accepting Inputs

Remember that in our `UserItemComponent` we had set `this.name = 'Felipe';` in the `constructor` of that component. Now we need to change this component to accept a value for this property.

Here’s what we need to change our `UserItemComponent` to:

**code/first_app/angular2_hello_world/src/app/user-item/user-item.component.ts**

```
import {
  Component,
  OnInit,
  Input       // <--- added this
} from '@angular/core';

@Component({
  selector: 'app-user-item',
  templateUrl: './user-item.component.html',
  styleUrls: ['./user-item.component.css']
})
export class UserItemComponent implements OnInit {
  @Input() name: string; // <-- added Input annotation

  constructor() {
    // removed setting name
  }

  ngOnInit() {
  }

}
```

Notice that we changed the `name` property to have an *annotation* of `@Input`. We talk a lot more about `Inputs` (and `Outputs`) in the next chapter, but for now, just know that this syntax allows us to pass in a value *from the parent template.*

In order to use `Input` we also had to add it to the list of constants in `import`.

Lastly, we don’t want to set a default value for `name` so we remove that from the ‘constructor.

So now that we have a `name` `Input`, how do we actually use it?

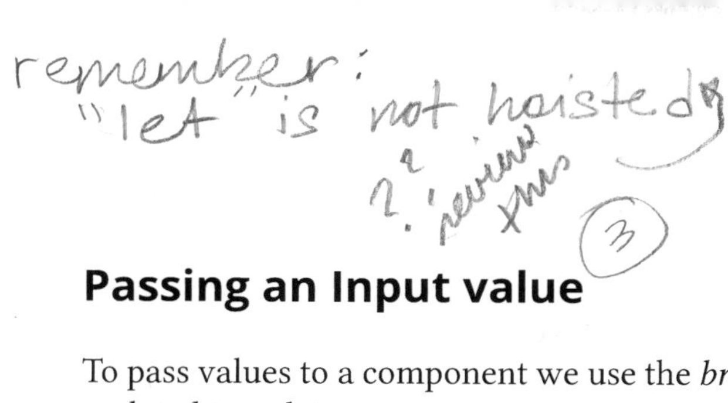
remember:
"let" is not hoisted

## Passing an Input value

To pass values to a component we use the *bracket* [] syntax in our template - let's take a look at our updated template:

**code/first_app/angular2_hello_world/src/app/user-list/user-list.component.html**

```
1 <ul>
2   <app-user-item
3     *ngFor="let name of names"
4     [name]="name">
5   </app-user-item>
6 </ul>
```

Notice that we've added a new attribute on our `app-user-item` tag: `[name]="name"` – in Angular when we add an attribute in brackets like `[foo]` we're saying we want to pass a value to the *input* named `foo` on that component.

In this case notice that the `name` on the right-hand side comes from the `let name ...` statement in `ngFor`. That is, consider if we had this instead:

```
1   <app-user-item
2     *ngFor="let individualUserName of names"
3     [name]="individualUserName">
4   </app-user-item>
```

The `[name]` part designates the `Input` on the `UserItemComponent`. Notice that we're *not* passing the literal string `"individualUserName"` instead we're passing the *value* of `individualUserName`, which is each element of `names`.

We talk more about inputs and outputs in detail in the next chapter. For now, know that we're:

1. Iterating over `names`
2. Creating a new `UserItemComponent` for each element in `names` and
3. Passing the value of that name into the `name` `Input` property on the `UserItemComponent`

Now rendering our list of names is working!

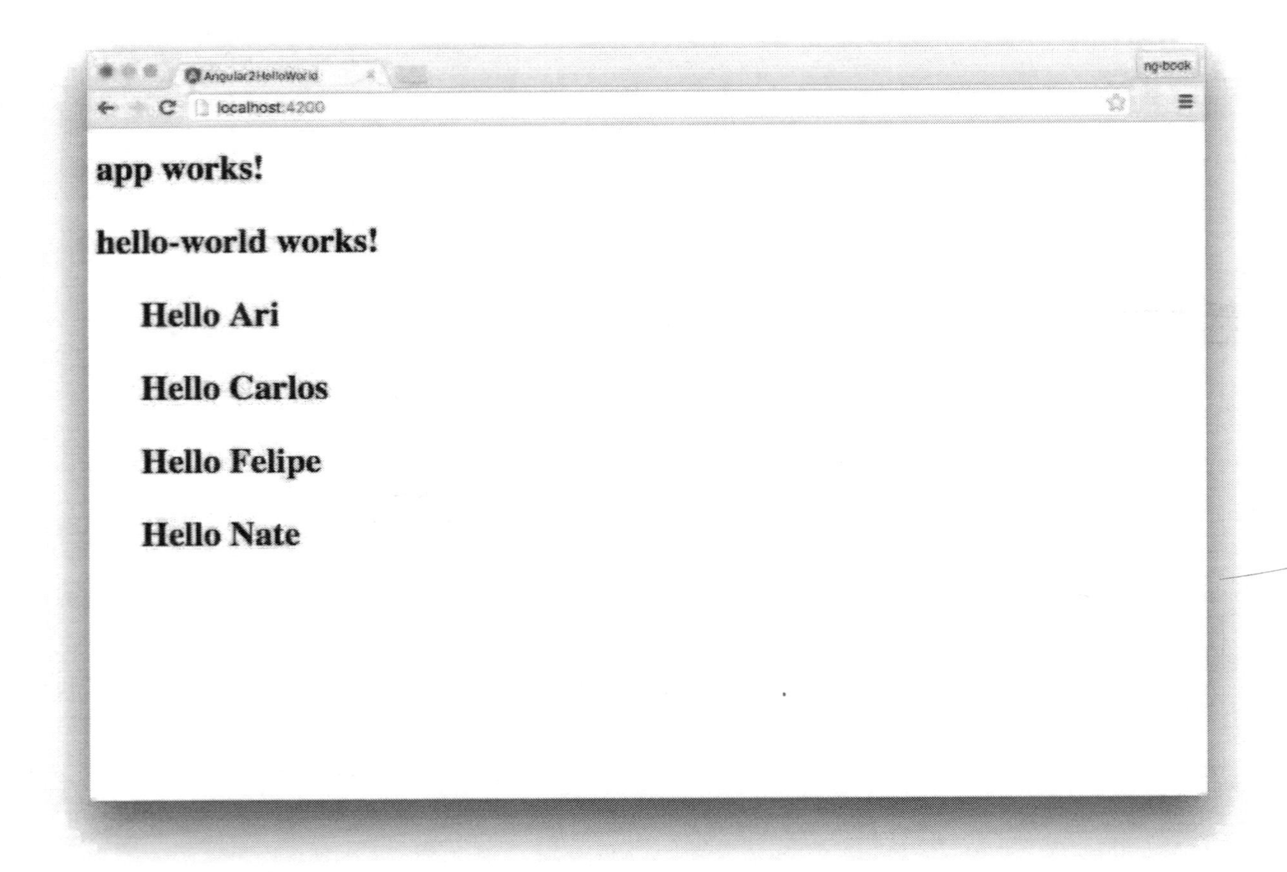


**Application with Names Working**

Congratulations! You've built your first Angular app with components!

Of course, this app is very simple and we'd like to build much more sophisticated applications. Don't worry, in this book we'll show you how to become an expert writing Angular apps. In fact, in this chapter we're going to build a voting-app (think Reddit or Product Hunt). This app will feature user interaction, and even more components!

But before we start building a new app, let's take a closer look at how Angular apps are bootstrapped.

## Bootstrapping Crash Course

Every app has a main entry point. This application was built using `angular-cli` which is built on a tool called webpack. You don't have to understand webpack to use Angular, but it is helpful to understand the flow of how your application boots.

We run this app by calling the command:

```
1   ng serve
```

ng will look at the file `angular-cli.json` to find the entry point to our app. Let's trace how ng finds the components we just built.

At a high level, it looks like this:

- `angular-cli.json` specifies a `"main"` file, which in this case is `main.ts`
- `main.ts` is the entry-point for our app and it *bootstraps* our application
- The bootstrap process boots **an Angular module** – we haven't talked about modules yet, but we will in a minute
- We use the `AppModule` to bootstrap the app. `AppModule` is specified in `src/app/app.module.ts`
- `AppModule` specifies which *component* to use as the top-level component. In this case it is `AppComponent`
- `AppComponent` has `<app-user-list>` tags in the template and this renders our list of users.

We'll talk about this process more later in the book, but for now the thing I want to focus on is the Angular module system: `NgModules`.

Angular also has a powerful concept of *modules*. When you boot an Angular app, you're not booting a component directly, but instead you create an `NgModule` which points to the component you want to load.

Let's a look at the code:

**code/first_app/angular2_hello_world/src/app/app.module.ts**

```
11  @NgModule({
12    declarations: [
13      AppComponent,
14      HelloWorldComponent,
15      UserItemComponent,
16      UserListComponent
17    ],
18    imports: [
19      BrowserModule,
20      FormsModule,
21      HttpModule
22    ],
23    providers: [],
24    bootstrap: [AppComponent]
25  })
26  export class AppModule { }
```

The first thing we see is an `@NgModule` annotation. Like all annotations, this `@NgModule( ... )` code adds metadata to the class immediately following (`AppModule`).

Our `@NgModule` annotation has three keys: `declarations`, `imports`, and `bootstrap`.

`declarations` specifies the components that are **defined in this module**. You may have noticed that when we used `ng generate` it automatically added our components to this list! This is an important idea in Angular:

**You have to declare components in a `NgModule` before you can use them in your templates.**

`imports` describes which *dependencies* this module has. We're creating a browser app, so we want to import the `BrowserModule`.

`bootstrap` tells Angular that when this module is used to bootstrap an app, we need to load the `AppComponent` component as the top-level component.

We talk more about `NgModules` in the section on `NgModules`

## Expanding our Application

Now that we know how to create a basic application, let's build our Reddit clone. Before we start coding, it's a good idea to look over our app and break it down into its logical components.

**Application with Data**

We're going to make two components in this app:

1. The overall application, which contains the form used to submit new articles (marked in magenta in the picture).
2. Each article (marked in mint green).

In a larger application, the form for submitting articles would probably become its own component. However, having the form be its own component makes the data passing more complex, so we're going to simplify in this chapter and only have two components.

For now, we'll just make two components, but we'll learn how to deal with more sophisticated data architectures in later chapters of this book.

But first thing's first, let's generate a new application by running the same **ng new** command we ran before to create a new application passing it the name of the app we want to create (here, we'll create an application called `angular2_reddit`):

```
1   ng new angular2_reddit
```

We provide a completed version of our `angular2_reddit` in the example code download

## Adding CSS

First thing we want to do is add some CSS styling so that our app isn't completely unstyled.

If you're building your app from scratch, you'll want to copy over a few files from our completed example in the `first_app/angular2_reddit` folder.

Copy:

- `src/index.html`
- `src/styles.css`
- `src/app/vendor`
- `src/assets/images`

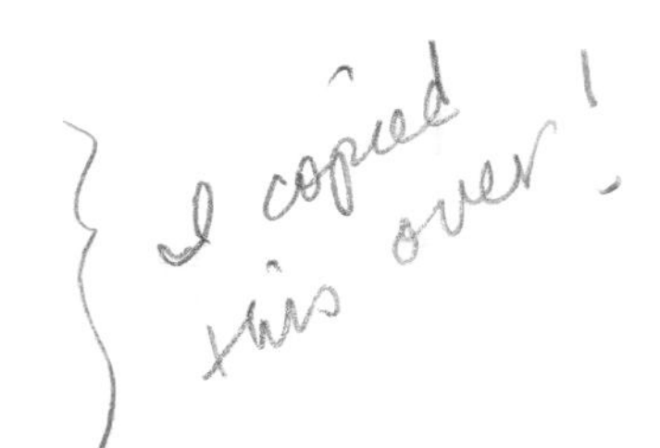

into your application's folder.

For this project we're going to be using Semantic-UI[12] to help with the styling. Semantic-UI is a CSS framework, similar to Zurb Foundation[13] or Twitter Bootstrap[14]. We've included it in the sample code download so all you need to do is copy over the files specified above.

## The Application Component

Let's now build a new component which will:

1. store our current list of articles
2. contain the form for submitting new articles.

We can find the main application component on the `src/app/app.component.ts` file. Let's open this file. Again, we'll see the same initial contents we saw previously.

[12]http://semantic-ui.com/
[13]http://foundation.zurb.com
[14]http://getbootstrap.com

**code/first_app/angular2_reddit/src/app/app.component.ts**

```
import { Component } from '@angular/core';

@Component({
  selector: 'app-root',
  templateUrl: './app.component.html',
  styleUrls: ['./app.component.css']
})
export class AppComponent {
  title = 'app works!';
}
```

Let's change the template a bit to include a form for adding links. We'll use a bit of styling from the `semantic-ui` package to make the form look a bit nicer:

**code/first_app/angular2_reddit/src/app/app.component.html**

```
<form class="ui large form segment">
  <h3 class="ui header">Add a Link</h3>

  <div class="field">
    <label for="title">Title:</label>
    <input name="title">
  </div>
  <div class="field">
    <label for="link">Link:</label>
    <input name="link">
  </div>
</form>
```

We're creating a `template` that defines two `input` tags: one for the `title` of the article and the other for the `link` URL.

When we load the browser you should see the rendered form:

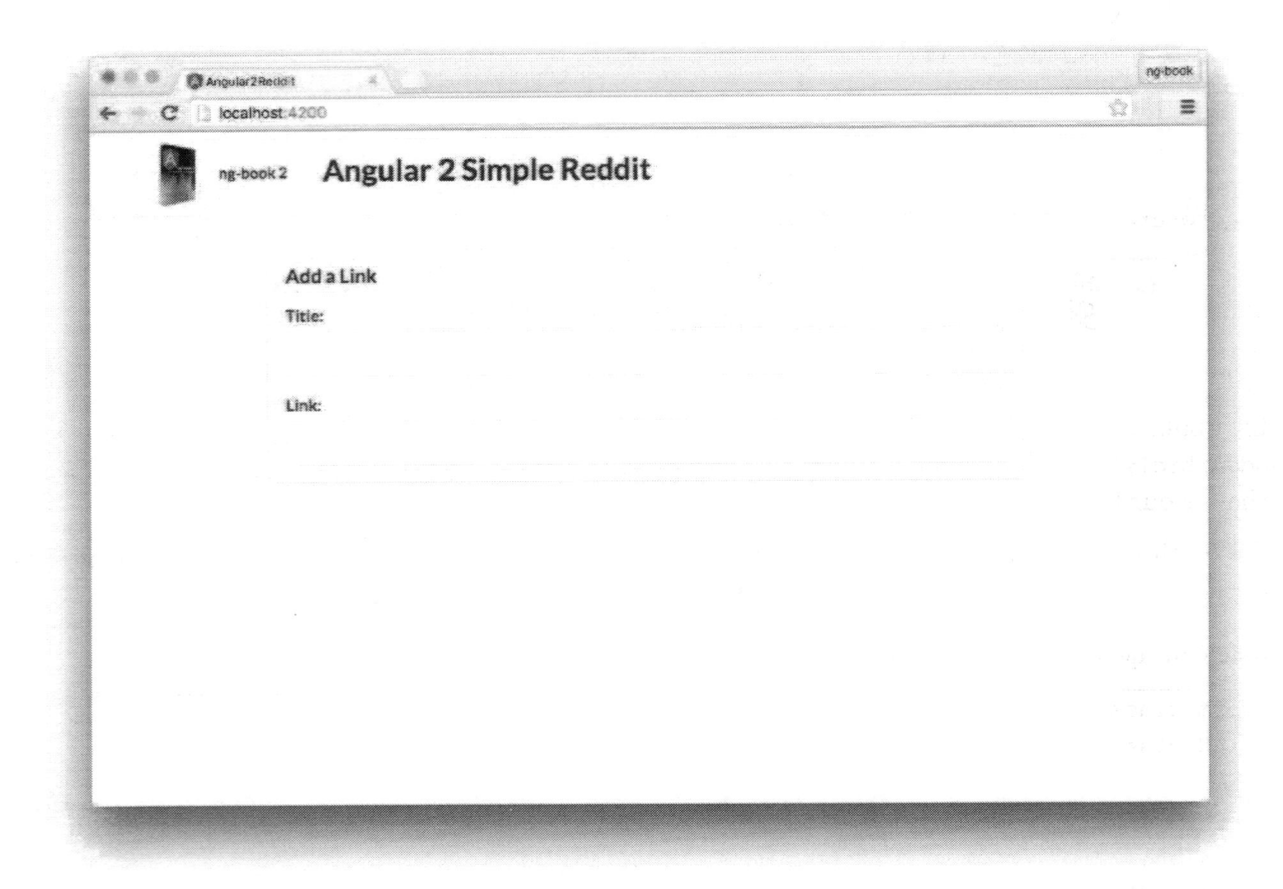


Form

## Adding Interaction

Now we have the form with input tags but we don't have any way to submit the data. Let's add some interaction by adding a submit button to our form.

When the form is submitted, we'll want to call a function to create and add a link. We can do this by adding an interaction event on the `<button />` element.

We tell Angular we want to respond to an event by surrounding the event name in parenthesis `()`. For instance, to add a function call to the `<button />` `onClick` event, we can pass it through like so:

```
1 <button (click)="addArticle()"
2         class="ui positive right floated button">
3   Submit link
4 </button>
```

Now, when the button is clicked, it will call a function called `addArticle()`, which we need to define on the `AppComponent` class. Let's do that now:

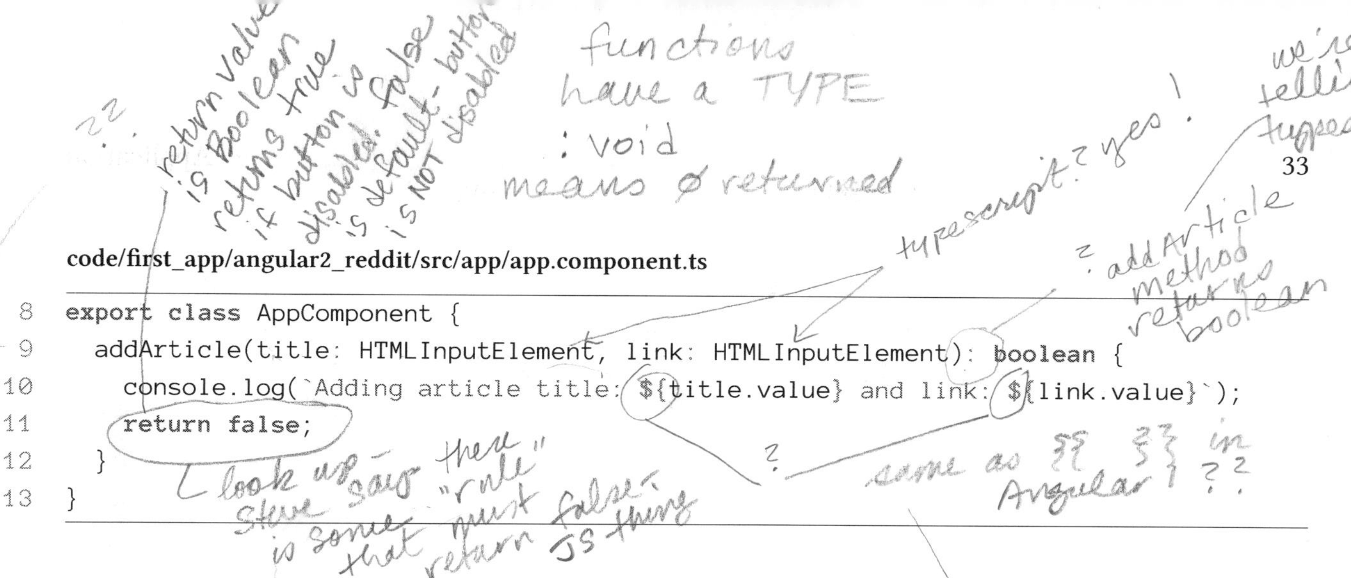

**code/first_app/angular2_reddit/src/app/app.component.ts**

```
 8  export class AppComponent {
 9    addArticle(title: HTMLInputElement, link: HTMLInputElement): boolean {
10      console.log(`Adding article title: ${title.value} and link: ${link.value}`);
11      return false;
12    }
13  }
```

With the `addArticle()` function added to the `AppComponent` and the `(click)` event added to the `<button />` element, this function will be called when the button is clicked. Notice that the `addArticle()` function can accept two arguments: the `title` and the `link` arguments. We need to change our template button to pass those into the call to the `addArticle()`.

We do this by populating a *template variable* by adding a special syntax to the `input` elements on our form. Here's what our template will look like:

**code/first_app/angular2_reddit/src/app/app.component.html**

```
 1  <form class="ui large form segment">
 2    <h3 class="ui header">Add a Link</h3>
 3
 4    <div class="field">
 5      <label for="title">Title:</label>
 6      <input name="title" #newtitle>  <!-- changed -->
 7    </div>
 8    <div class="field">
 9      <label for="link">Link:</label>
10      <input name="link" #newlink> <!-- changed -->
11    </div>
12
13    <!-- added this button -->
14    <button (click)="addArticle(newtitle, newlink)"
15            class="ui positive right floated button">
16      Submit link
17    </button>
18
19  </form>
```

Notice that in the `input` tags we used the `#` (hash) to tell Angular to assign those tags to *a local variable*. By adding the `#title` and `#link` to the appropriate `<input />` elements, we can **pass them as variables** into the `addArticle()` function on the button!

To recap what we've done, we've made **four** changes:

1. Created a `button` tag in our markup that shows the user where to click
2. We created a function named `addArticle` that defines what we want to do when the button is clicked
3. We added a `(click)` attribute on the `button` that says "call the function `addArticle` when this button is pressed".
4. We added the attribute `#newtitle` and `#newlink` to the `<input>` tags

Let's cover each one of these steps in reverse order:

## Binding `inputs` to values

Notice in our first input tag we have the following:

```
1 <input name="title" #newtitle>
```

This markup tells Angular to *bind* this `<input>` to the variable `newtitle`. The `#newtitle` syntax is called a *resolve*. The effect is that this makes the variable `newtitle` available to the expressions within this view.

`newtitle` is now an **object** that represents this `input` DOM element (specifically, the type is `HTMLInputElement`). Because `newtitle` is an object, that means we get the value of the input tag using `newtitle.value`.

Similarly we add `#newlink` to the other `<input>` tag, so that we'll be able to extract the value from it as well.

## Binding actions to events

On our `button` tag we add the attribute `(click)` to define what should happen when the button is clicked on. When the `(click)` event happens we call `addArticle` with two arguments: `newtitle` and `newlink`. Where did this function and two arguments come from?

1. `addArticle` is a function on our component definition class `AppComponent`
2. `newtitle` comes from the resolve (`#newtitle`) on our `<input>` tag named `title`
3. `newlink` comes from the resolve (`#newlink`) on our `<input>` tag named `link`

All together:

```
<button (click)="addArticle(newtitle, newlink)"
        class="ui positive right floated button">
  Submit link
</button>
```

The markup `class="ui positive right floated button"` comes from Semantic UI and it gives the button the pleasant green color.

## Defining the Action Logic

On our `class AppComponent` we define a new function called `addArticle`. It takes two arguments: `title` and `link`. Again, it's important to realize that `title` and `link` are both **objects** of type `HTMLInputElement` and *not the input values directly*. To get the value from the `input` we have to call `title.value`. For now, we're just going to `console.log` out those arguments.

**code/first_app/angular2_reddit/src/app/app.component.ts**

```
  addArticle(title: HTMLInputElement, link: HTMLInputElement): boolean {
    console.log(`Adding article title: ${title.value} and link: ${link.value}`);
    return false;
  }
```

Notice that we're using backtick strings again. This is a really handy feature of ES6: backtick strings will expand template variables!

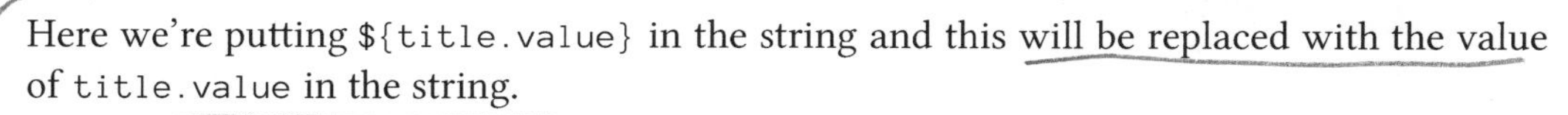

Here we're putting `${title.value}` in the string and this will be replaced with the value of `title.value` in the string.

## Try it out!

Now when you click the submit button, you can see that the message is printed on the console:

**Clicking the Button**

## Adding the Article Component

Now we have a form to submit new articles, but we aren't showing the new articles anywhere. Because every article submitted is going to be displayed as a list on the page, this is the perfect candidate for a new component.

Let's create a new component to represent the individual submitted articles.

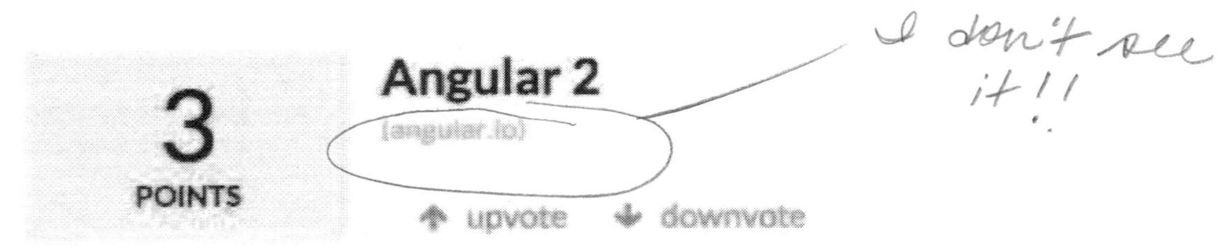


**A reddit-article**

For that, let's use the `ng` tool to generate a new component:

```
1 ng generate component article
```

We have three parts to defining this new component:

1. Define the `ArticleComponent` view in the template
2. Define the `ArticleComponent` properties by annotating the class with `@Component`
3. Define a component-definition class (`ArticleComponent`) which houses our component logic

Let's talk through each part in detail:

#### Creating the `ArticleComponent` template

We define the template using the file `article.component.html`:

**code/first_app/angular2_reddit/src/app/article/article.component.html**

```
1  <div class="four wide column center aligned votes">
2    <div class="ui statistic">
3      <div class="value">
4        {{ votes }}
5      </div>
6      <div class="label">
7        Points
8      </div>
9    </div>
10 </div>
11 <div class="twelve wide column">
12   <a class="ui large header" href="{{ link }}">
13     {{ title }}
14   </a>
15   <ul class="ui big horizontal list voters">
16     <li class="item">
17       <a href (click)="voteUp()">
18         <i class="arrow up icon"></i>
19           upvote
20         </a>
21     </li>
22     <li class="item">
23       <a href (click)="voteDown()">
24         <i class="arrow down icon"></i>
25         downvote
26       </a>
```

```
27      </li>
28    </ul>
29  </div>
```

There's a lot of markup here, so let's break it down :

**A Single reddit-article Row**

We have two columns:

1. the number of votes on the left and
2. the article information on the right.

We specify these columns with the CSS classes `four wide column` and `twelve wide column` respectively (remember that these come from SemanticUI's CSS).

We're showing `votes` and the `title` with the template expansion strings `{{ votes }}` and `{{ title }}`. The values come from the value of `votes` and `title` property of the `ArticleComponent` class, which we'll define in a minute.

Notice that we can use template strings in **attribute values**, as in the `href` of the `a` tag: `href="{{ link }}"`. In this case, the value of the `href` will be dynamically populated with the value of `link` from the component class

On our upvote/downvote links we have an action. We use `(click)` to bind `voteUp()`/`voteDown()` to their respective buttons. When the upvote button is pressed, the `voteUp()` function will be called on the `ArticleComponent` class (similarly with downvote and `voteDown()`).

#### Creating the `ArticleComponent`

**code/first_app/angular2_reddit/src/app/article/article.component.ts**

```
 3  @Component({
 4    selector: 'app-article',
 5    templateUrl: './article.component.html',
 6    styleUrls: ['./article.component.css'],
 7    host: {
 8      class: 'row'
 9    }
10  })
```

First, we define a new Component with `@Component`. The `selector` says that this component is placed on the page by using the tag `<app-article>` (i.e. the selector is a tag name).

So the most essential way to use this component would be to place the following tag in our markup:

```
1  <app-article>
2  </app-article>
```

These tags will remain in our view when the page is rendered.

We want each `app-article` to be on its own row. We're using Semantic UI, and Semantic provides a CSS class for rows[15] called `row`.

In Angular, a component *host* is **the element this component is attached to**. You'll notice on our `@Component` we're passing the option: `host: { class: 'row' }`. This tells Angular that on the **host element** (the `app-article` tag) we want to set the `class` attribute to have "`row`".

Using the `host` option is nice because it means we can encapsulate the `app-article` markup *within* our component. That is, we don't have to both use a `app-article` tag **and** require a `class="row"` in the markup of the parent view. By using the `host` option, we're able to configure our host element from *within* the component.

### Creating the `ArticleComponent` Definition Class

Finally, we create the `ArticleComponent` definition class:

[15] http://semantic-ui.com/collections/grid.html

**code/first_app/angular2_reddit/src/app/article/article.component.ts**

```
11 export class ArticleComponent implements OnInit {
12   votes: number;
13   title: string;
14   link: string;
15
16   constructor() {
17     this.title = 'Angular 2';
18     this.link = 'http://angular.io';
19     this.votes = 10;
20   }
21
22   voteUp() {
23     this.votes += 1;
24   }
25
26   voteDown() {
27     this.votes -= 1;
28   }
29
30   ngOnInit() {
31   }
32
33 }
```

Here we create three properties on `ArticleComponent`:

1. `votes` - a `number` representing the sum of all upvotes, minus the downvotes
2. `title` - a `string` holding the title of the article
3. `link` - a `string` holding the URL of the article

In the `constructor()` we set some default attributes:

**code/first_app/angular2_reddit/src/app/article/article.component.ts**

```
16    constructor() {
17      this.title = 'Angular 2';
18      this.link = 'http://angular.io';
19      this.votes = 10;
20    }
```

And we define two functions for voting, one for voting up `voteUp` and one for voting down `voteDown`:

**code/first_app/angular2_reddit/src/app/article/article.component.ts**

```
22    voteUp() {
23      this.votes += 1;
24    }
25
26    voteDown() {
27      this.votes -= 1;
28    }
```

In `voteUp` we increment `this.votes` by one. Similarly we decrement for `voteDown`.

### Using the `app-article` Component

In order to use this component and make the data visible, we have to add a `<app-article></app-article>` tag somewhere in our markup.

In this case, we want the `AppComponent` to render this new component, so let's update the code in that component. Add the `<app-article>` tag to the `AppComponent`'s template right after the closing `</form>` tag:

```
1    <button (click)="addArticle(newtitle, newlink)"
2            class="ui positive right floated button">
3      Submit link
4    </button>
5  </form>
6
7  <div class="ui grid posts">
8    <app-article>
9    </app-article>
10 </div>
```

If we reload the browser now, we will see that the `<app-article>` tag wasn't compiled. Oh no!

Whenever hitting a problem like this, the first thing to do is open up your browser's developer console. If we inspect our markup (see screenshot below), we can see that the `app-article` tag is on our page, but it hasn't been compiled into markup. Why not?

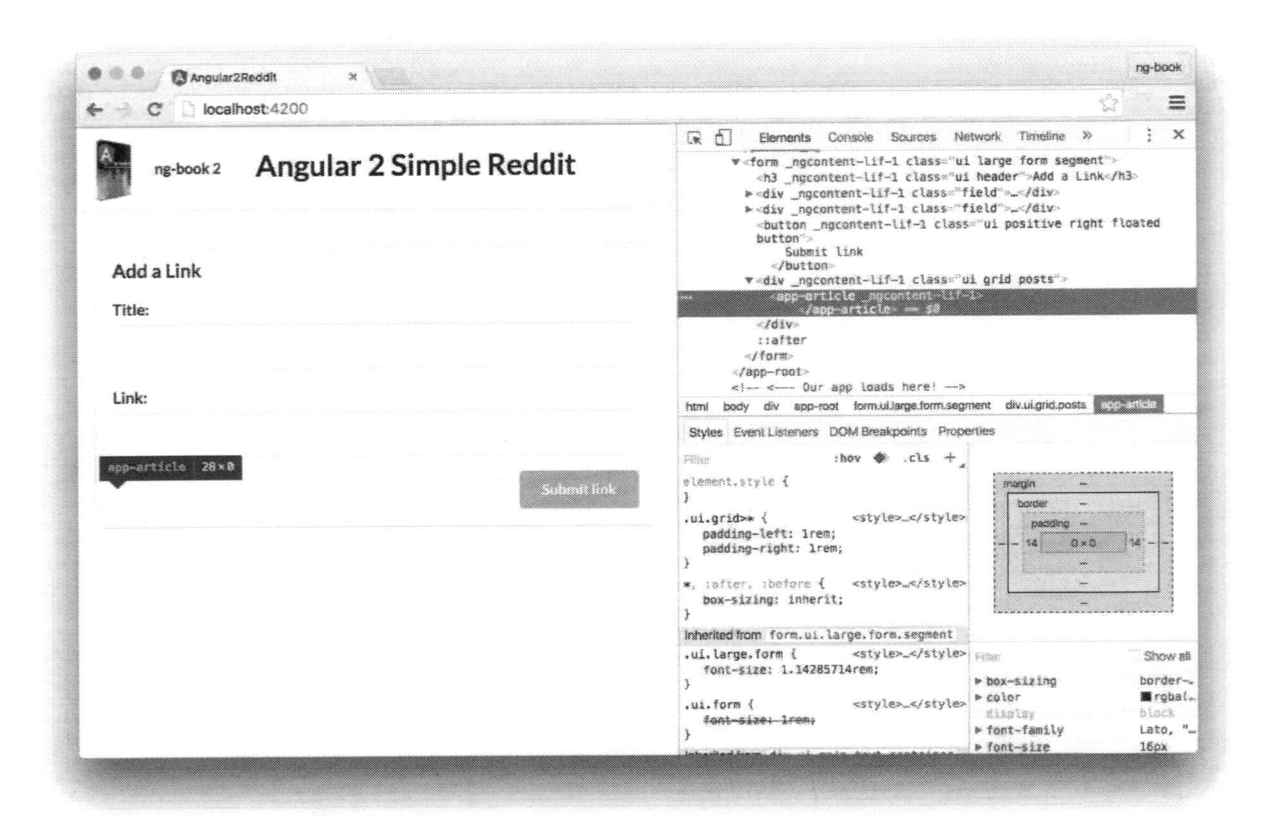


**Unexpanded tag when inspecting the DOM**

This happens because the `AppComponent` component **doesn't know about the `ArticleComponent` component** yet.

> **Angular 1 Note**: If you've used Angular 1 it might be surprising that our app doesn't know about our new `app-article` component. This is because in Angular 1, directives match globally. However, in Angular you need to explicitly specify which components (and therefore, which selectors) you want to use.
>
> On the one hand, this requires a little more configuration. On the other hand, it's great for building scalable apps because it means we don't have to share our directive selectors in a global namespace.

In order to tell our `AppComponent` about our new `ArticleComponent` component, we need to **add the `ArticleComponent` to the list of `declarations` in this `NgModule`**.

We add `ArticleComponent` to our `declarations` because `ArticleComponent` is part of this module (`RedditAppModule`). However, if `ArticleComponent` were part of a *different* module, then we might import it with `imports`.

We'll discuss more about `NgModules` later on, but for now, know that when you create a new component, you have to put in a `declarations` in `NgModules`.

**code/first_app/angular2_reddit/src/app/app.module.ts**

```
 6  import { AppComponent } from './app.component';
 7  import { ArticleComponent } from './article/article.component.ts';
 8
 9  @NgModule({
10    declarations: [
11      AppComponent,
12      ArticleComponent // <-- added this
13    ],
```

See here that we are:

1. `importing` `ArticleComponent` and then
2. Adding `ArticleComponent` to the list of `declarations`

After you've added `ArticleComponent` to `declarations` in the `NgModule`, if we reload the browser we should see the article properly rendered:

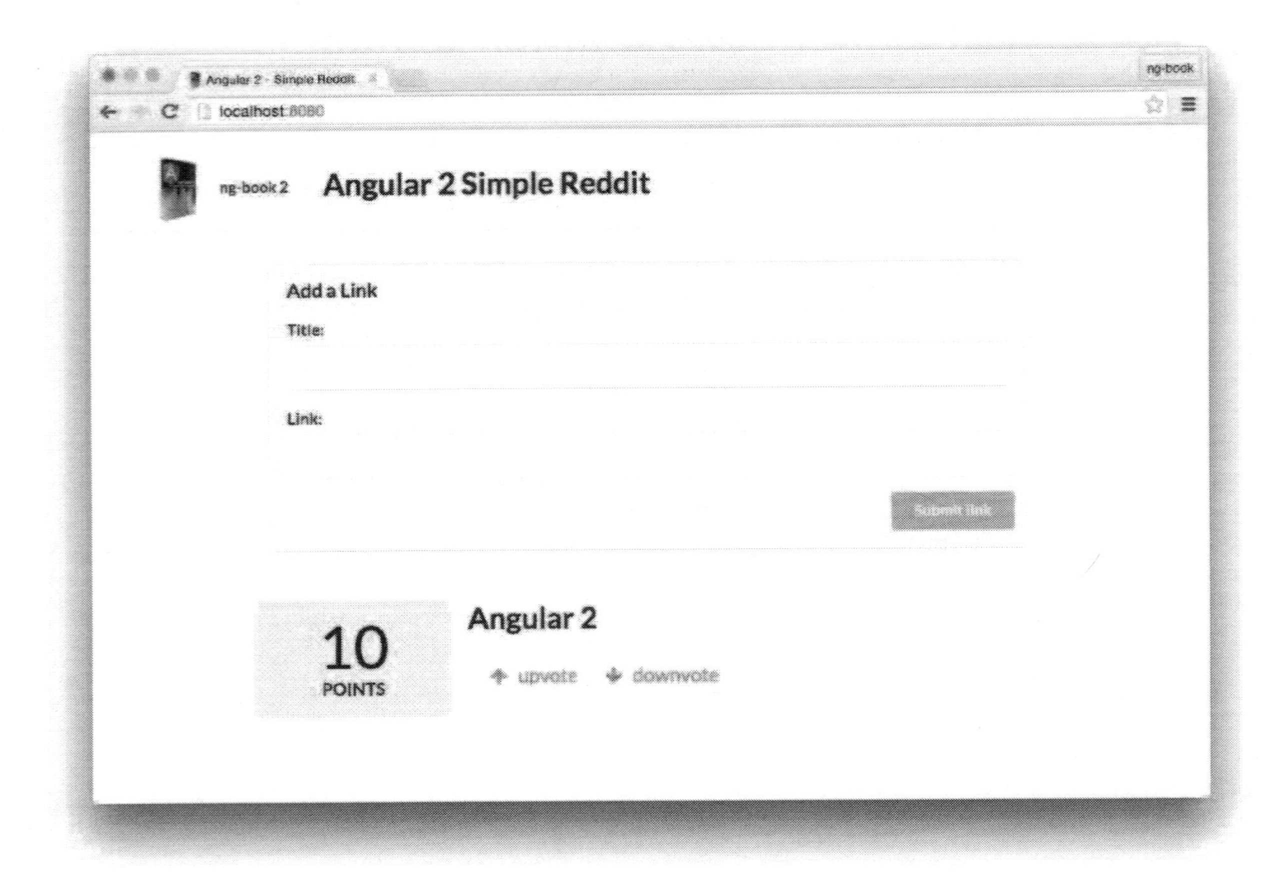


**Rendered ArticleComponent component**

However, clicking on the **vote up** or **vote down** links will **cause the page to reload** instead of updating the article list.

JavaScript, by default, **propagates the `click` event to all the parent components**. Because the `click` event is propagated to parents, our browser is trying to follow the empty link, which tells the browser to reload.

To fix that, we need to make the click event handler to return `false`. This will ensure the browser won't try to refresh the page. Let's update our code so that each of the functions `voteUp()` and `voteDown()` return a boolean value of `false` (tells the browser *not* to propagate the event upwards):

```
1 voteDown(): boolean {
2   this.votes -= 1;
3   return false;
4 }
5 // and similarly with `voteUp()`
```

Now when we click the links we'll see that the votes increase and decrease properly without a page refresh.

# Rendering Multiple Rows

Right now we only have one article on the page and there's no way to render more, unless we paste another <app-article> tag. And even if we did that all the articles would have the same content, so it wouldn't be very interesting.

## Creating an Article class

A good practice when writing Angular code is to try to isolate the data structures we are using from the component code. To do this, let's create a data structure that represents a single article. Let's add a new file `article.model.ts` to define an `Article` class that we can use.

**code/first_app/angular2_reddit/src/app/article/article.model.ts**

```
export class Article {
  title: string;
  link: string;
  votes: number;

  constructor(title: string, link: string, votes?: number) {
    this.title = title;
    this.link = link;
    this.votes = votes || 0;
  }
}
```

Here we are creating a new class that represents an `Article`. Note that this is a **plain class and not an Angular component.** In the Model-View-Controller pattern this would be the **Model**.

Each article has a `title`, a `link`, and a total for the `votes`. When creating a new article we need the `title` and the `link`. The `votes` parameter is optional (denoted by the `?` at the end of the name) and defaults to zero.

Now let's update the `ArticleComponent` code to use our new `Article` class. Instead of storing the properties directly on the `ArticleComponent` component let's **store the properties on an instance of the `Article` class.**

First let's import the class: {lang=javascript,crop-start-line=2,crop-end-line=2,starting-line-number=2} <<code/first_app/angular2_reddit/src/app/article/article.component.ts[16]

Then let's use it: {lang=javascript,crop-start-line=12,crop-end-line=35,starting-line-number=12} <<code/-first_app/angular2_reddit/src/app/article/article.component.ts[17]

[16]code/first_app/angular2_reddit/src/app/article/article.component.2.ts

[17]code/first_app/angular2_reddit/src/app/article/article.component.2.ts

Notice what we've changed: instead of storing the `title`, `link`, and `votes` properties directly on the component, we're storing a reference to an `article`. What's neat is that we've defined the type of `article` to be our new `Article` class.

When it comes to `voteUp` (and `voteDown`), we don't increment `votes` on the component, but rather, we need to increment the `votes` on the `article`.

However, this refactoring introduces another change: we need to update our view to get the template variables from the right location. To do that, we need to change our template tags to read from `article`. That is, where before we had `{{ votes }}`, we need to change it to `{{ article.votes }}`, and same with `title` and `link`:

**code/first_app/angular2_reddit/src/app/article/article.component.html**

```
1  <div class="four wide column center aligned votes">
2    <div class="ui statistic">
3      <div class="value">
4        {{ article.votes }}
5      </div>
6      <div class="label">
7        Points
8      </div>
9    </div>
10 </div>
11 <div class="twelve wide column">
12   <a class="ui large header" href="{{ article.link }}">
13     {{ article.title }}
14   </a>
15   <ul class="ui big horizontal list voters">
16     <li class="item">
17       <a href (click)="voteUp()">
18         <i class="arrow up icon"></i>
19           upvote
20         </a>
21     </li>
22     <li class="item">
23       <a href (click)="voteDown()">
24         <i class="arrow down icon"></i>
25         downvote
26       </a>
27     </li>
28   </ul>
29 </div>
```

Reload the browser and everything still works.

This situation is better but something in our code is still off: our `voteUp` and `voteDown` methods break the encapsulation of the `Article` class by changing the article's internal properties directly.

`voteUp` and `voteDown` currently break the Law of Demeter[18] which says that a given object should assume as little as possible about the structure or properties of other objects.

The problem is that our `ArticleComponent` component knows too much about the `Article` class internals. To fix that, let's add `voteUp` and `voteDown` methods on the `Article` class.

**code/first_app/angular2_reddit/src/app/article/article.model.ts**

```
export class Article {
  title: string;
  link: string;
  votes: number;

  constructor(title: string, link: string, votes?: number) {
    this.title = title;
    this.link = link;
    this.votes = votes || 0;
  }

  voteUp(): void {
    this.votes += 1;
  }

  voteDown(): void {
    this.votes -= 1;
  }

  domain(): string {
    try {
      const link: string = this.link.split('//')[1];
      return link.split('/')[0];
    } catch (err) {
      return null;
    }
  }
}
```

We can then change `ArticleComponent` to call these methods:

[18] http://en.wikipedia.org/wiki/Law_of_Demeter

**code/first_app/angular2_reddit/src/app/article/article.component.ts**

```
12 export class ArticleComponent implements OnInit {
13   article: Article;
14 
15   constructor() {
16     this.article = new Article(
17       'Angular 2',
18       'http://angular.io',
19       10);
20   }
21 
22   voteUp(): boolean {
23     this.article.voteUp();
24     return false;
25   }
26 
27   voteDown(): boolean {
28     this.article.voteDown();
29     return false;
30   }
31 
32   ngOnInit() {
33   }
34 
35 }
```

**Why do we have a `voteUp` function in both the model and the component?**

The reason we have a `voteUp()` and a `voteDown()` on both classes is because each function does a slightly different thing. The idea is that the `voteUp()` on the `ArticleComponent` relates to the **component view**, whereas the `Article` model `voteUp()` defines what *mutations* happen **in the model**.

That is, it allows the `Article` class to encapsulate what functionality should happen **to a model** when voting happens. In a "real" app, the internals of the `Article` model would probably be more complicated, e.g. make an API request to a webserver, and you wouldn't want to have that sort of model-specific code in your component controller.

Similarly, in the `ArticleComponent` we `return false;` as a way to say "don't propagate the event" - this is a view-specific piece of logic and we shouldn't allow the `Article` model's `voteUp()` function to have to knowledge about that sort of view-specific API. That is, the `Article` model should allow voting apart from the specific view.

After reloading our browser, we'll notice everything works the same way, but we now have clearer, simpler code.

Checkout our `ArticleComponent` component definition now: it's so short! We've moved a lot of logic **out** of our component and into our models. The corresponding MVC guideline here might be Fat Models, Skinny Controllers[19]. The idea is that we want to move most of our logic to our models so that our components do the minimum work possible.

## Storing Multiple `Article`s

Let's write the code that allows us to have a list of multiple `Articles`.

Let's start by changing `AppComponent` to have a collection of articles:

**code/first_app/angular2_reddit/src/app/app.component.ts**

```
 9  export class AppComponent {
10    articles: Article[];
11
12    constructor() {
13      this.articles = [
14        new Article('Angular 2', 'http://angular.io', 3),
15        new Article('Fullstack', 'http://fullstack.io', 2),
16        new Article('Angular Homepage', 'http://angular.io', 1),
17      ];
18    }
19
20    addArticle(title: HTMLInputElement, link: HTMLInputElement): boolean {
21      console.log(`Adding article title: ${title.value} and link: ${link.value}`);
22      this.articles.push(new Article(title.value, link.value, 0));
23      title.value = '';
24      link.value = '';
25      return false;
26    }
27  }
```

Notice that our `AppComponent` has the line:

```
1   articles: Article[];
```

[19] http://weblog.jamisbuck.org/2006/10/18/skinny-controller-fat-model

The `Article[]` might look a little unfamiliar. We're saying here that `articles` is an `Array` of `Articles`. Another way this could be written is `Array<Article>`. The word for this pattern is *generics*. It's a concept seen in Java, C#, and other languages. The idea is that our collection (the `Array`) is typed. That is, the `Array` is a collection that will only hold objects of type `Article`.

We populate this `Array` by setting `this.articles` in the constructor:

**code/first_app/angular2_reddit/src/app/app.component.ts**

```
12   constructor() {
13     this.articles = [
14       new Article('Angular 2', 'http://angular.io', 3),
15       new Article('Fullstack', 'http://fullstack.io', 2),
16       new Article('Angular Homepage', 'http://angular.io', 1),
17     ];
18   }
```

## Configuring the `ArticleComponent` with `inputs`

Now that we have a list of `Article` *models*, how can we pass them to our `ArticleComponent` *component*?

Here again we use `Inputs`. Previously we had our `ArticleComponent` class defined like this:

**code/first_app/angular2_reddit/src/app/article/article.component.ts**

```
12 export class ArticleComponent implements OnInit {
13   article: Article;
14 
15   constructor() {
16     this.article = new Article(
17       'Angular 2',
18       'http://angular.io',
19       10);
20   }
```

The problem here is that we've hard coded a particular `Article` in the constructor. The point of making components is not only encapsulation, but also reusability.

What we would really like to do is to configure the `Article` we want to display. If, for instance, we had two articles, `article1` and `article2`, we would like to be able to reuse the `app-article` component by passing an `Article` as a "parameter" to the component like this:

```
1 <app-article [article]="article1"></app-article>
2 <app-article [article]="article2"></app-article>
```

Angular allows us to do this by using the `Input` annotation on a property of a `Component`:

```
1 class ArticleComponent {
2   @Input() article: Article;
3   // ...
```

Now if we have an `Article` in a variable `myArticle` we could pass it to our `ArticleComponent` in our view. Remember, we can pass a variable in an element by surrounding it in square brackets `[variableName]`, like so:

```
1 <app-article [article]="myArticle"></app-article>
```

Notice the syntax here: we put the name of the input in brackets as in: `[article]` and the value of the attribute is what we want to pass in to that input.

Then, and this is important, the `this.article` on the `ArticleComponent` instance will be set to `myArticle`. We can think about the variable `myArticle` as being passed as a *parameter* (i.e. input) to our components.

Here's what our `ArticleComponent` component now looks like using `@Input`:

**code/first_app/angular2_reddit/src/app/article/article.component.ts**

```
16 export class ArticleComponent implements OnInit {
17   @Input() article: Article;
18 
19   voteUp(): boolean {
20     this.article.voteUp();
21     return false;
22   }
23 
24   voteDown(): boolean {
25     this.article.voteDown();
26     return false;
27   }
28 
29   ngOnInit() {
30   }
31 
32 }
```

## Rendering a List of Articles

Earlier we configured our `AppComponent` to store an array of `articles`. Now let's configure `AppComponent` to *render* all the `articles`. To do so, instead of having the `<app-article>` tag alone, we are going to use the `NgFor` directive to iterate over the list of `articles` and render a `app-article` for each one:

Let's add this in the `template` of the `AppComponent` `@Component`, just below the closing `<form>` tag:

```
1      Submit link
2    </button>
3  </form>
4
5  <!-- start adding here -->
6  <div class="ui grid posts">
7    <app-article
8      *ngFor="let article of articles"
9      [article]="article">
10   </app-article>
11 </div>
12 <!-- end adding here -->
```

Remember when we rendered a list of names as a bullet list using the `NgFor` directive earlier in the chapter? This syntax also works for rendering multiple components.

The `*ngFor="let article of articles"` syntax will iterate through the list of `articles` and create the local variable `article` (for each item in the list).

To specify the `article` input on a component, we are using the `[inputName]="inputValue"` expression. In this case, we're saying that we want to set the `article` input to the value of the local variable `article` set by `ngFor`.

We are using the variable `article` many times in that previous code snippet, it's (potentially) clearer if we rename the temporary variable created by `NgFor` to `foobar`:

```
1   <app-article
2     *ngFor="let foobar of articles"
3     [article]="foobar">
4   </app-article>
```

So here we have three variables:

1. `articles` which is an `Array` of `Articles`, defined on the `RedditApp` component
2. `foobar` which is a single element of `articles` (an `Article`), defined by `NgFor`
3. `article` which is the name of the field defined on `inputs` of the `ArticleComponent`

Basically, `NgFor` generates a temporary variable `foobar` and then we're passing it in to `app-article`

Reloading our browser now, we will see all articles will be rendered:

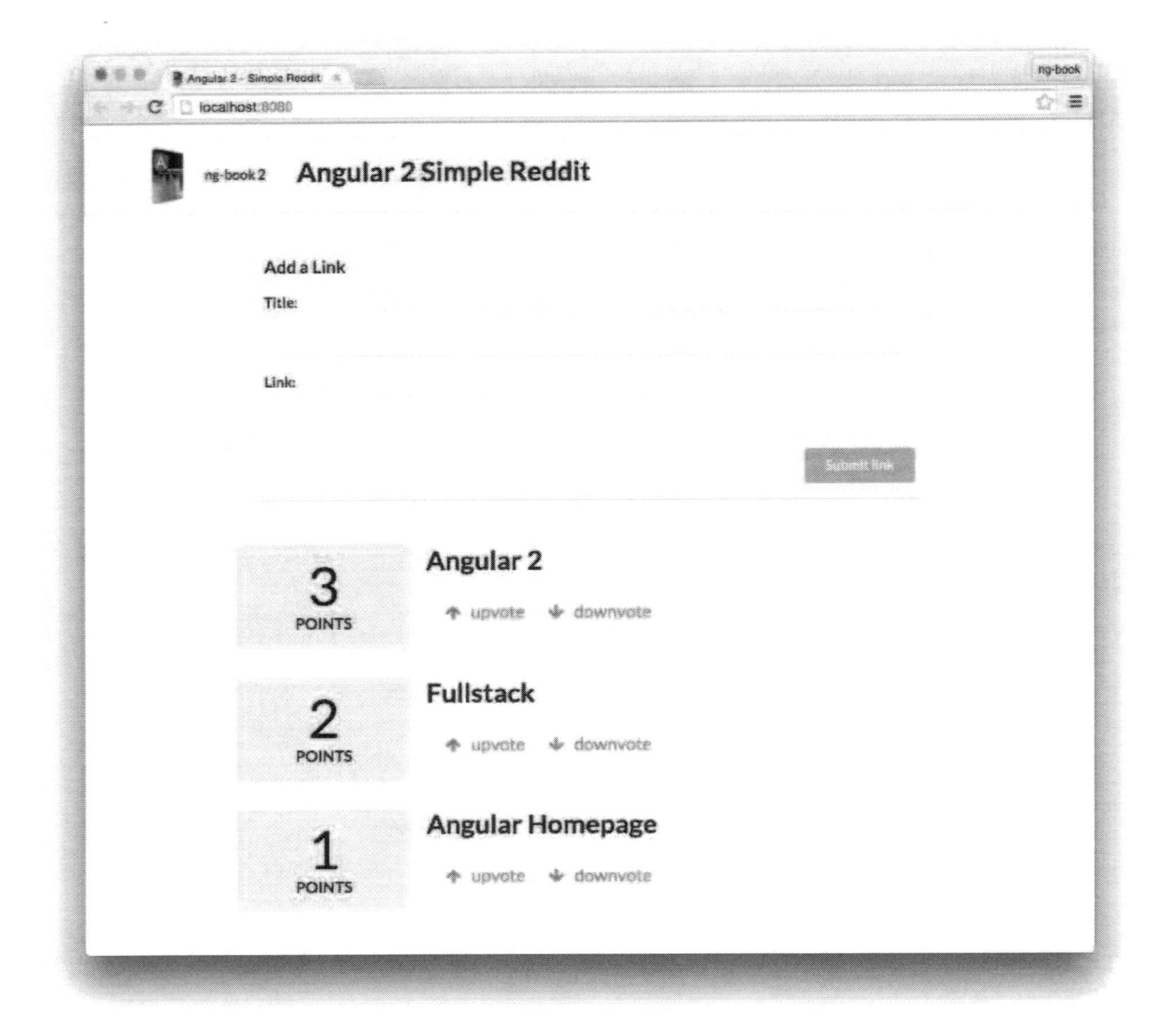


**Multiple articles being rendered**

## Adding New `Article`s

Now we need to change `addArticle` to actually add new articles when the button is pressed. Change the `addArticle` method to match the following:

**code/first_app/angular2_reddit/src/app/app.component.ts**

```
20    addArticle(title: HTMLInputElement, link: HTMLInputElement): boolean {
21      console.log(`Adding article title: ${title.value} and link: ${link.value}`);
22      this.articles.push(new Article(title.value, link.value, 0));
23      title.value = '';
24      link.value = '';
25      return false;
26    }
```

This will:

1. create a new `Article` instance with the submitted title and URL
2. add it to the array of `Articles` and
3. clear the `input` field values

How are we clearing the `input` field values? Well, if you recall, `title` and `link` are `HTMLInputElement` *objects*. That means we can set their properties. When we change the `value` property , the `input` tag on our page changes.

After adding a new article in our input fields and clicking the **Submit Link** we will see the new article added!

# Finishing Touches

## Displaying the Article Domain

As a nice touch, let's add a hint next to the link that shows the domain where the user will be redirected to when the link is clicked.

Let's add a `domain` method to the `Article` class:

**code/first_app/angular2_reddit/src/app/article/article.model.ts**

```
20    domain(): string {
21      try {
22        const link: string = this.link.split('//')[1];
23        return link.split('/')[0];
24      } catch (err) {
25        return null;
26      }
27    }
```

Let's add a call to this function on the `ArticleComponent`'s template:

```
1  <div class="twelve wide column">
2    <a class="ui large header" href="{{ article.link }}">
3      {{ article.title }}
4    </a>
5    <!-- right here -->
6    <div class="meta">({{ article.domain() }})</div>
7    <ul class="ui big horizontal list voters">
8      <li class="item">
9        <a href (click)="voteUp()">
```

And now when we reload the browser, we will see the domain name of each URL (note: URL must include *http://*).

## Re-sorting Based on Score

Clicking and voting on articles, we'll see that something doesn't feel quite right: our articles don't sort based on score! We definitely want to see the highest-rated items on top and the lower ranking ones sink to the bottom.

We're storing the `articles` in an `Array` in our `AppComponent` class, but that `Array` is unsorted. An easy way to handle this is to create a new method `sortedArticles` on `AppComponent`:

**code/first_app/angular2_reddit/src/app/app.component.ts**

```
28    sortedArticles(): Article[] {
29      return this.articles.sort((a: Article, b: Article) => b.votes - a.votes);
30    }
```

In our `ngFor` we can iterate over `sortedArticles()` (instead of `articles` directly):

app.component.html



```
1 <div class="ui grid posts">
2   <app-article
3     *ngFor="let article of sortedArticles()"
4     [article]="article">
5   </app-article>
6 </div>
```

## Full Code Listing

We've been exploring many small pieces of code for this chapter. You can find all of the files and the complete TypeScript code for our app in the example code download included with this book.

## Wrapping Up

We did it! We've created our first Angular 2 App. That wasn't so bad, was it? There's lots more to learn: understanding data flow, making AJAX requests, built-in directives, routing, manipulating the DOM etc.

But for now, bask in our success! Much of writing Angular apps is just as we did above:

1. Split your app into components
2. Create the views
3. Define your models
4. Display your models
5. Add interaction

In the future chapters of this book we'll cover everything you need to write sophisticated apps with Angular.

## Getting Help

Did you have any trouble with this chapter? Did you find a bug or have trouble getting the code running? We'd love to hear from you!

- Come join our (free!) community and chat with us on Gitter[20]
- Email us directly at us@fullstack.io[21]

Onward!

[20]https://gitter.im/ng-book/ng-book
[21]mailto:us@fullstack.io

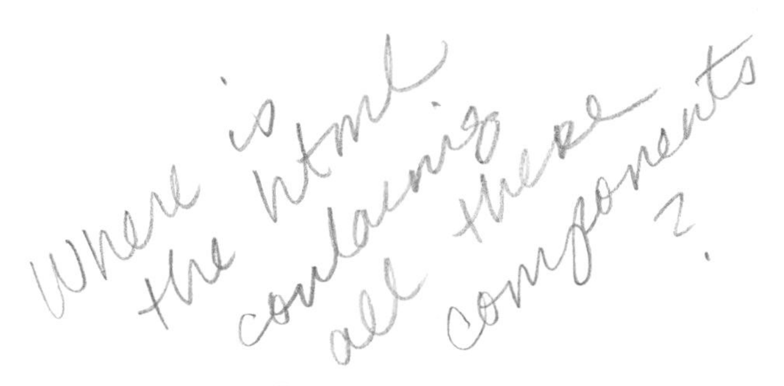

Where is the html containing all these components?


# How Angular Works

In this chapter, we're going to talk about the high-level concepts of Angular 2. We're going to take a step back so that we can see how all the pieces fit together.

> If you've used Angular 1, you'll notice that Angular 2 has a new mental-model for building applications. Don't panic! As Angular 1 users we've found Angular 2 to be both straightforward and familiar. A little later in this book we're going to talk specifically about how to convert your Angular 1 apps to Angular 2.

In the chapters that follow, we'll be taking a deep dive into each concept, but here we're just going to give an overview and explain the foundational ideas.

The first big idea is that an Angular 2 application is made up of *Components*. One way to think of Components is a way to teach the browser new tags. If you have an Angular 1 background, Components are analogous to *directives* in Angular 1 (it turns out, Angular 2 has directives too, but we'll talk more about this distinction later on).

However, Angular 2 Components have some significant advantages over Angular 1 directives and we'll talk about that below. First, let's start at the top: the Application.

## Application

An Angular 2 Application is nothing more than a tree of Components.

At the root of that tree, the top level Component is the application itself. And that's what the browser will render when "booting" (a.k.a *bootstrapping*) the app.

One of the great things about Components is that they're **composable**. This means that we can build up larger Components from smaller ones. The Application is simply a Component that renders other Components.

Because Components are structured in a parent/child tree, when each Component renders, it recursively renders its children Components.

For example, let's create a simple inventory management application that is represented by the following page mockup:

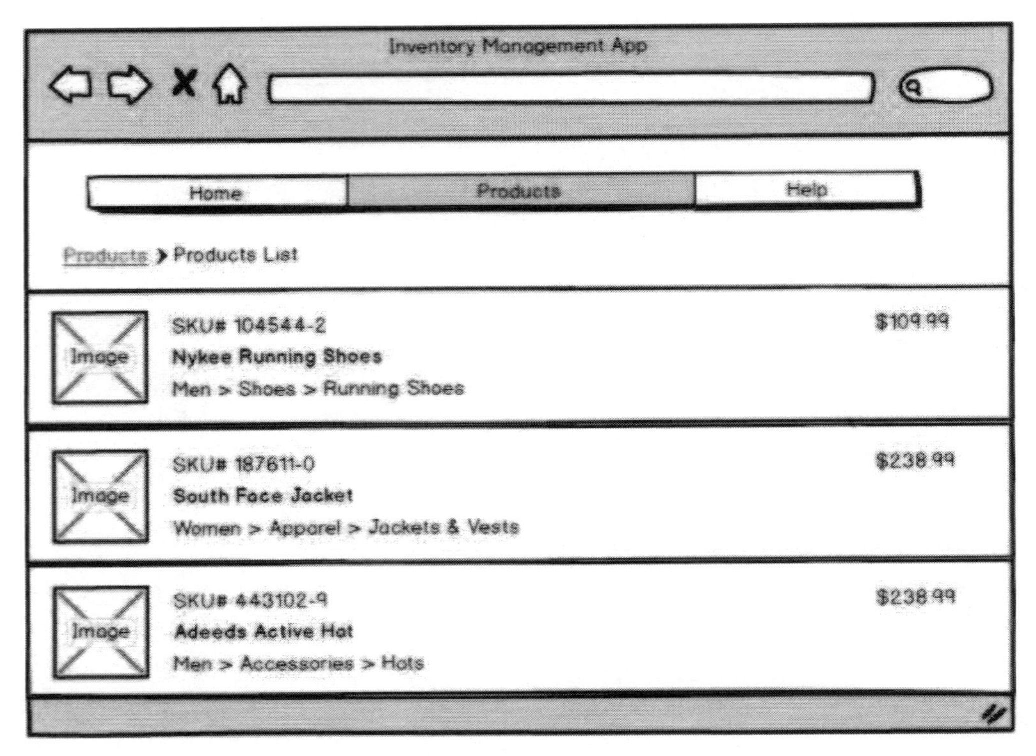


**Inventory Management App**

Given this mockup, to write this application the first thing we want to do is split it into components.

In this example, we could group the page into three high level components

1. The Navigation Component
2. The Breadcrumbs Component
3. The Product Info Component

## The Navigation Component

This component would render the navigation section. This would allow the user to visit other areas of the application.

**Navigation Component**

## The Breadcrumbs Component

This would render a hierarchical representation of where in the application the user currently is.

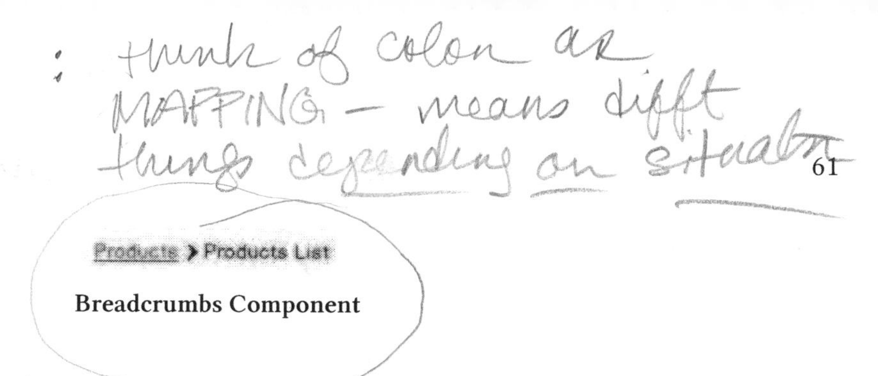


Breadcrumbs Component

## The Product List Component

The Products List component would be a representation of a collection of products.

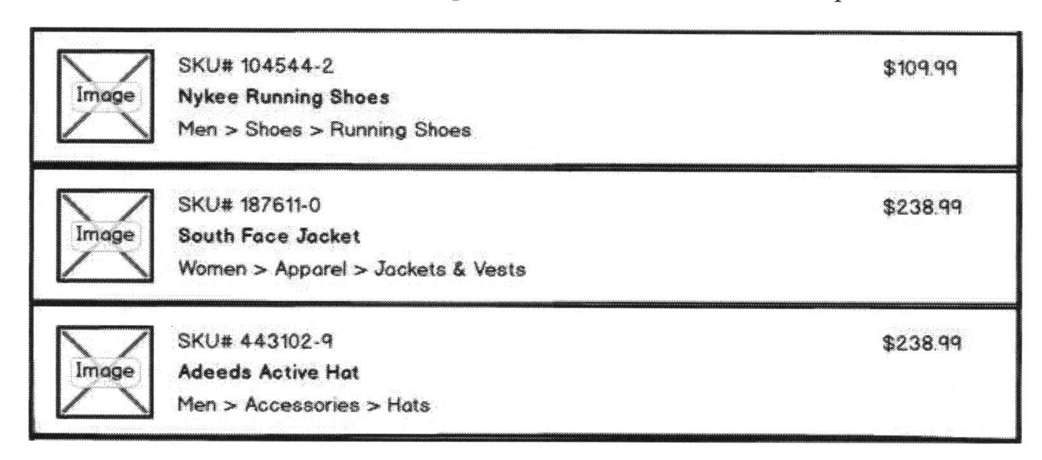


**Product List Component**

Breaking this component down into the next level of smaller components, we could say that the Product List is composed of multiple Product Rows.

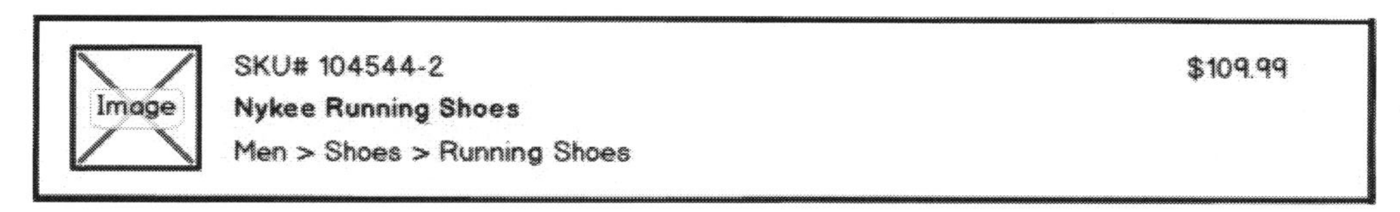


**Product Row Component**

And of course, we could continue one step further, breaking each Product Row into smaller pieces:

- the **Product Image** component would be responsible for rendering a product image, given its image name
- the **Product Department** component would render the department tree, like *Men > Shoes > Running Shoes*
- the **Price Display** component would render the price. Imagine that our implementation customizes the pricing if the user is logged in to include system-wide tier discounts or include shipping for instance. We could implement all this behavior into this component.

Finally, putting it all together into a tree representation, we end up with the following diagram:

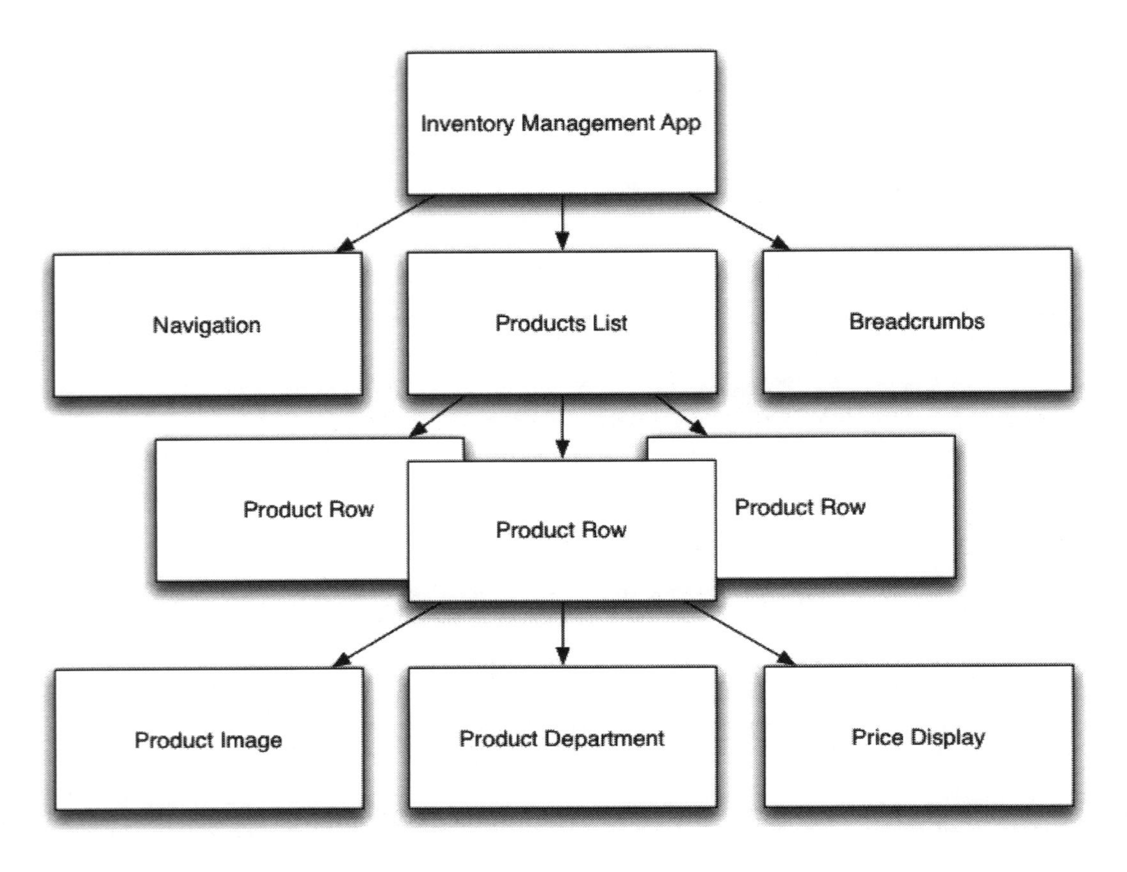


**App Tree Diagram**

At the top we see **Inventory Management App**: that's our application.

Under the application we have the Navigation, the Breadcrumb and the Products List components.

The Products List component has Product Rows, one for each product.

And the Product Row uses three components itself: one for the image, the department, and the price.

Let's work together to build this application.

You can find the full code listing for this chapter in the downloads under `how_angular_-works/inventory_app`.

Here's a screenshot of what our app will look like when we're done:

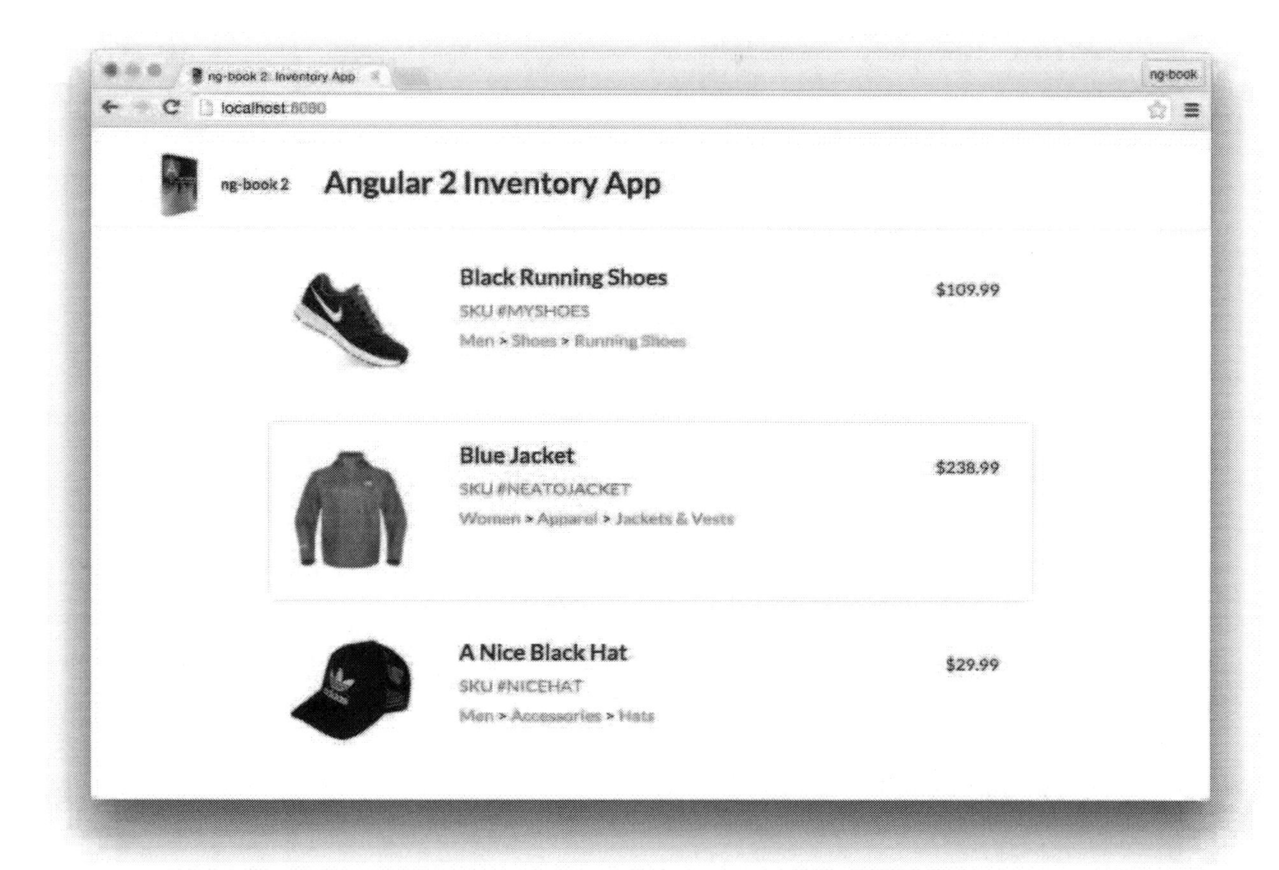


**Completed Inventory App**

# Product Model

One of the key things to realize about Angular is that it **doesn't prescribe a particular model library**.

Angular is flexible enough to be support many different kinds of models (and data architectures). However, this means the choice is left to you as the user to determine how to implement these things.

We'll have **a lot** to say about data architectures in future chapters. For now, though, we're going to have our models be plain JavaScript objects.

**code/how_angular_works/inventory_app/app.ts**

```
18  /**
19   * Provides a `Product` object
20   */
21  class Product {
22    constructor(
23      public sku: string,
24      public name: string,
25      public imageUrl: string,
26      public department: string[],
27      public price: number) {
28    }
29  }
```

If you're new to ES6/TypeScript this syntax might be a bit unfamiliar.

We're creating a new `Product` class and the `constructor` takes 5 arguments. When we write `public sku: string`, we're saying two things:

- there is a `public` variable on instances of this class called `sku`
- `sku` is of type `string`.

If you're already familiar with JavaScript, you can quickly catch up on some of the differences, including the `public constructor` shorthand, here at learnxinyminutes[22]

This `Product` class doesn't have any dependencies on Angular, it's just a model that we'll use in our app.

## Components

As we mentioned before, Components are the fundamental building block of Angular 2 applications. The "application" itself is just the top-level Component. Then we break our application into smaller child Components.

TIP: When building a new Angular application, mockup the design and then break it down into Components.

We'll be using Components a lot, so it's worth looking at them more closely.

Each components is composed of three parts:

[22] https://learnxinyminutes.com/docs/typescript/

- Component *Decorator*
- A View
- A Controller

To illustrate the key concepts we need to understand about components, we'll start with the top level Inventory App and then focus on the **Products List** and child components:

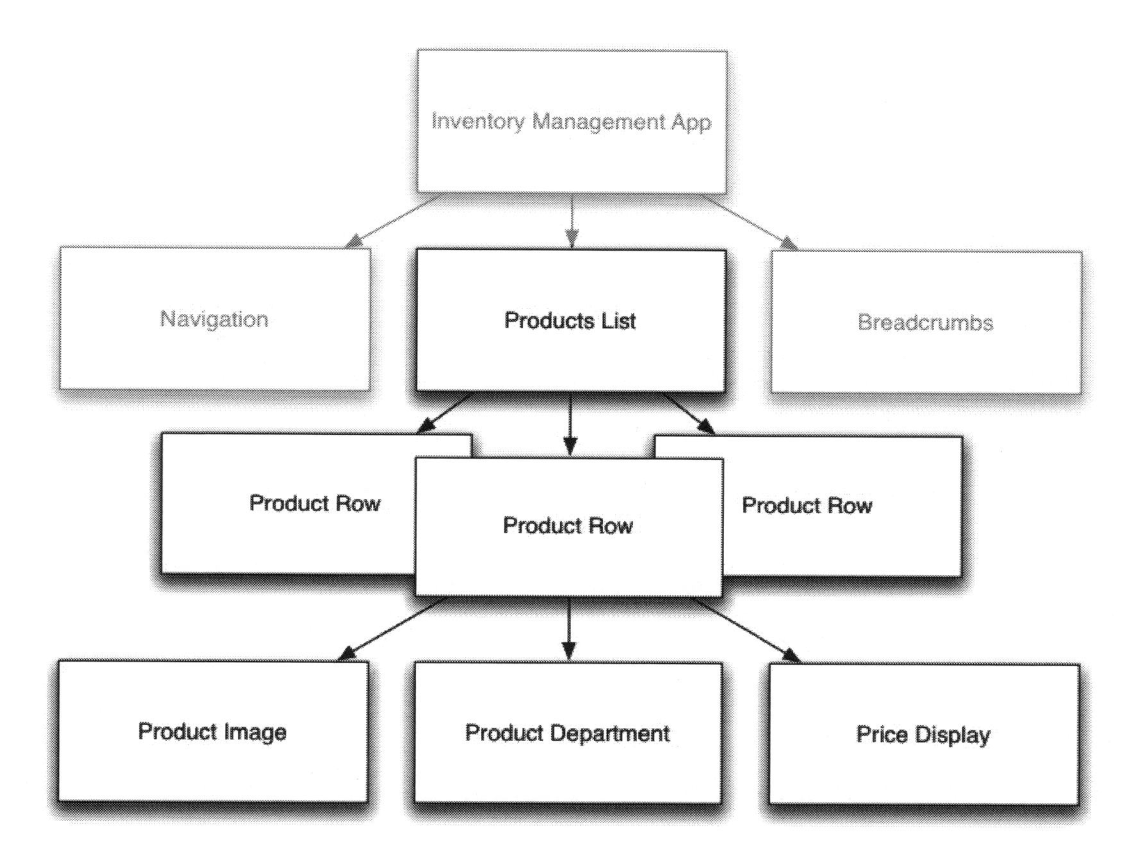


**Products List Component**

Here's what a basic, top-level `InventoryApp` looks like:

```
1  @Component({
2    selector: 'inventory-app',
3    template: `
4    <div class="inventory-app">
5       (Products will go here soon)
6    </div>
7      `
8  })
9  class InventoryApp {
```

```
10     // Inventory logic here
11 }
12
13 // module boot here...
```

If you've been using Angular 1 the syntax might look pretty foreign! But the ideas are pretty similar, so let's take them step by step:

The `@Component` is called a **decorator**. It adds metadata to the class that follows it (`InventoryApp`). The `@Component` annotation specifies:

- a `selector`, which tells Angular what element to match
- a `template`, which defines the view

The Component **controller** is defined by a `class`, the `InventoryApp` class, in this case.

Let's take a look into each part now in more detail.

## Component Decorator

The `@Component` decorator is where you configure your component. Primarily, `@Component` will configure how the outside world will interact with your component.

There are lots of options available to configure a component (many of which we cover in the Advanced Components Chapter). In this chapter we're just going to touch on some of the basics.

### Component `selector`

With the `selector` key, you indicate how your component will be recognized when rendering HTML templates. The idea is similar to CSS or XPath selectors. The `selector` is a way to define what elements in the HTML will match this component. In this case, by saying `selector: 'inventory-app'`, we're saying that in our HTML we want to match the `inventory-app` tag, that is, we're defining a new tag that has new functionality whenever we use it. E.g. when we put this in our HTML:

```
1 <inventory-app></inventory-app>
```

Angular will use the `InventoryApp` component to implement the functionality.

Alternatively, with this selector, we can also use a regular `div` and specify the component as an attribute:

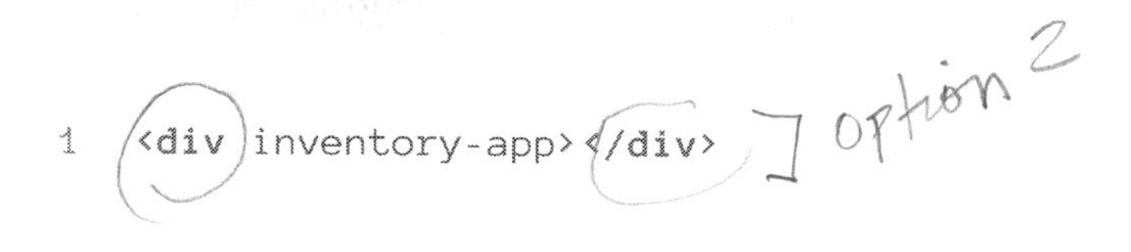

```
<div inventory-app></div>
```

## Component `template`

The view is the visual part of the component. By using the `template` option on `@Component`, we declare the HTML template that the component will have.

```
@Component({
  selector: 'inventory-app',
  template: `
  <div class="inventory-app">
     (Products will go here soon)
  </div>
  `
})
```

For this `template`, notice that we're using TypeScript's backtick multi-line string syntax. Our template so far is pretty sparse: just a `div` with some placeholder text.

We could use the option `templateUrl` instead of `template` if we wanted to put the template in a separate file.

## Adding A `Product`

Our app isn't very interesting without `Products` to view. Let's add some now.

We can create a new `Product` like this:

```
let newProduct = new Product(
      'NICEHAT',                                 // sku
      'A Nice Black Hat',                        // name
      '/resources/images/products/black-hat.jpg', // imageUrl
      ['Men', 'Accessories', 'Hats'],            // department
      29.99);                                    // price
```

Our constructor for `Product` takes 5 arguments. We can create a new `Product` by using the `new` keyword.

> Normally, I probably wouldn't pass more than 5 arguments to a function. Another option here is to configure the `Product` class to take an Object in the constructor, then if we wouldn't have to remember the order of the arguments. That is, `Product` could be changed to do something like this:
>
> ```
> new Product({sku: "MYHAT", name: "A green hat"})
> ```
>
> But for now, a 5 argument constructor is fine.

We want to be able to show this `Product` in the view. In order to make properties accessible to our template **we add them as instance variables to the Component**.

For instance, if we want to access `newProduct` in our view we would write:

```
class InventoryApp {
  product: Product;

  constructor() {
    let newProduct = new Product(
          'NICEHAT',
          'A Nice Black Hat',
          '/resources/images/products/black-hat.jpg',
          ['Men', 'Accessories', 'Hats'],
          29.99);

    this.product = newProduct;
  }
}
```

or more concisely:

```
class InventoryApp {
  product: Product;

  constructor() {
    this.product = new Product(
          'NICEHAT',
          'A Nice Black Hat',
          '/resources/images/products/black-hat.jpg',
          ['Men', 'Accessories', 'Hats'],
          29.99);
  }
}
```

Notice that we did three things here:

1. **We added a `constructor`** - When Angular creates a new instance of this Component, it calls the `constructor` function. This is where we can put setup for this Component.
2. **We described an instance variable** - On `InventoryApp`, when we write: `product: Product`, we're specifying that the `InventoryApp` instances have a property `product` which is a `Product` object.
3. **We assigned a `Product` to `product`** - In the `constructor` we create an instance of `Product` and assigned it to the instance variable

## Viewing the `Product` with Template Binding

Now that we have `product` assigned, we can use that variable in our view. Let's change our template to the following:

```
@Component({
  selector: 'inventory-app',
  template: `
  <div class="inventory-app">
     <h1>{{ product.name }}</h1>
     <span>{{ product.sku }}</span>
  </div>
  `
})
```

Using the `{{…}}` syntax is called *template binding*. It tells the view we want to use the value of the expression inside the brackets at this location in our template.

So in this case, we have two bindings:

- `{{ product.name }}`
- `{{ product.sku }}`

The `product` variable comes from the instance variable `product` on our Component instance of `InventoryApp`.

What's neat about template binding is that the code inside the brackets is *an expression*. That means you can do things like this:

- `{{ count + 1 }}`
- `{{ myFunction(myArguments) }}`

In the first case, we're using an operator to change the displayed value of `count`. In the second case, we're able to replace the tags with the value of the function `myFunction(myArguments)`. Using template binding tags is the main way that you'll show data in your Angular applications.

## Adding More Products

We actually don't want to show only a single product in our app - we actually want to show a whole list of products. So let's change our `InventoryApp` to store an array of `Products` rather than a single `Product`:

```
class InventoryApp {
  products: Product[];

  constructor() {
    this.products = [];
  }
}
```

Notice that we've renamed the variable `product` to `products`, and we've changed the type to `Product[]`. The `[]` characters at the end mean we want `products` to be an `Array` of `Products`. We also could have written this as: `Array<Product>`.

Now that our `InventoryApp` holds an array of `Products`. Let's create some `Products` in the constructor:

**code/how_angular_works/inventory_app/app.ts**

```
class InventoryApp {
  products: Product[];

  constructor() {
    this.products = [
      new Product(
        'MYSHOES',
        'Black Running Shoes',
        '/resources/images/products/black-shoes.jpg',
        ['Men', 'Shoes', 'Running Shoes'],
        109.99),
      new Product(
        'NEATOJACKET',
        'Blue Jacket',
        '/resources/images/products/blue-jacket.jpg',
        ['Women', 'Apparel', 'Jackets & Vests'],
        238.99),
      new Product(
        'NICEHAT',
        'A Nice Black Hat',
        '/resources/images/products/black-hat.jpg',
```

```
196             ['Men', 'Accessories', 'Hats'],
197             29.99)
198         ];
199     }
```

This code will give us some `Products` to work with in our app.

## Selecting a `Product`

We want to support user interaction in our app. For instance, the user might *select* a particular product to view more information about the product, add it to the cart, etc.

Let's add some functionality here in our `InventoryApp` to handle what happens when a new `Product` is selected. To do that, let's define a new function, `productWasSelected`:

**code/how_angular_works/inventory_app/app.ts**

```
201   productWasSelected(product: Product): void {
202     console.log('Product clicked: ', product);
203   }
```

## Listing products using `<products-list>`

Now that we have our top-level `InventoryApp` component, we need to add a new component for rendering a list of products. In the next section we'll create the implementation of a `ProductsList` component that matches the selector `products-list`. Before we dive into the implementation details, here's how we will *use* this new component:

**code/how_angular_works/inventory_app/app.ts**

```
164 @Component({
165   selector: 'inventory-app',
166   template: `
167   <div class="inventory-app">
168     <products-list
169       [productList]="products"
170       (onProductSelected)="productWasSelected($event)">
171     </products-list>
172   </div>
173   `
174 })
175 class InventoryApp {
```

There's some new syntax and options here, so let's talk about each of them:

## Inputs and Outputs

When we use `products-list` we're using a key feature of Angular components: inputs and outputs:

```
1     <products-list
2       [productList]="products"                                   <!-- input -->
3       (onProductSelected)="productWasSelected($event)">          <!-- output -->
4     </products-list>
```

The `[squareBrackets]` pass inputs and the `(parenthesis)` handle outputs.

Data flows *in* to your component via *input bindings* and events flow *out* of your component through *output bindings*.

Think of the set of input + output bindings as defining the **public API** of your component.

### `[squareBrackets]` pass inputs

In Angular, you pass data into child components via *inputs*.

In our code where we show:

```
1     <products-list
2       [productList]="products"
```

We're using an *input* of the `ProductList` component.

It can be tricky to understand where `products/productList` are coming from. There are two sides to this attribute:

- `[productList]` (the left-hand side) and
- `"products"` (the right-hand side)

The left-hand side `[productList]` says we want to use the `productList` *input* of the `products-list` component

The right-hand side `"products"` says we want to send the *value of the expression* `products`. That is, the array `this.products` in the `InventoryApp` class.

> You might ask, "how would I know that `productList` is a valid input to the `products-list` component? The answer is: you'd read the docs for that component. The `inputs` (and `outputs`) are part of the "public API" of a component.
>
> You'd know the inputs for a component that you're using in the same way that you'd know what the arguments are for a function that you're using.

### `(parens)` handle outputs

In Angular, you send data out of components via *outputs*.

In our code where we show:

```
1      <products-list
2        ...
3        (onProductSelected)="productWasSelected($event)">
```

We're saying that we want to listen to the `onProductSelected` *output* from the `ProductsList` component.

That is:

- `(onProductSelected)`, the left-hand side is the name of the `output` we want to "listen" on
- `"productWasSelected"`, the right-hand side is the function we want to call when something new is on this output
- `$event` is a special variable here that represents the thing emitted on the output.

Now, we haven't talked about how to define inputs or outputs on our own components yet, but we will shortly when we define the `ProductsList` component.

### Full `InventoryApp` Listing

Here's the full code listing of our `InventoryApp` component:

**code/how_angular_works/inventory_app/app.ts**

```
164  @Component({
165    selector: 'inventory-app',
166    template: `
167    <div class="inventory-app">
168      <products-list
169        [productList]="products"
170        (onProductSelected)="productWasSelected($event)">
171      </products-list>
172    </div>
173    `
174  })
175  class InventoryApp {
176    products: Product[];
177
178    constructor() {
```

```
179     this.products = [
180       new Product(
181         'MYSHOES',
182         'Black Running Shoes',
183         '/resources/images/products/black-shoes.jpg',
184         ['Men', 'Shoes', 'Running Shoes'],
185         109.99),
186       new Product(
187         'NEATOJACKET',
188         'Blue Jacket',
189         '/resources/images/products/blue-jacket.jpg',
190         ['Women', 'Apparel', 'Jackets & Vests'],
191         238.99),
192       new Product(
193         'NICEHAT',
194         'A Nice Black Hat',
195         '/resources/images/products/black-hat.jpg',
196         ['Men', 'Accessories', 'Hats'],
197         29.99)
198       ];
199   }
200
201   productWasSelected(product: Product): void {
202     console.log('Product clicked: ', product);
203   }
204 }
```

## The `ProductsList` Component

Now that we have our top-level application component, let's write the `ProductsList` component, which will render a list of product rows.

We want to allow the user to select **one** `Product` and we want to keep track of which `Product` is the currently selected one. The `ProductsList` component is a great place to do this because it "knows" all of the `Products` at the same time.

Let's write the `ProductsList` Component in three steps:

- Configuring the `ProductsList` `@Component` options
- Writing the `ProductsList` controller class
- Writing the `ProductsList` view template

## Configuring the `ProductsList` `@Component` Options

Let's take a look at the `@Component` configuration for `ProductsList`:

**code/how_angular_works/inventory_app/app.ts**

```
/**
 * @ProductsList: A component for rendering all ProductRows and
 * storing the currently selected Product
 */
@Component({
  selector: 'products-list',
  inputs: ['productList'],
  outputs: ['onProductSelected'],
  template: `
```

We start our `ProductsList` Component with a familiar option: `selector`. This selector means we can place our `ProductsList` component with the tag `<products-list>`.

There are two new options though: `inputs` and `outputs`.

## Component `inputs`

With the `inputs` option, we're specifying the parameters we expect our component to receive. `inputs` takes an array of strings which specify the input keys.

When we specify that a Component takes an input, it is expected that the definition class **will have an instance variable** that will receive the value. For example, say we have the following code:

```
@Component({
  selector: 'my-component',
  inputs: ['name', 'age']
})
class MyComponent {
  name: string;
  age: number;
}
```

The `name` and `age` inputs map to the `name` and `age` properties on instances of the `MyComponent` class.

If we want to use `MyComponent` from another template, we write something like: `<my-component [name]="myName" [age]="myAge"></my-component>`.

Notice that the attribute `name` matches the input `name`, which in turn matches the `MyComponent` property `name`. They don't always have to match.

For instance, say we wanted our attribute key and instance property to differ. That is, we want to use our component like this:

```
<my-component [shortName]="myName" [oldAge]="myAge"></my-component>
```

To do this, we would change the format of the string in the `inputs` option:

```
@Component({
  selector: 'my-component',
  inputs: ['name: shortName', 'age: oldAge']
})
class MyComponent {
  name: string;
  age: number;
}
```

More generally, `inputs` strings can have the format `'componentProperty: exposedProperty'`.

For instance we could have a different component that looks like this:

```
@Component({
  //...
  inputs: ['name', 'age', 'enabled']
  //...
})
class MyComponent {
  name: string;
  age: number;
  enabled: boolean;
}
```

However, if we wanted to represent the exposed property `enabled` in my component as `isEnabled`, we could use the alternative notation, like this:

```
@Component({
  //...
  inputs: [
    'name: name',
    'age: age',
    'isEnabled: enabled'
  ]
  //...
})
class MyComponent {
  name: string;
```

```
12        age: number;
13        isEnabled: boolean;
14      }
```

And going a little further, since the only property that requires an explicit mapping is `enabled` to `isEnabled`, we could even simplify and write it like this:

```
1     @Component({
2       // ...
3       inputs: ['name', 'age', 'isEnabled: enabled']
4       // ...
5     })
6     class MyComponent {
7       name: string;
8       age: number;
9       isEnabled: boolean;
10    }
```

In the `inputs` array, when the strings are in the `key: value` format, each have a specific meaning:

- The **key** (`name`, `age` and `isEnabled`) represent how that incoming property will be **visible ("bound") in the controller**.
- The **value** (`name`, `age` and `enabled`) configures how the property is **visible to the outside world**.

### Passing `products` through via the `inputs`

If you recall, in our `InventoryApp`, we passed `products` to our `products-list` via the `[productList]` input:

**code/how_angular_works/inventory_app/app.ts**

```
161 /**
162  * @InventoryApp: the top-level component for our application
163  */
164 @Component({
165   selector: 'inventory-app',
166   template: `
167   <div class="inventory-app">
168     <products-list
169       [productList]="products"
170       (onProductSelected)="productWasSelected($event)">
```

```
171       </products-list>
172     </div>
173     `
174 })
175 class InventoryApp {
176   products: Product[];
177 
178   constructor() {
179     this.products = [
```

Hopefully this now makes a bit more sense: we're passing `this.products` in via an input on `ProductsList`.

## Component outputs

When you want to send data from your component to the outside world, you use *output bindings*.

Let's say a component we're writing has a button and we need to do something when that button is clicked.

The way to do this is by binding the *click* output of the button to a method declared on our component's controller. You do that using the `(output)="action"` notation.

Here's an example where we keep a counter and increment (or decrement) based on which button is pressed:

```
1  @Component({
2    selector: 'counter',
3    template: `
4      {{ value }}
5      <button (click)="increase()">Increase</button>
6      <button (click)="decrease()">Decrease</button>
7    `
8  })
9  class Counter {
10   value: number;
11 
12   constructor() {
13     this.value = 1;
14   }
15 
16   increase() {
17     this.value = this.value + 1;
```

```
18        return false;
19      }
20
21      decrease() {
22        this.value = this.value - 1;
23        return false;
24      }
25    }
```

In this example we're saying that every time the first button is clicked, we want the `increase()` method on our controller to be invoked. And, similarly, when the second button clicked, we want to call the `decrease()` method.

The parentheses attribute syntax looks like this: `(output)="action"`. In this case, the output we're listening for is `click` event on this button. There are many other built-in events you can listen to: `mousedown`, `mousemove`, `dbl-click`, etc.

In this example, the event is internal to the component. When creating our own components we can also expose "public events" (component `outputs`) that allow the component to talk to the outside world.

The key thing to understand here is that in a view, we can listen to an event by using the `(output)="action"` syntax.

## Emitting Custom Events

Let's say we want to create a component that emits a custom event, like `click` or `mousedown` above. To create a custom output event we do three things:

1. Specify `outputs` in the `@Component` configuration
2. Attach an `EventEmitter` to the output property
3. Emit an event from the `EventEmitter`, at the right time

Perhaps `EventEmitter` is unfamiliar to you. Don't panic! It's not too hard.

An `EventEmitter` is simply an object that helps you implement the Observer Pattern[23]. That is, it's an object that can maintain a list of subscribers and publish events to them. That's it.

Here's a short and sweet example of how you can use `EventEmitter`

```
1 let ee = new EventEmitter();
2 ee.subscribe((name: string) => console.log(`Hello ${name}`));
3 ee.emit("Nate");
4
5 // -> "Hello Nate"
```

When we assign an `EventEmitter` to an output *Angular automatically subscribes* for us. You don't need to do the subscription yourself (necessarily, though you can add your own subscriptions if you want to).

Here's a code example of how we write a component that has `outputs`:

```
1  @Component({
2    selector: 'single-component',
3    outputs: ['putRingOnIt'],
4    template: `
5      <button (click)="liked()">Like it?</button>
6    `
7  })
8  class SingleComponent {
9    putRingOnIt: EventEmitter<string>;
10
11   constructor() {
12     this.putRingOnIt = new EventEmitter();
13   }
14
15   liked(): void {
16     this.putRingOnIt.emit("oh oh oh");
17   }
18 }
```

Notice that we did all three steps: 1. specified `outputs`, 2. created an `EventEmitter` that we attached to the output property `putRingOnIt` and 3. Emitted an event when `liked` is called.

If we wanted to use this output in a parent component we could do something like this:

[23] https://en.wikipedia.org/wiki/Observer_pattern

```
@Component({
  selector: 'club',
  template: `
    <div>
      <single-component
        (putRingOnIt)="ringWasPlaced($event)"
        ></single-component>
    </div>
  `
})
class ClubComponent {
  ringWasPlaced(message: string) {
    console.log(`Put your hands up: ${message}`);
  }
}

// logged -> "Put your hands up: oh oh oh"
```

Again, notice that:

- `putRingOnIt` comes from the `outputs` of `SingleComponent`
- `ringWasPlaced` is a function on the `ClubComponent`
- `$event` contains the thing that was emitted, in this case a `string`

## Writing the `ProductsList` Controller Class

Back to our store example, our `ProductsList` controller class needs three instance variables:

- One to hold the list of Products (that come from the `productList` input)
- One to output events (that emit from the `onProductSelected` output)
- One to hold a reference to the currently selected product

Here's how we define those in code:

**code/how_angular_works/inventory_app/app.ts**

```
125 class ProductsList {
126   /**
127    * @input productList - the Product[] passed to us
128    */
129   productList: Product[];
130 
131   /**
132    * @output onProductSelected - outputs the current
133    *          Product whenever a new Product is selected
134    */
135   onProductSelected: EventEmitter<Product>;
136 
137   /**
138    * @property currentProduct - local state containing
139    *             the currently selected `Product`
140    */
141   private currentProduct: Product;
142 
143   constructor() {
144     this.onProductSelected = new EventEmitter();
145   }
```

Notice that our `productList` is an Array of `Products` - this comes in from the `inputs`.

`onProductSelected` is our output.

`currentProduct` is a property internal to `ProductsList`. You might also hear this being referred to as “local component state”. It’s only used here within the component.

## Writing the `ProductsList` View Template

Here’s the `template` for our `products-list` component:

**code/how_angular_works/inventory_app/app.ts**

```
114   template: `
115   <div class="ui items">
116     <product-row
117       *ngFor="let myProduct of productList"
118       [product]="myProduct"
119       (click)='clicked(myProduct)'
120       [class.selected]="isSelected(myProduct)">
121     </product-row>
122   </div>
123   `
```

Here we're using the `product-row` tag, which comes from the `ProductRow` component, which we'll define in a minute.

We're using `ngFor` to iterate over each `Product` in `productList`. We've talked about `ngFor` before in this book, but just as a reminder the `let thing of things` syntax says, "iterate over `things` and create a copy of this element for each item, and assign each item to the variable `thing`".

So in this case, we're iterating over the `Products` in `productList` and generating a local variable `myProduct` for each one.

Stylistically, I probably wouldn't call this variable `myProduct` in a real app. I'd probably just call it `product`, or even `p`. But I want to be explicit about what we're passing around, and so `myProduct` is slightly clearer.

The interesting thing to note about this `myProduct` variable is that we can now use it *even on the same tag*. As you can see, we do this on the following three lines.

The line that reads `[product]="myProduct"` says that we want to pass `myProduct` (the local variable) to the input `product` of the `product-row`. (We'll define this input when we define the `ProductRow` component below.)

The `(click)='clicked(myProduct)'` line describes what we want to do when this element is clicked. `click` is a built-in event that is triggered when the host element is clicked on. In this case, we want to call the component function `clicked` on `ProductsList` whenever this element is clicked on.

The line `[class.selected]="isSelected(myProduct)"` is a fun one: Angular allows us to set `classes` conditionally on an element using this syntax. This syntax says "add the CSS class `selected` if `isSelected(myProduct)` returns `true`." This is a really handy way for us to mark the currently selected product.

You may have noticed that we didn't define `clicked` nor `isSelected` yet, so let's do that now (in `ProductsList`):

**`clicked`**

**code/how_angular_works/inventory_app/app.ts**

```
147   clicked(product: Product): void {
148     this.currentProduct = product;
149     this.onProductSelected.emit(product);
150   }
```

This function does two things:

1. Set `this.currentProduct` to the `Product` that was passed in.
2. Emit the `Product` that was clicked on our output

#### `isSelected`

**code/how_angular_works/inventory_app/app.ts**

```
152   isSelected(product: Product): boolean {
153     if (!product || !this.currentProduct) {
154       return false;
155     }
156     return product.sku === this.currentProduct.sku;
157   }
```

This function accepts a `Product` and returns `true` if `product`'s `sku` matches the `currentProduct`'s `sku`. It returns `false` otherwise.

## The Full `ProductsList` Component

Here's the full code listing so we can see everything in context:

**code/how_angular_works/inventory_app/app.ts**

```
106 /**
107  * @ProductsList: A component for rendering all ProductRows and
108  * storing the currently selected Product
109  */
110 @Component({
111   selector: 'products-list',
112   inputs: ['productList'],
113   outputs: ['onProductSelected'],
114   template: `
115   <div class="ui items">
```

```
116       <product-row
117         *ngFor="let myProduct of productList"
118         [product]="myProduct"
119         (click)='clicked(myProduct)'
120         [class.selected]="isSelected(myProduct)">
121       </product-row>
122     </div>
123     `
124 })
125 class ProductsList {
126   /**
127    * @input productList - the Product[] passed to us
128    */
129   productList: Product[];
130 
131   /**
132    * @output onProductSelected - outputs the current
133    *          Product whenever a new Product is selected
134    */
135   onProductSelected: EventEmitter<Product>;
136 
137   /**
138    * @property currentProduct - local state containing
139    *             the currently selected `Product`
140    */
141   private currentProduct: Product;
142 
143   constructor() {
144     this.onProductSelected = new EventEmitter();
145   }
146 
147   clicked(product: Product): void {
148     this.currentProduct = product;
149     this.onProductSelected.emit(product);
150   }
151 
152   isSelected(product: Product): boolean {
153     if (!product || !this.currentProduct) {
154       return false;
155     }
156     return product.sku === this.currentProduct.sku;
157   }
```

```
158
159 }
```

## The `ProductRow` Component

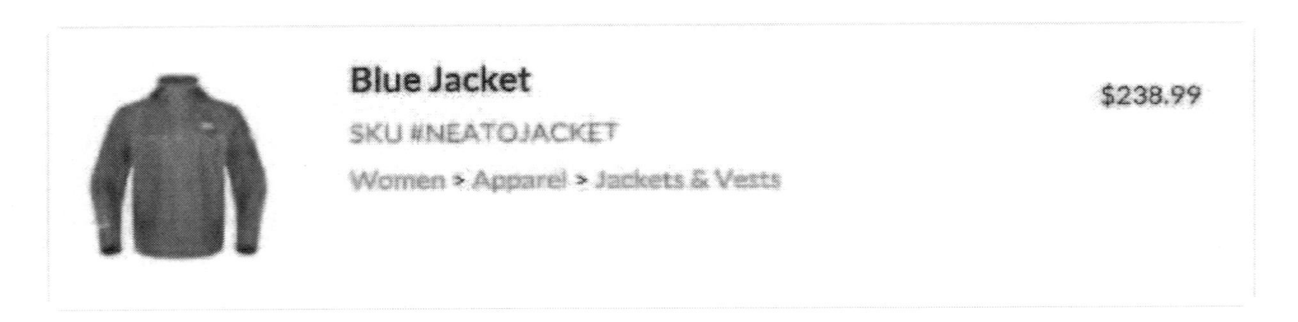


**A Selected Product Row Component**

Our `ProductRow` displays our `Product`. `ProductRow` will have its own template, but will also be split up into three smaller Components:

- `ProductImage` - for the image
- `ProductDepartment` - for the department "breadcrumbs"
- `PriceDisplay` - for showing the product's price

Here's a visual of the three Components that will be used within the `ProductRow`:

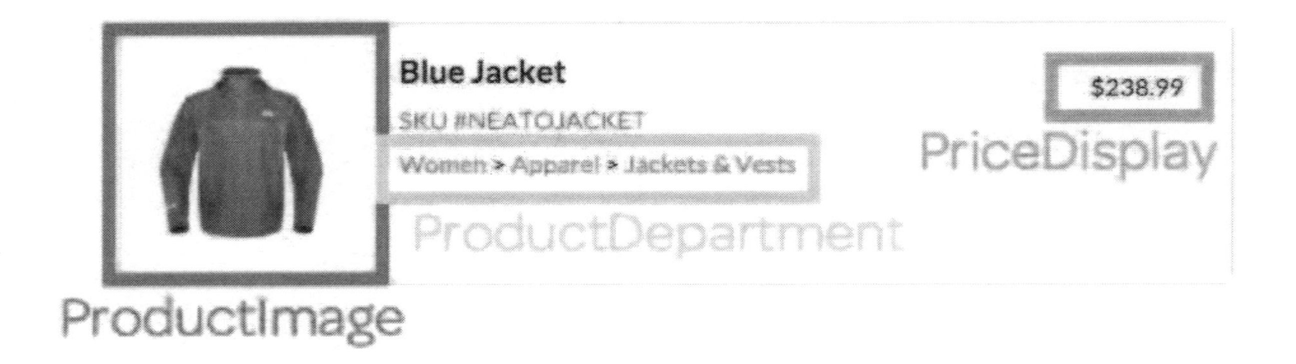


**ProductRow's Sub-components**

Let's take a look at the `ProductRow`'s Component configuration, definition class, and template:

### `ProductRow` Component Configuration

**code/how_angular_works/inventory_app/app.ts**

```
81  /**
82   * @ProductRow: A component for the view of single Product
83   */
84  @Component({
85    selector: 'product-row',
86    inputs: ['product'],
87    host: {'class': 'item'},
88    template: `
```

We start by defining the `selector` of `product-row`. We've seen this several times now - this defines that this component will match the tag `product-row`.

Next we define that this row takes an input of `product`. This will be the `Product` that was passed in from our parent Component.

The `host` option lets us set attributes on the host element. In this case, we're using the Semantic UI `item` class[24]. Here when we say `host: {'class': 'item'}` we're saying that we want to attach the CSS class "item" to the host element.

Using `host` is nice because it means we can configure our host element from *within* the component. This is great because otherwise we'd require the host element to specify the CSS tag and that is bad because we would then make assigning a CSS class part of the requirement to using the Component.

We'll talk about the `template` in a minute.

## `ProductRow` Component Definition Class

The `ProductRow` Component definition class is straightforward:

**code/how_angular_works/inventory_app/app.ts**

```
102  class ProductRow {
103    product: Product;
104  }
```

Here we're specifying that the `ProductRow` will have an instance variable `product`. Because we specified an input of `product`, when Angular creates an instance of this Component, it will automatically assign the `product` for us. We don't need to do it manually, and we don't need a `constructor`.

[24]http://semantic-ui.com/views/item.html

### `ProductRow template`

Now let's take a look at the `template`:

**code/how_angular_works/inventory_app/app.ts**

```
88   template: `
89   <product-image [product]="product"></product-image>
90   <div class="content">
91     <div class="header">{{ product.name }}</div>
92     <div class="meta">
93       <div class="product-sku">SKU #{{ product.sku }}</div>
94     </div>
95     <div class="description">
96       <product-department [product]="product"></product-department>
97     </div>
98   </div>
99   <price-display [price]="product.price"></price-display>
100  `
```

Our template doesn't have anything conceptually new.

In the first line we use our `product-image` directive and we pass our `product` to the `product` input of the `ProductImage` component. We use the `product-department` directive in the same way.

We use the `price-display` directive slightly differently in that we pass the `product.price`, instead of the `product` directly.

The rest of the template is standard HTML elements with custom CSS classes and some template bindings.

## `ProductRow` Full Listing

Here's the `ProductRow` component all together:

**code/how_angular_works/inventory_app/app.ts**

```
81  /**
82   * @ProductRow: A component for the view of single Product
83   */
84  @Component({
85    selector: 'product-row',
86    inputs: ['product'],
87    host: {'class': 'item'},
88    template: `
```

```
89     <product-image [product]="product"></product-image>
90     <div class="content">
91       <div class="header">{{ product.name }}</div>
92       <div class="meta">
93         <div class="product-sku">SKU #{{ product.sku }}</div>
94       </div>
95       <div class="description">
96         <product-department [product]="product"></product-department>
97       </div>
98     </div>
99     <price-display [price]="product.price"></price-display>
100    `
101  })
102  class ProductRow {
103    product: Product;
104  }
```

Now let's talk about the three components we used. They're pretty short.

## The `ProductImage` Component

**code/how_angular_works/inventory_app/app.ts**

```
31  /**
32   * @ProductImage: A component to show a single Product's image
33   */
34  @Component({
35    selector: 'product-image',
36    host: {class: 'ui small image'},
37    inputs: ['product'],
38    template: `
39    <img class="product-image" [src]="product.imageUrl">
40    `
41  })
42  class ProductImage {
43    product: Product;
44  }
```

The one thing to note here is in the `img` tag, notice how we use the `[src]` input to `img`.

We could have written this tag this way:

```
1 <!-- wrong, don't do it this way -->
2 <img src="{{ product.imageUrl }}">
```

Why is that wrong? Well, because in the case where your browser loads that template before Angular has run, your browser will try to load the image with the literal string `{{ product.imageUrl }}` and then it will get a 404 not found, which can show a broken image on your page until Angular runs.

By using the `[src]` attribute, we're telling Angular that we want to use the `[src]` *input* on this `img` tag. Angular will then replace the value of the `src` attribute once the expression is resolved.

## The `PriceDisplay` Component

Next, let's look at `PriceDisplay`:

**code/how_angular_works/inventory_app/app.ts**

```
66 /**
67  * @PriceDisplay: A component to show the price of a
68  * Product
69  */
70 @Component({
71   selector: 'price-display',
72   inputs: ['price'],
73   template: `
74   <div class="price-display">\${{ price }}</div>
75   `
76 })
77 class PriceDisplay {
78   price: number;
79 }
```

It's pretty straightforward, but one thing to note is that we're escaping the dollar sign `$` because this is a backtick string and the dollar sign is used for template variables.

## The `ProductDepartment` Component

**code/how_angular_works/inventory_app/app.ts**

```
46  /**
47   * @ProductDepartment: A component to show the breadcrumbs to a
48   * Product's department
49   */
50  @Component({
51    selector: 'product-department',
52    inputs: ['product'],
53    template: `
54    <div class="product-department">
55      <span *ngFor="let name of product.department; let i=index">
56        <a href="#">{{ name }}</a>
57        <span>{{i < (product.department.length-1) ? '>' : ''}}</span>
58      </span>
59    </div>
60    `
61  })
62  class ProductDepartment {
63    product: Product;
64  }
```

The thing to note about the `ProductDepartment` Component is the `ngFor` and the `span` tag.

Our `ngFor` loops over `product.department` and assigns each department string to `name`. The new part is the second expression that says: `let i=index`. This is how you get the iteration number out of `ngFor`.

In the `span` tag, we use the `i` variable to determine if we should show the greater-than `>` symbol.

The idea is that given a department, we want to show the department string like:

```
1  Women > Apparel > Jackets & Vests
```

The expression `{{i < (product.department.length-1) ? '>' : ''}}` says that we only want to use the `'>'` character if we're not the last department. On the last department just show an empty string `''`.

This format: `test ? valueIfTrue : valueIfFalse` is called the *ternary operator*.

## `NgModule` and Booting the App

The final thing we have to do is create the `NgModule` for this app and boot it up:

**code/how_angular_works/inventory_app/app.ts**

```
206 @NgModule({
207   declarations: [
208     InventoryApp,
209     ProductImage,
210     ProductDepartment,
211     PriceDisplay,
212     ProductRow,
213     ProductsList
214   ],
215   imports: [ BrowserModule ],
216   bootstrap: [ InventoryApp ]
217 })
218 class InventoryAppModule {}
```

Angular provides a *module* system that helps organize our code. Unlike Angular 1, where all directives are essentially globals, in Angular 2 you must specifically say which components you're going to be using in your app.

While it is a bit more configuration to do it this way, it's a lifesaver for larger apps.

When you create new components in Angular, in order to use them they must be *accessible* from the current module. That is, if we want to use the `ProductsList` component with the `products-list` selector in the `InventoryApp template`, then we need to make sure that the `InventoryApp`'s module either:

1. is in the same module as the `ProductsList` component or
2. The `InventoryApp`'s module `imports` the module that contains `ProductsList`

Remember **every** component you write must be declared in one `NgModule` before it can be used in a template.

In this case, we're putting `InventoryApp`, `ProductsList`, and **all** the other components for this app in one module. This is easy and it means they can all "see" each other.

Notice that we tell `NgModule` that we want to `bootstrap` with `InventoryApp`. This says that `InventoryApp` will be the top-level component.

Because we are writing a browser app, we also put `BrowserModule` in the `imports` of the `NgModule`.

To learn more about `NgModule` checkout the section on `NgModule` later in the book

## Booting the app

We're writing a browser app with no "ahead-of-time" compilation (more on this later in the book). So to bootstrap we do this:

**code/how_angular_works/inventory_app/app.ts**

```
219 platformBrowserDynamic().bootstrapModule(InventoryAppModule);
```

# The Completed Project

Now we have all the pieces we need for the working project!

Here's what it will look like when we're done:

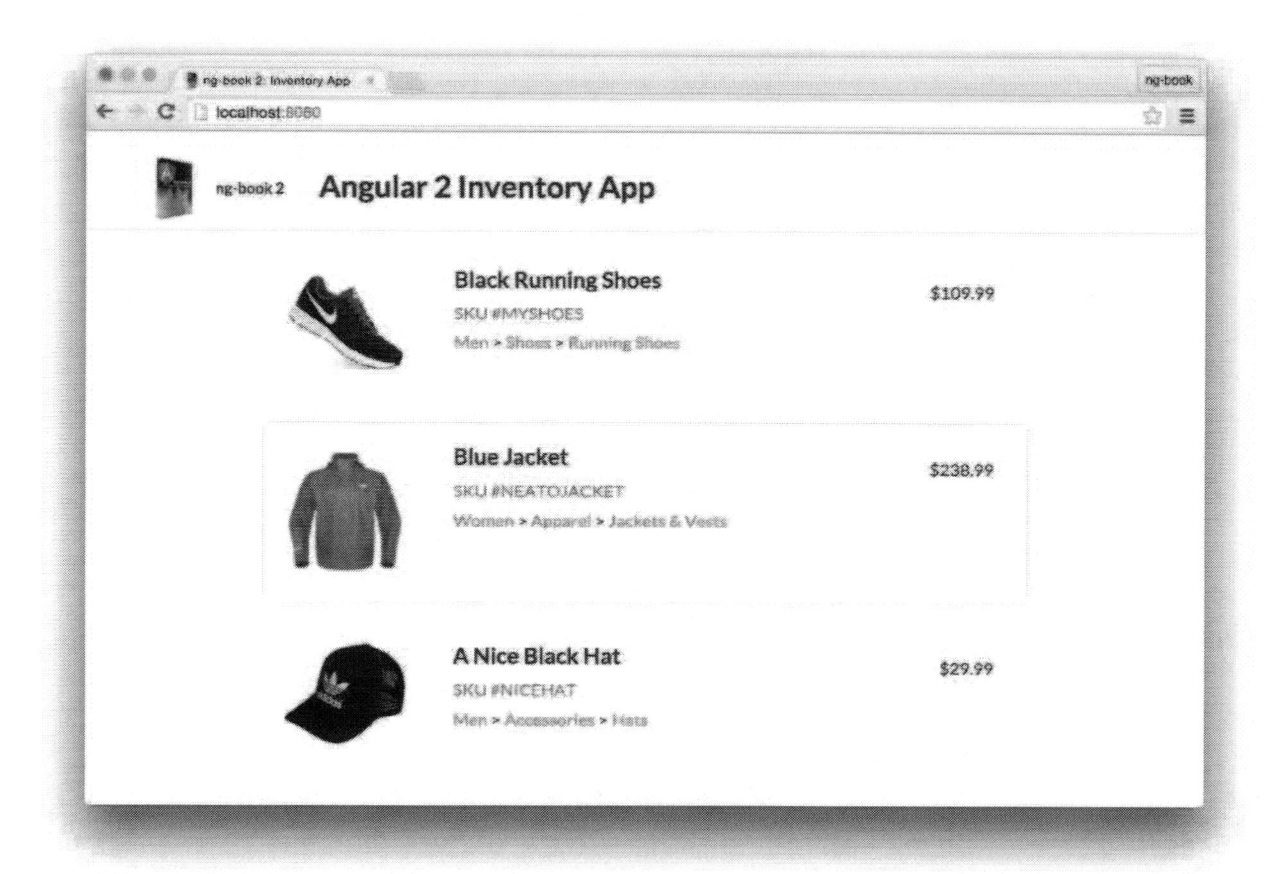


**Completed Inventory App**

You can run the code example in `how_angular_works/inventory_app`. See the `README` there.

Now you can click to select a particular product and have it render a nice purple outline when selected. If you add new `Products` in your code, you'll see them rendered.

# A Word on Data Architecture

You might be wondering at this point how we would manage the data flow if we started adding more functionality to this app.

For instance, say we wanted to add a shopping cart view and then we would add items to our cart. How could we implement this?

The only tools we've talked about are emitting output events. When we click add-to-cart do we simply bubble up an `addedToCart` event and handle at the root component? That feels a bit awkward.

Data architecture is a large topic with many opinions. Thankfully, Angular is flexible enough to handle a wide variety of data architectures, but that means that you have to decide for yourself which to use.

In Angular 1, the default option was two-way data binding. Two-way data binding is super easy to get started: your controllers have data, your forms manipulate that data directly, and your views show the data.

The problem with two-way data binding is that it often causes cascading effects throughout your application and makes it really difficult to trace data flow as your project grows.

Another problem with two-way data binding is that because you're passing data down through components it often forces your "data layout tree" to match your "dom view tree". In practice, these two things should really be separate.

One way you might handle this scenario would be to create a `ShoppingCartService`, which would be a singleton that would hold the list of the current items in the cart. This service could notify any interested objects when an item in the cart changes.

The idea is easy enough, but in practice there are a lot of details to be worked out.

The recommended way in Angular 2, and in many modern web frameworks (such as React), is to adopt a pattern of **one-way data binding**. That is, your data flows only **down** through components. If you need to make changes, you emit events that cause changes to happen "at the top" which then trickle down.

One-way data binding can seem like it adds some overhead in the beginning but it saves *a lot* of complication around change detection and it makes your systems easier to reason about.

Thankfully there are two major contenders for managing your data architecture:

1. Use an Observables-based architecture like RxJS
2. Use a Flux-based architecture

Later in this book we'll talk about how to implement a scalable data architecture for your app. For now, bask in the joy of your new Component-based application!

# Built-in Directives

## Introduction

Angular 2 provides a number of built-in directives. In this chapter, we're going to cover each built-in directive and show you examples of how to use them.

The built-in directives are imported and made available to your components automatically, so you don't need to inject them as a directive like you would do with your own components.

### NgIf

The `ngIf` directive is used when you want to display or hide an element based on a condition. The condition is determined by the result of the *expression* that you pass into the directive.

If the result of the expression returns a false value, the element will be removed from the DOM.

Some examples are:

```
<div *ngIf="false"></div>             <!-- never displayed -->
<div *ngIf="a > b"></div>             <!-- displayed if a is more than b -->
<div *ngIf="str == 'yes'"></div>      <!-- displayed if str holds the string "yes" -\
->
<div *ngIf="myFunc()"></div>          <!-- displayed if myFunc returns a true value \
-->
```

If you have experience with Angular 1, you probably used `ngIf` directive before. You can think of the Angular 2 version as a direct substitute. On the other hand, Angular 2 offers no built-in alternative for `ng-show`. So, if your goal is to just change the CSS visibility of an element, you should look into either the `ng_style` or the `class` directives, described later in this chapter.

### NgSwitch

Sometimes you need to render different elements depending on a given condition.

When you run into this situation, you could use `ngIf` several times like this:

```
1 <div class="container">
2   <div *ngIf="myVar == 'A'">Var is A</div>
3   <div *ngIf="myVar == 'B'">Var is B</div>
4   <div *ngIf="myVar != 'A' && myVar != 'B'">Var is something else</div>
5 </div>
```

But as you can see, the scenario where `myVar` is neither `A` nor `B` is pretty verbose, all we're really trying to express is an `else`. And as we add more values the last `ngIf` condition will become more complex.

To illustrate this growth in complexity, let's say we wanted to handle a new hypothetical `C` value.

In order to do that, we'd have to not only add the new element with `ngIf`, but also change the last case:

```
1 <div class="container">
2   <div *ngIf="myVar == 'A'">Var is A</div>
3   <div *ngIf="myVar == 'B'">Var is B</div>
4   <div *ngIf="myVar == 'C'">Var is C</div>
5   <div *ngIf="myVar != 'A' && myVar != 'B' && myVar != 'C'">Var is something els\
6 e</div>
7 </div>
```

For cases like this, Angular 2 introduces the `ngSwitch` directive.

If you're familiar with the `switch` statement then you'll feel very at home.

The idea behind this directive is the same: allow a single evaluation of an expression, and then display nested elements based on the value that resulted from that evaluation.

Once we have the result then we can:

- Describe the known results, using the `ngSwitchCase` directive
- Handle all the other unknown cases with `ngSwitchDefault`

Let's rewrite our example using this new set of directives:

```
1 <div class="container" [ngSwitch]="myVar">
2   <div *ngSwitchCase="'A'">Var is A</div>
3   <div *ngSwitchCase="'B'">Var is B</div>
4   <div *ngSwitchDefault>Var is something else</div>
5 </div>
```

Then if we want to handle the new value `C` we insert a single line:

```
1  <div class="container" [ngSwitch]="myVar">
2    <div *ngSwitchCase="'A'">Var is A</div>
3    <div *ngSwitchCase="'B'">Var is B</div>
4    <div *ngSwitchCase="'C'">Var is C</div>
5    <div *ngSwitchDefault>Var is something else</div>
6  </div>
```

And we don't have to touch the default (i.e. *fallback*) condition.

Having the `ngSwitchDefault` element is optional. If we leave it out, nothing will be rendered when `myVar` fails to match any of the expected values.

You can also declare the same `*ngSwitchCase` value for different elements, so you're not limited to matching only a single time. Here's an example:

**code/built_in_directives/app/ts/ng_switch/ng_switch.ts**

```
7    template: `
8      <h4 class="ui horizontal divider header">
9        Current choice is {{ choice }}
10     </h4>
11
12     <div class="ui raised segment">
13       <ul [ngSwitch]="choice">
14         <li *ngSwitchCase="1">First choice</li>
15         <li *ngSwitchCase="2">Second choice</li>
16         <li *ngSwitchCase="3">Third choice</li>
17         <li *ngSwitchCase="4">Fourth choice</li>
18         <li *ngSwitchCase="2">Second choice, again</li>
19         <li *ngSwitchDefault>Default choice</li>
20       </ul>
21     </div>
22
23     <div style="margin-top: 20px;">
24       <button class="ui primary button" (click)="nextChoice()">
25         Next choice
26       </button>
27     </div>
28   `
```

In the example above when the `choice` is `2`, both the second and fifth `li`s will be rendered.

## NgStyle

With the `NgStyle` directive, you can set a given DOM element CSS properties from Angular expressions.

The simplest way to use this directive is by doing `[style.<cssproperty>]="value"`. For example:

**code/built_in_directives/app/ts/ng_style/ng_style.ts**

```
12      <div [style.background-color]="'yellow'">
13        Uses fixed yellow background
14      </div>
```

This snippet is using the `NgStyle` directive to set the `background-color` CSS property to the literal string `'yellow'`.

Another way to set fixed values is by using the `NgStyle` attribute and using key value pairs for each property you want to set, like this:

**code/built_in_directives/app/ts/ng_style/ng_style.ts**

```
20      <div [ngStyle]="{color: 'white', 'background-color': 'blue'}">
21        Uses fixed white text on blue background
22      </div>
```

> Notice that in the `ng-style` specification we have single quotes around `background-color` but not around color. Why is that? Well, the argument to `ng-style` is a Javascript object and `color` is a valid key, without quotes. With `background-color`, however, the dash character isn't allowed in an object key, unless it's a string so we have to quote it.
>
> Generally I'd leave out quoting as much as possible in object keys and only quote keys when we have to.

Here we are setting both the `color` and the `background-color` properties.

But the real power of the `NgStyle` directive comes with using dynamic values.

In our example, we are defining two input boxes with an apply settings button:

**code/built_in_directives/app/ts/ng_style/ng_style.ts**

```
63      <div class="ui input">
64        <input type="text" name="color" value="{{color}}" #colorinput>
65      </div>
66
67      <div class="ui input">
68        <input type="text" name="fontSize" value="{{fontSize}}" #fontinput>
69      </div>
70
71      <button class="ui primary button" (click)="apply(colorinput.value, fontinput\
72  .value)">
73        Apply settings
74      </button>
```

And then using their values to set the CSS properties for three elements.

On the first one, we're setting the font size based on the input value:

**code/built_in_directives/app/ts/ng_style/ng_style.ts**

```
28      <div>
29        <span [ngStyle]="{color: 'red'}" [style.font-size.px]="fontSize">
30          red text
31        </span>
32      </div>
```

It's important to note that we have to specify units where appropriate. For instance, it isn't valid CSS to set a `font-size` of `12` - we have to specify a unit such as `12px` or `1.2em`. Angular provides a handy syntax for specifying units: here we used the notation `[style.font-size.px]`.

The `.px` suffix indicates that we're setting the `font-size` property value in pixels. You could easily replace that by `[style.font-size.em]` to express the font size in ems or even in percentage using `[style.font-size.%]`.

The other two elements use the `#colorinput` to set the text and background colors:

**code/built_in_directives/app/ts/ng_style/ng_style.ts**

```
40      <h4 class="ui horizontal divider header">
41        ngStyle with object property from variable
42      </h4>
43
44      <div>
45        <span [ngStyle]="{color: color}">
46          {{ color }} text
47        </span>
48      </div>
49
50      <h4 class="ui horizontal divider header">
51        style from variable
52      </h4>
53
54      <div [style.background-color]="color"
55           style="color: white;">
56        {{ color }} background
57      </div>
```

This way, when we click the **Apply settings** button, we call a method that sets the new values:

**code/built_in_directives/app/ts/ng_style/ng_style.ts**

```
98    apply(color: string, fontSize: number) {
99      this.color = color;
100     this.fontSize = fontSize;
101   }
```

And with that, both the color and the font size will be applied to the elements using the `NgStyle` directive.

## NgClass

The `NgClass` directive, represented by a `ngClass` attribute in your HTML template, allows you to dynamically set and change the CSS classes for a given DOM element.

If you're coming from Angular 1, the `NgClass` directive will feel very similar to what `ngClass` used to do in Angular 1.

The first way to use this directive is by passing in an object literal. The object is expected to have the keys as the class names and the values should be a truthy/falsy value to indicate whether the class should be applied or not.

Let's assume we have a CSS class called `bordered` that adds a dashed black border to an element:

**code/built_in_directives/app/css/styles.scss**

```
11  .bordered {
12    border: 1px dashed black;
13    background-color: #eee;
14  }
```

Let's add two `div` elements: one always having the `bordered` class (and therefore always having the border) and another one never having it:

**code/built_in_directives/app/ts/ng_class/ng_class.ts**

```
9    <div [ngClass]="{bordered: false}">This is never bordered</div>
10   <div [ngClass]="{bordered: true}">This is always bordered</div>
```

As expected, this is how those two `divs` would be rendered :

**Simple class directive usage**

Of course, it's a lot more useful to use the `NgClass` directive to make class assignments dynamic.

To make it dynamic we add a variable as the value for the object value, like this:

**code/built_in_directives/app/ts/ng_class/ng_class.ts**

```
12   <div [ngClass]="{bordered: isBordered}">
13    Using object literal. Border {{ isBordered ? "ON" : "OFF" }}
14   </div>
```

Alternatively, we can define the object in our component:

**code/built_in_directives/app/ts/ng_class/ng_class.ts**

```
50 export class NgClassSampleApp {
51   isBordered: boolean;
52   classesObj: Object;
53   classList: string[];
```

And use the object directly:

**code/built_in_directives/app/ts/ng_class/ng_class.ts**

```
16   <div [ngClass]="classesObj">
17     Using object var. Border {{ classesObj.bordered ? "ON" : "OFF" }}
18   </div>
```

Again, be careful when you have class names that contains dashes, like `bordered-box`. JavaScript objects don't allow literal keys to have dashes. If you need to use them, you must make the key a string like this:

```
1 <div [ngClass]="{'bordered-box': false}">...</div>
```

We can also use a list of class names to specify which class names should be added to the element. For that, we can either pass in an array literal:

**code/built_in_directives/app/ts/ng_class/ng_class.ts**

```
37   <div class="base" [ngClass]="['blue', 'round']">
38     This will always have a blue background and
39     round corners
40   </div>
```

Or declare an array variable in our component:

```
1     this.classList = ['blue', 'round'];
```

And passing it in:

**code/built_in_directives/app/ts/ng_class/ng_class.ts**

```
42    <div class="base" [ngClass]="classList">
43      This is {{ classList.indexOf('blue') > -1 ? "" : "NOT" }} blue
44      and {{ classList.indexOf('round') > -1 ? "" : "NOT" }} round
45    </div>
```

In this last example, the `[ngClass]` assignment works alongside existing values assigned by the HTML `class` attribute.

The resulting classes added to the element will always be the set of the classes provided by usual `class` HTML attribute and the result of the evaluation of the `[class]` directive.

In this example:

**code/built_in_directives/app/ts/ng_class/ng_class.ts**

```
37    <div class="base" [ngClass]="['blue', 'round']">
38      This will always have a blue background and
39      round corners
40    </div>
```

The element will have all three classes: `base` from the `class` HTML attribute and also `blue` and `round` from the `[class]` assigment:

**Classes from both the attribute and directive**

## NgFor

The role of this directive is to **repeat a given DOM element** (or a collection of DOM elements), each time passing it a different value from an array.

This directive is the successor of ng1's `ng-repeat`.

The syntax is `*ngFor="let item of items"`.

- The `let item` syntax specifies a (template) variable that's receiving each element of the `items` array;
- The `items` is the collection of items from your controller.

To illustrate, we can take a look at the code example. We declare an array of cities on our component controller:

```
1 this.cities = ['Miami', 'Sao Paulo', 'New York'];
```

And then, in our template we can have the following HTML snippet:

**code/built_in_directives/app/ts/ng_for/ng_for.ts**

```
 8   <h4 class="ui horizontal divider header">
 9     Simple list of strings
10   </h4>
11 
12   <div class="ui list" *ngFor="let c of cities">
13     <div class="item">{{ c }}</div>
14   </div>
```

And it will render each city inside the `div` as you would expect:

**Result of the ng_for directive usage**

We can also iterate through an array of objects like these:

**code/built_in_directives/app/ts/ng_for/ng_for.ts**

```
72     this.people = [
73       { name: 'Anderson', age: 35, city: 'Sao Paulo' },
74       { name: 'John', age: 12, city: 'Miami' },
75       { name: 'Peter', age: 22, city: 'New York' }
76     ];
```

And then render a table based on each row of data:

**code/built_in_directives/app/ts/ng_for/ng_for.ts**

```
16   <h4 class="ui horizontal divider header">
17     List of objects
18   </h4>
19 
20   <table class="ui celled table">
21     <thead>
22       <tr>
23         <th>Name</th>
24         <th>Age</th>
25         <th>City</th>
```

```
26        </tr>
27      </thead>
28      <tr *ngFor="let p of people">
29        <td>{{ p.name }}</td>
30        <td>{{ p.age }}</td>
31        <td>{{ p.city }}</td>
32      </tr>
33    </table>
```

Getting the following result:

**Rendering array of objects**

We can also work with nested arrays. If we wanted to have the same table as above, broken down by city, we could easily declare a new array of objects:

**code/built_in_directives/app/ts/ng_for/ng_for.ts**

```
77      this.peopleByCity = [
78        {
79          city: 'Miami',
80          people: [
81            { name: 'John', age: 12 },
82            { name: 'Angel', age: 22 }
83          ]
84        },
85        {
86          city: 'Sao Paulo',
87          people: [
88            { name: 'Anderson', age: 35 },
89            { name: 'Felipe', age: 36 }
90          ]
91        }
92      ];
93    };
```

And then we could use `NgFor` to render one `h2` for each city :

**code/built_in_directives/app/ts/ng_for/ng_for.ts**

```
39    <div *ngFor="let item of peopleByCity">
40      <h2 class="ui header">{{ item.city }}</h2>
```

And use a nested directive to iterate through the people for a given city :

**code/built_in_directives/app/ts/ng_for/ng_for.ts**

```
42      <table class="ui celled table">
43        <thead>
44          <tr>
45            <th>Name</th>
46            <th>Age</th>
47          </tr>
48        </thead>
49        <tr *ngFor="let p of item.people">
50          <td>{{ p.name }}</td>
51          <td>{{ p.age }}</td>
52        </tr>
53      </table>
```

Resulting in the following template code:

**code/built_in_directives/app/ts/ng_for/ng_for.ts**

```
35    <h4 class="ui horizontal divider header">
36      Nested data
37    </h4>
38
39    <div *ngFor="let item of peopleByCity">
40      <h2 class="ui header">{{ item.city }}</h2>
41
42      <table class="ui celled table">
43        <thead>
44          <tr>
45            <th>Name</th>
46            <th>Age</th>
47          </tr>
48        </thead>
49        <tr *ngFor="let p of item.people">
50          <td>{{ p.name }}</td>
51          <td>{{ p.age }}</td>
```

```
52        </tr>
53      </table>
54    </div>
```

And it would render one table for each city:

Rendering nested arrays

## Getting an index

There are times that we need the index of each item when we're iterating an array.

We can get the index by appending the syntax `let idx = index` to the value of our `ngFor` directive, separated by a semi-colon. When we do this, ng2 will assign the current index into the variable we provide (in this case, the variable `idx`).

Note that, like JavaScript, the index is always zero based. So the index for first element is 0, 1 for the second and so on...

Making some changes to our first example, adding the `let num = index` snippet like below:

**code/built_in_directives/app/ts/ng_for/ng_for.ts**

```
60    <div class="ui list" *ngFor="let c of cities; let num = index">
61      <div class="item">{{ num+1 }} - {{ c }}</div>
62    </div>
```

It will add the position of the city before the name, like this:

Using an index

## NgNonBindable

We use `ngNonBindable` when we want tell Angular **not** to compile or bind a particular section of our page.

Let's say we want to render the literal text `{{ content }}` in our template. Normally that text will be *bound* to the value of the `content` variable because we're using the `{{ }}` template syntax.

So how can we render the exact text `{{ content }}`? We use the `ngNonBindable` directive.

Let's say we want to have a `div` that renders the contents of that `content` variable and right after we want to point that out by outputting *<- this is what {{ content }} rendered* next to the actual value of the variable.

To do that, here's the template we'd have to use:

**code/built_in_directives/app/ts/ng_non_bindable/ng_non_bindable.ts**

```
7    template: `
8    <div class='ngNonBindableDemo'>
9      <span class="bordered">{{ content }}</span>
10     <span class="pre" ngNonBindable>
11       &larr; This is what {{ content }} rendered
12     </span>
13   </div>
14   `
```

And with that `ngNonBindable` attribute, ng2 will not compile within that second `span`'s context, leaving it intact:

**Result of using ngNonBindable**

# Conclusion

Angular 2 has only a few core directives, but we can combine these simple pieces to create dynamic apps.

# Forms in Angular 2

## Forms are Crucial, Forms are Complex

Forms are probably the most crucial aspect of your web application. While we often get events from clicking on links or moving the mouse, it's through *forms* where we get the majority of our rich data input from users.

On the surface, forms seem straightforward: you make an `input` tag, the user fills it out, and hits submit. How hard could it be?

It turns out, forms can end up being really complex. Here's a few reasons why:

- Form inputs are meant to modify data, both on the page and the server
- Changes often need to be reflected elsewhere on the page
- Users have a lot of leeway in what they enter, so you need to validate values
- The UI needs to clearly state expectations and errors, if any
- Dependent fields can have complex logic
- We want to be able to test our forms, without relying on DOM selectors

Thankfully, Angular 2 has tools to help with all of these things.

- **`FormControls`** encapsulate the inputs in our forms and give us objects to work with them
- **`Validators`** give us the ability to validate inputs, any way we'd like
- **Observers** let us watch our form for changes and respond accordingly

In this chapter we're going to walk through building forms, step by step. We'll start with some simple forms and build up to more complicated logic.

## `FormControl`s and `FormGroup`s

The two fundamental objects in Angular 2 forms are `FormControl` and `FormGroup`.

### `FormControl`

A `FormControl` represents a single input field - it is the smallest unit of an Angular form.

`FormControls` encapsulate the field's value, and states such as being valid, dirty (changed), or has errors.

For instance, here's how we might use a `FormControl` in TypeScript:

```
// create a new FormControl with the value "Nate"
let nameControl = new FormControl("Nate");

let name = nameControl.value; // -> Nate

// now we can query this control for certain values:
nameControl.errors // -> StringMap<string, any> of errors
nameControl.dirty  // -> false
nameControl.valid  // -> true
// etc.
```

To build up forms we create `FormControl`s (and groups of `FormControl`s) and then attach metadata and logic to them.

Like many things in Angular, we have a class (`FormControl`, in this case) that we attach to the DOM with an attribute (`formControl`, in this case). For instance, we might have the following in our form:

```
<!-- part of some bigger form -->
<input type="text" [formControl]="name" />
```

This will create a new `FormControl` object within the context of our `form`. We'll talk more about how that works below.

### `FormGroup`

Most forms have more than one field, so we need a way to manage multiple `FormControl`s. If we wanted to check the validity of our form, it's cumbersome to iterate over an array of `FormControl`s and check each `FormControl` for validity. `FormGroups` solve this issue by providing a wrapper interface around a collection of `FormControl`s.

Here's how you create a `FormGroup`:

```
let personInfo = new FormGroup({
    firstName: new FormControl("Nate"),
    lastName: new FormControl("Murray"),
    zip: new FormControl("90210")
})
```

`FormGroup` and `FormControl` have a common ancestor (`AbstractControl`[25]). That means we can check the `status` or `value` of `personInfo` just as easily as a single `FormControl`:

[25] https://github.com/angular/angular/blob/master/modules/@angular/forms/src/model.ts

```
personInfo.value; // -> {
//   firstName: "Nate",
//   lastName: "Murray",
//   zip: "90210"
// }

// now we can query this control group for certain values, which have sensible
// values depending on the children FormControl's values:
personInfo.errors // -> StringMap<string, any> of errors
personInfo.dirty  // -> false
personInfo.valid  // -> true
// etc.
```

Notice that when we tried to get the `value` from the `FormGroup` we received an **object** with key-value pairs. This is a really handy way to get the full set of values from our form without having to iterate over each `FormControl` individually.

## Our First Form

There are lots of moving pieces to create a form, and several important ones we haven't touched on. Let's jump in to a full example and I'll explain each piece as we go along.

You can find the full code listing for this section in the code download under `forms/`

Here's a screenshot of the very first form we're going to build:

**Demo Form with Sku: Simple Version**

In our imaginary application we're creating an e-commerce-type site where we're listing products for sale. In this app we need to store the product's SKU, so let's create a simple form that takes the SKU as the only input field.

SKU is an abbreviation for "stockkeeping unit". It's a term for a unique id for a product that is going to be tracked in inventory. When we talk about a SKU, we're talking about a human-readable item ID.

Our form is super simple: we have a single input for `sku` (with a label) and a submit button.

Let's turn this form into a Component. If you recall, there are three parts to defining a component:

- Configure the `@Component()` annotation
- Create the template
- Implement custom functionality in the component definition class

Let's take these in turn:

## Loading the `FormsModule`

In order to use the new forms library we need to first make sure we import the forms library in our `NgModule`.

There are two ways of using forms in Angular and we'll talk about them both in this chapter: using `FormsModule` or using `ReactiveFormsModule`. Since we'll use both, we'll import them both into our module. To do this we do the following in our `app.ts` where we bootstrap the app:

```
 1  import {
 2    FormsModule,
 3    ReactiveFormsModule
 4  } from '@angular/forms';
 5
 6  // farther down...
 7
 8  @NgModule({
 9    declarations: [
10      FormsDemoApp,
11      DemoFormSku,
12      // ... our declarations here
13    ],
14    imports: [
15      BrowserModule,
16      FormsModule,          // <-- add this
17      ReactiveFormsModule   // <-- and this
18    ],
19    bootstrap: [ FormsDemoApp ]
20  })
21  class FormsDemoAppModule {}
```

This ensures that we're able to use the form directives in our views. At the risk of jumping ahead, the `FormsModule` gives us *template driven* directives such as:

- `ngModel` and
- `NgForm`

Whereas `ReactiveFormsModule` gives us directives like

- `formControl` and
- `ngFormGroup`

... and several more. We haven't talked about how to use these directives or what they do, but we will shortly. For now, just know that by importing `FormsModule` and `ReactiveFormsModule` into our `NgModule` means we can *use any of the directives in that list* in our view template or *inject any of their respective providers* into our components.

## Simple SKU Form: @Component Annotation

Now we can start creating our component:

**code/forms/app/forms/demo_form_sku.ts**

```
1 import { Component } from '@angular/core';
2
3 @Component({
4   selector: 'demo-form-sku',
```

Here we define a `selector` of `demo-form-sku`. If you recall, `selector` tells Angular what elements this component will bind to. In this case we can use this component by having a `demo-form-sku` tag like so:

```
1 <demo-form-sku></demo-form-sku>
```

## Simple SKU Form: `template`

Let's look at our `template`:

**code/forms/app/ts/forms/demo_form_sku.ts**

```
6     template: `
7     <div class="ui raised segment">
8       <h2 class="ui header">Demo Form: Sku</h2>
9       <form #f="ngForm"
10            (ngSubmit)="onSubmit(f.value)"
11            class="ui form">
12
13        <div class="field">
14          <label for="skuInput">SKU</label>
15          <input type="text"
16                 id="skuInput"
17                 placeholder="SKU"
18                 name="sku" ngModel>
19        </div>
20
21        <button type="submit" class="ui button">Submit</button>
22      </form>
23    </div>
24    `
```

#### `form` & `NgForm`

Now things get interesting: because we imported `FormsModule`, that makes `NgForm` available to our view. Remember that whenever we make directives available to our view, they will **get attached to any element that matches their `selector`**.

`NgForm` does something handy but **non-obvious**: it includes the `form` tag in its selector (instead of requiring you to explicitly add `ngForm` as an attribute). What this means is that if you import `FormsModule`, `NgForm` will get *automatically* attached to any `<form>` tags you have in your view. This is really useful but potentially confusing because it happens behind the scenes.

There are two important pieces of functionality that `NgForm` gives us:

1. A `FormGroup` named `ngForm`
2. A **`(ngSubmit)`** output

You can see that we use both of these in the `<form>` tag in our view:

**code/forms/app/ts/forms/demo_form_sku.ts**

```
9      <form #f="ngForm"
10           (ngSubmit)="onSubmit(f.value)"
```

First we have `#f="ngForm"`. The `#v=thing` syntax says that we want to create a local variable for this view.

Here we're creating an alias to `ngForm`, for this view, bound to the variable `#f`. Where did `ngForm` come from in the first place? It came from the `NgForm` directive.

And what type of object is `ngForm`? It is a `FormGroup`. That means we can use `f` as a `FormGroup` in our view. And that's exactly what we do in the `(ngSubmit)` output.

Astute readers might notice that I just said above that `NgForm` is automatically attached to `<form>` tags (because of the default `NgForm` selector), which means we don't have to add an `ngForm` attribute to use `NgForm`. But here we're putting `ngForm` in an attribute (value) tag. Is this a typo?

No, it's not a typo. If `ngForm` were the *key* of the attribute then we would be telling Angular that we want to use `NgForm` on this attribute. In this case, we're using `ngForm` as the *attribute* `when we're assigning a _reference_. That is, we're saying the value of the evaluated expression` ngForm `should be assigned to a local template variable` f'.

`ngForm` is already on this element and you can think of it as if we are "exporting" this `FormGroup` so that we can reference it elsewhere in our view.

We bind to the `ngSubmit` action of our form by using the syntax: `(ngSubmit)="onSubmit(f.value)"`.

- `(ngSubmit)` - comes from `NgForm`
- `onSubmit()` - will be implemented in our component definition class (below)
- `f.value` - `f` is the `FormGroup` that we specified above. And `.value` will return the key/value pairs of this `FormGroup`

Put it all together and that line says "when I submit the form, call `onSubmit` on my component instance, passing the value of the form as the arguments".

#### `input` & `NgModel`

Our `input` tag has a few things we should touch on before we talk about `NgModel`:

**code/forms/app/ts/forms/demo_form_sku.ts**

```
 9    <form #f="ngForm"
10          (ngSubmit)="onSubmit(f.value)"
11          class="ui form">
12
13     <div class="field">
14       <label for="skuInput">SKU</label>
15       <input type="text"
16              id="skuInput"
17              placeholder="SKU"
18              name="sku" ngModel>
19     </div>
```

- `class="ui form"` and `class="field"` - these two classes are totally optional. They come from the CSS framework Semantic UI[26]. I've added them in some of our examples just to give them a nice coat of CSS but they're not part of Angular.
- The `label` "`for`" attribute and the `input` "`id`" attribute are to match, as per W3C standard[27]
- We set a `placeholder` of "SKU", which is just a hint to the user for what this `input` should say when it is blank

The `NgModel` directive specifies a `selector` of `ngModel`. This means we can attach it to our `input` tag by adding this sort of attribute: `ngModel="whatever"`. In this case, we specify `ngModel` with no attribute value.

There are a couple of different ways to specify `ngModel` in your templates and this is the first. When we use `ngModel` with no attribute value we are specifying:

1. a *one-way* data binding
2. we want to create a `FormControl` on this form with the name `sku` (because of the `name` attribute on the `input` tag)

`NgModel` **creates a new `FormControl`** that is **automatically added** to the parent `FormGroup` (in this case, on the form) and then binds a DOM element to that new `FormControl`. That is, it sets up an association between the `input` tag in our view and the `FormControl` and the association is matched by a name, in this case `"sku"`.

[26]http://semantic-ui.com/
[27]http://www.w3.org/TR/WCAG20-TECHS/H44.html

**`NgModel` vs. `ngModel`**: what's the difference? Generally, when we use PascalCase, like `NgModel`, we're specifying the *class* and referring to the object as it's defined in code. The lower case (CamelCase), as in `ngModel`, comes from the `selector` of the directive and it's only used in the DOM / template.

It's also worth pointing out that `NgModel` and `FormControl` are separate objects. `NgModel` is the *directive* that you use in your view, whereas `FormControl` is the object used for representing the data and validations in your form.

Sometimes we want to do *two-way* binding with `ngModel` like we used to do in Angular 1. We'll look at how to do that towards the end of this chapter.

## Simple SKU Form: Component Definition Class

Now let's look at our class definition:

**code/forms/app/ts/forms/demo_form_sku.ts**

```
26 export class DemoFormSku {
27   onSubmit(form: any): void {
28     console.log('you submitted value:', form);
29   }
30 }
```

Here our class defines one function: `onSubmit`. This is the function that is called when the form is submitted. For now, we'll just `console.log` out the value that is passed in.

## Try it out!

Putting it all together, here's what our code listing looks like:

**code/forms/app/ts/forms/demo_form_sku.ts**

```
1 import { Component } from '@angular/core';
2
3 @Component({
4   selector: 'demo-form-sku',
5
6   template: `
7   <div class="ui raised segment">
8     <h2 class="ui header">Demo Form: Sku</h2>
```

```
 9       <form #f="ngForm"
10             (ngSubmit)="onSubmit(f.value)"
11             class="ui form">
12 
13         <div class="field">
14           <label for="skuInput">SKU</label>
15           <input type="text"
16                  id="skuInput"
17                  placeholder="SKU"
18                  name="sku" ngModel>
19         </div>
20 
21         <button type="submit" class="ui button">Submit</button>
22       </form>
23     </div>
24     `
25 })
26 export class DemoFormSku {
27   onSubmit(form: any): void {
28     console.log('you submitted value:', form);
29   }
30 }
```

If we try this out in our browser, here's what it looks like:

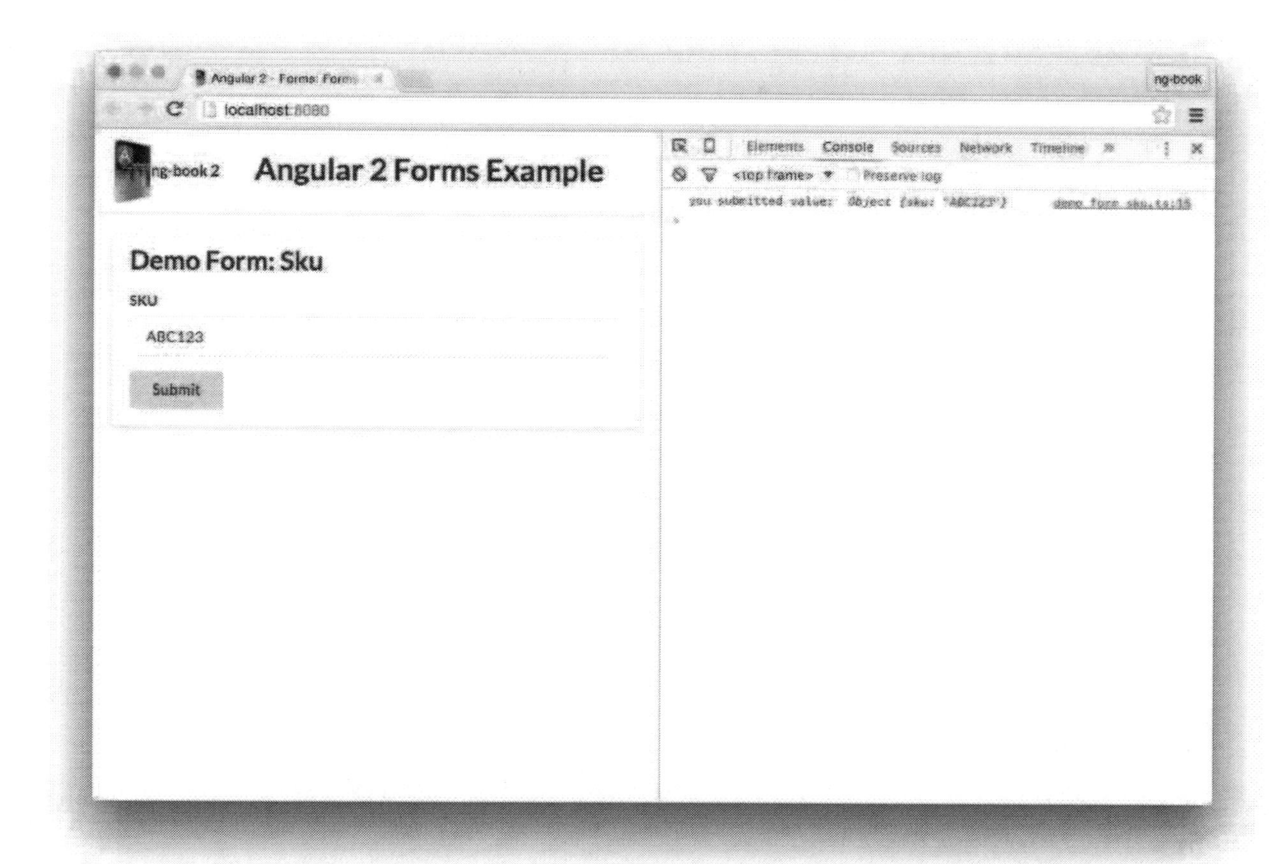


Demo Form with Sku: Simple Version, Submitted

## Using `FormBuilder`

Building our `FormControls` and `FormGroups` implicitly using `ngForm` and `ngControl` is convenient, but doesn't give us a lot of customization options. A more flexible and common way to configure forms is to use a `FormBuilder`.

`FormBuilder` is an aptly-named helper class that helps us build forms. As you recall, forms are made up of `FormControls` and `FormGroups` and the `FormBuilder` helps us make them (you can think of it as a "factory" object).

Let's add a `FormBuilder` to our previous example. Let's look at:

- how to use the `FormBuilder` in our component definition class
- how to use our custom `FormGroup` on a `form` in the view

# Reactive Forms with `FormBuilder`

For this component we're going to be using the `formGroup` and `formControl` directives which means we need to import the appropriate classes. We start by importing them like so:

**code/forms/app/ts/forms/demo_form_sku_with_builder.ts**

```
1  import { Component } from '@angular/core';
2  import {
3    FormBuilder,
4    FormGroup
5  } from '@angular/forms';
6
7  @Component({
8    selector: 'demo-form-sku-builder',
```

## Using `FormBuilder`

We inject `FormBuilder` by creating an argument in the `constructor` of our component class:

**What does `inject` mean?** We haven't talked much about dependency injection (DI) or how DI relates to the hierarchy tree, so that last sentence may not make a lot of sense. We talk a lot more about dependency injection in the Dependency Injection chapter, so go there if you'd like to learn more about it in depth.

At a high level, Dependency Injection is a way to tell Angular what dependencies this component needs to function properly.

**code/forms/app/ts/forms/demo_form_sku_with_builder.ts**

```
29 export class DemoFormSkuBuilder {
30   myForm: FormGroup;
31
32   constructor(fb: FormBuilder) {
33     this.myForm = fb.group({
34       'sku': ['ABC123']
35     });
36   }
37
38   onSubmit(value: string): void {
39     console.log('you submitted value: ', value);
40   }
41 }
```

During injection an instance of `FormBuilder` will be created and we assign it to the `fb` variable (in the constructor).

There are two main functions we'll use on `FormBuilder`:

- `control` - creates a new `FormControl`
- `group` - creates a new `FormGroup`

Notice that we've setup a new *instance variable* called `myForm` on this class. (We could have just as easily called it `form`, but I want to differentiate between our `FormGroup` and the `form` we had before.)

`myForm` is typed to be a `FormGroup`. We create a `FormGroup` by calling `fb.group()`. `.group` takes an object of key-value pairs that specify the `FormControls` in this group.

In this case, we're setting up one control `sku`, and the value is `["ABC123"]` - this says that the default value of this control is `"ABC123"`. (You'll notice that is an array. That's because we'll be adding more configuration options there later.)

Now that we have `myForm` we need to use that in the view (i.e. we need to *bind* it to our `form` element).

## Using `myForm` in the view

We want to change our `<form>` to use `myForm`. If you recall, in the last section we said that `ngForm` is applied for us automatically when we use `FormsModule`. We also mentioned that `ngForm` creates its own `FormGroup`. Well, in this case, we **don't** want to use an outside `FormGroup`. Instead we want to use our instance variable `myForm`, which we created with our `FormBuilder`. How can we do that?

Angular provides another directive that we use **when we have an existing `FormGroup`**: it's called `formGroup` and we use it like this:

**code/forms/app/ts/forms/demo_form_sku_with_builder.ts**

```
11      <h2 class="ui header">Demo Form: Sku with Builder</h2>
12      <form [formGroup]="myForm"
```

Here we're telling Angular that we want to use `myForm` as the `FormGroup` for this form.

Remember how earlier we said that when using `FormsModule` that `NgForm` will be automatically applied to a `<form>` element? There is an exception: `NgForm` won't be applied to a `<form>` that has `formGroup`.

If you're curious, the `selector` for `NgForm` is:

```
1   form:not([ngNoForm]):not([formGroup]),ngForm,[ngForm]
```

This means you *could* have a form that doesn't get `NgForm` applied by using the `ngNoForm` attribute.

We also need to change onSubmit to use myForm instead of f, because now it is myForm that has our configuration and values.

There's one last thing we need to do to make this work: bind our FormControl to the input tag. Remember that **ngControl creates a new FormControl object**, and attaches it to the parent FormGroup. But in this case, we used FormBuilder to create our own FormControls.

When we want to bind an **existing FormControl** to an input we use formControl:

**code/forms/app/ts/forms/demo_form_sku_with_builder.ts**

```
17          <label for="skuInput">SKU</label>
18          <input type="text"
19                 id="skuInput"
20                 placeholder="SKU"
21                 [formControl]="myForm.controls['sku']">
```

Here we are instructing the formControl directive to look at myForm.controls and use the existing sku FormControl for this input.

## Try it out!

Here's what it looks like all together:

**code/forms/app/ts/forms/demo_form_sku_with_builder.ts**

```
 1  import { Component } from '@angular/core';
 2  import {
 3    FormBuilder,
 4    FormGroup
 5  } from '@angular/forms';
 6
 7  @Component({
 8    selector: 'demo-form-sku-builder',
 9    template: `
10    <div class="ui raised segment">
11      <h2 class="ui header">Demo Form: Sku with Builder</h2>
12      <form [formGroup]="myForm"
13            (ngSubmit)="onSubmit(myForm.value)"
14            class="ui form">
15
16        <div class="field">
17          <label for="skuInput">SKU</label>
18          <input type="text"
```

```
19                  id="skuInput"
20                  placeholder="SKU"
21                  [formControl]="myForm.controls['sku']">
22        </div>
23
24      <button type="submit" class="ui button">Submit</button>
25      </form>
26    </div>
27    `
28  })
29  export class DemoFormSkuBuilder {
30    myForm: FormGroup;
31
32    constructor(fb: FormBuilder) {
33      this.myForm = fb.group({
34        'sku': ['ABC123']
35      });
36    }
37
38    onSubmit(value: string): void {
39      console.log('you submitted value: ', value);
40    }
41  }
```

Remember:

To create a new `FormGroup` and `FormControls` implicitly use:

- `ngForm` and
- `ngModel`

But to bind to an existing `FormGroup` and `FormControls` use:

- `formGroup` and
- `formControl`

# Adding Validations

Our users aren't always going to enter data in exactly the right format. If someone enters data in the wrong format, we want to give them feedback and not allow the form to be submitted. For this we use *validators*.

Validators are provided by the `Validators` module and the simplest validator is `Validators.required` which simply says that the designated field is required or else the `FormControl` will be considered invalid.

To use validators we need to do two things:

1. Assign a validator to the `FormControl` object
2. Check the status of the validator in the view and take action accordingly

To assign a validator to a `FormControl` object we simply pass it as the second argument to our `FormControl` constructor:

```
1 let control = new FormControl('sku', Validators.required);
```

Or in our case, because we're using `FormBuilder` we will use the following syntax:

**code/forms/app/ts/forms/demo_form_with_validations_explicit.ts**

```
44   constructor(fb: FormBuilder) {
45     this.myForm = fb.group({
46       'sku':  ['', Validators.required]
47     });
48 
49     this.sku = this.myForm.controls['sku'];
50   }
```

Now we need to use our validation in the view. There are two ways we can access the validation value in the view:

1. We can explicitly assign the `FormControl` `sku` to an instance variable of the class - which is more verbose, but gives us easy access to the `FormControl` in the view.
2. We can lookup the `FormControl` `sku` from `myForm` in the view. This requires less work in the component definition class, but is slightly more verbose in the view.

To make this difference clearer, let's look at this example both ways:

### Explicitly setting the `sku` `FormControl` as an instance variable

Here's a screenshot of what our form is going to look like with validations:

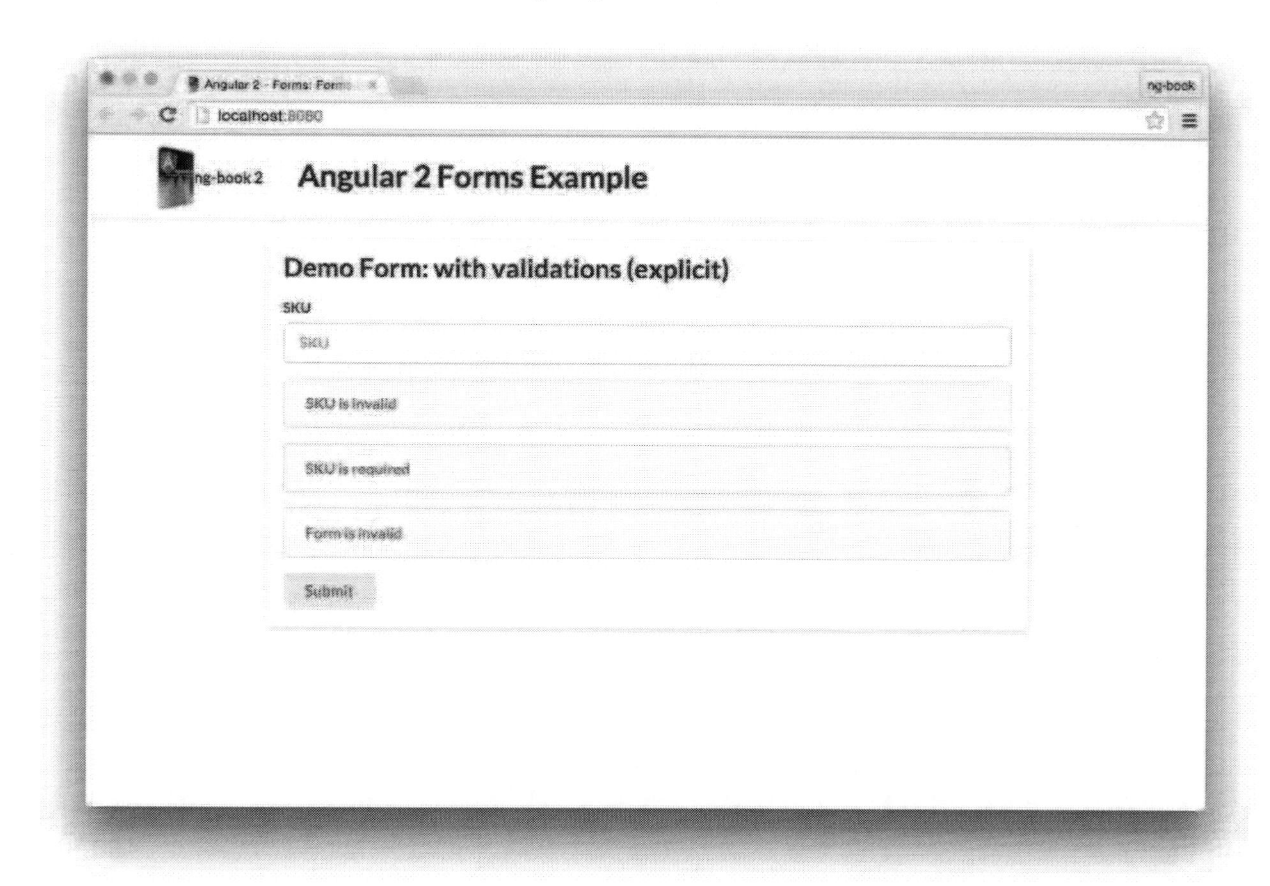


**Demo Form with Validations**

The most flexible way to deal with individual `FormControls` in your view is to set each `FormControl` up as an instance variable in your component definition class. Here's how we could setup `sku` in our class:

**code/forms/app/ts/forms/demo_form_with_validations_explicit.ts**

```
40  export class DemoFormWithValidationsExplicit {
41    myForm: FormGroup;
42    sku: AbstractControl;
43
44    constructor(fb: FormBuilder) {
45      this.myForm = fb.group({
46        'sku':  ['', Validators.required]
47      });
```

```
48
49      this.sku = this.myForm.controls['sku'];
50    }
51
52    onSubmit(value: string): void {
53      console.log('you submitted value: ', value);
54    }
55  }
```

Notice that:

1. We setup `sku: AbstractControl` at the top of the class and
2. We assign `this.sku` after we've created `myForm` with the `FormBuilder`

This is great because it means we can reference `sku` anywhere in our component view. The downside is that by doing it this way, we'd have to setup an instance variable **for every field in our form**. For large forms, this can get pretty verbose.

Now that we have our `sku` being validated, I want to look at four different ways we can use it in our view:

1. Checking the validity of our whole form and displaying a message
2. Checking the validity of our individual field and displaying a message
3. Checking the validity of our individual field and coloring the field red if it's invalid
4. Checking the validity of our individual field on a particular requirement and displaying a message

### Form message

We can check the validity of our whole form by looking at `myForm.valid`:

**code/forms/app/ts/forms/demo_form_with_validations_explicit.ts**

```
26          <div *ngIf="!sku.valid"
27            class="ui error message">SKU is invalid</div>
```

Remember, `myForm` is a `FormGroup` and a `FormGroup` is valid if all of the children `FormControls` are also valid.

### Field message

We can also display a message for the specific field if that field's `FormControl` is invalid:

**code/forms/app/ts/forms/demo_form_with_validations_explicit.ts**

```
26          <div *ngIf="!sku.valid"
27            class="ui error message">SKU is invalid</div>
28          <div *ngIf="sku.hasError('required')"
29            class="ui error message">SKU is required</div>
```

## Field coloring

I'm using the Semantic UI CSS Framework's CSS class `.error`, which means if I add the class `error` to the `<div class= "field">` it will show the input tag with a red border.

To do this, we can use the property syntax to set conditional classes:

**code/forms/app/ts/forms/demo_form_with_validations_explicit.ts**

```
19        <div class="field"
20            [class.error]="!sku.valid && sku.touched">
```

Notice here that we have two conditions for setting the `.error` class: We're checking for `!sku.valid` and `sku.touched`. The idea here is that we only want to show the error state if the user has tried editing the form ("touched" it) and it's now invalid.

To try this out, enter some data into the `input` tag and then delete the contents of the field.

## Specific validation

A form field can be invalid for many reasons. We often want to show a different message depending on the reason for a failed validation.

To look up a specific validation failure we use the `hasError` method:

**code/forms/app/ts/forms/demo_form_with_validations_explicit.ts**

```
28          <div *ngIf="sku.hasError('required')"
29            class="ui error message">SKU is required</div>
```

Note that `hasError` is defined on both `FormControl` and `FormGroup`. This means you can pass a second argument of `path` to lookup a specific field from `FormGroup`. For example, we could have written the previous example as:

```
1       <div *ngIf="myForm.hasError('required', 'sku')"
2         class="error">SKU is required</div>
```

## Putting it together

Here's the full code listing of our form with validations with the `FormControl` set as an instance variable:

**code/forms/app/ts/forms/demo_form_with_validations_explicit.ts**

```
1  /* tslint:disable:no-string-literal */
2  import { Component } from '@angular/core';
3  import {
4    FormBuilder,
5    FormGroup,
6    Validators,
7    AbstractControl
8  } from '@angular/forms';
9
10 @Component({
11   selector: 'demo-form-with-validations-explicit',
12   template: `
13   <div class="ui raised segment">
14     <h2 class="ui header">Demo Form: with validations (explicit)</h2>
15     <form [formGroup]="myForm"
16           (ngSubmit)="onSubmit(myForm.value)"
17           class="ui form">
18
19       <div class="field"
20           [class.error]="!sku.valid && sku.touched">
21         <label for="skuInput">SKU</label>
22         <input type="text"
23                id="skuInput"
24                placeholder="SKU"
25                [formControl]="sku">
26          <div *ngIf="!sku.valid"
27            class="ui error message">SKU is invalid</div>
28          <div *ngIf="sku.hasError('required')"
29            class="ui error message">SKU is required</div>
30       </div>
31
32       <div *ngIf="!myForm.valid"
33         class="ui error message">Form is invalid</div>
```

```
34
35         <button type="submit" class="ui button">Submit</button>
36       </form>
37     </div>
38     `
39 })
40 export class DemoFormWithValidationsExplicit {
41   myForm: FormGroup;
42   sku: AbstractControl;
43
44   constructor(fb: FormBuilder) {
45     this.myForm = fb.group({
46       'sku':  ['', Validators.required]
47     });
48
49     this.sku = this.myForm.controls['sku'];
50   }
51
52   onSubmit(value: string): void {
53     console.log('you submitted value: ', value);
54   }
55 }
```

## Removing the `sku` instance variable

In the example above we set `sku: AbstractControl` as an instance variable. We often wont want to create an instance variable for each `AbstractControl`, so how would we reference this `FormControl` in our view without an instance variable?

Instead we can use the `myForm.controls` property as in:

**code/forms/app/ts/forms/demo_form_with_validations_shorthand.ts**

```
20         <input type="text"
21                id="skuInput"
22                placeholder="SKU"
23                [formControl]="myForm.controls['sku']">
24          <div *ngIf="!myForm.controls['sku'].valid"
25            class="ui error message">SKU is invalid</div>
26          <div *ngIf="myForm.controls['sku'].hasError('required')"
27            class="ui error message">SKU is required</div>
```

In this way we can access the `sku` control without being forced to explicitly add it as an instance variable on the component class.

## Custom Validations

We often are going to want to write our own custom validations. Let's take a look at how to do that.

To see how validators are implemented, let's look at `Validators.required` from the Angular core source:

```
1 export class Validators {
2   static required(c: FormControl): StringMap<string, boolean> {
3     return isBlank(c.value) || c.value == "" ? {"required": true} : null;
4   }
```

A validator: - Takes a `FormControl` as its input and - Returns a `StringMap<string, boolean>` where the key is "error code" and the value is `true` if it fails

### Writing the Validator

Let's say we have specific requirements for our `sku`. For example, say our `sku` needs to begin with `123`. We could write a validator like so:

**code/forms/app/ts/forms/demo_form_with_custom_validations.ts**

```
19 function skuValidator(control: FormControl): { [s: string]: boolean } {
20   if (!control.value.match(/^123/)) {
21     return {invalidSku: true};
22   }
23 }
```

This validator will return an error code `invalidSku` if the input (the `control.value`) does not begin with `123`.

### Assigning the Validator to the `FormControl`

Now we need to add the validator to our `FormControl`. However, there's one small problem: we already have a validator on `sku`. How can we add multiple validators to a single field?

For that, we use `Validators.compose`:

**code/forms/app/ts/forms/demo_form_with_custom_validations.ts**

```
60    constructor(fb: FormBuilder) {
61      this.myForm = fb.group({
62        'sku':  ['', Validators.compose([
63          Validators.required, skuValidator])]
64      });
```

`Validators.compose` wraps our two validators and lets us assign them both to the `FormControl`. The `FormControl` is not `valid` unless both validations are valid.

Now we can use our new validator in the view:

**code/forms/app/ts/forms/demo_form_with_custom_validations.ts**

```
45          <div *ngIf="sku.hasError('invalidSku')"
46            class="ui error message">SKU must begin with <span>123</span></div>
```

Note that in this section, I'm using "explicit" notation of adding an instance variable for each `FormControl`. That means that in the view in this section, `sku` refers to a `FormControl`.

If you run the sample code, one neat thing you'll notice is that if you type something in to the field, the `required` validation will be fulfilled, but the `invalidSku` validation may not. This is great - it means we can partially-validate our fields and show the appropriate messages.

# Watching For Changes

So far we've only extracted the value from our form by calling `onSubmit` when the form is submitted. But often we want to watch for any value changes on a control.

Both `FormGroup` and `FormControl` have an `EventEmitter` that we can use to observe changes.

`EventEmitter` is an *Observable*, which means it conforms to a defined specification for watching for changes. If you're interested in the Observable spec, you can find it here[28]

To watch for changes on a control we:

1. get access to the `EventEmitter` by calling `control.valueChanges`. Then we
2. add an *observer* using the `.subscribe` method

Here's an example:

[28] https://github.com/jhusain/observable-spec

**code/forms/app/ts/forms/demo_form_with_events.ts**

```
44   constructor(fb: FormBuilder) {
45     this.myForm = fb.group({
46       'sku':  ['', Validators.required]
47     });
48 
49     this.sku = this.myForm.controls['sku'];
50 
51     this.sku.valueChanges.subscribe(
52       (value: string) => {
53         console.log('sku changed to:', value);
54       }
55     );
56 
57     this.myForm.valueChanges.subscribe(
58       (form: any) => {
59         console.log('form changed to:', form);
60       }
61     );
62 
63   }
```

Here we're observing two separate events: changes on the sku field and changes on the form as a whole.

The observable that we pass in is an object with a single key: `next` (there are other keys you can pass in, but we're not going to worry about those now). `next` is the function we want to call with the new value whenever the value changes.

If we type '`kj`' into the text box we will see in our console:

```
1 sku changed to:  k
2 form changed to:  Object {sku: "k"}
3 sku changed to:  kj
4 form changed to:  Object {sku: "kj"}
```

As you can see each keystroke causes the control to change, so our observable is triggered. When we observe the individual `FormControl` we receive a value (e.g. `kj`), but when we observe the whole form, we get an object of key-value pairs (e.g. `{sku: "kj"}`).

# ngModel

`NgModel` is a special directive: it binds a model to a form. `ngModel` is special in that it **implements two-way data binding**. Two-way data binding is almost always more complicated and difficult to reason about vs. one-way data binding. Angular 2 is built to generally have data flow one-way: top-down. However, when it comes to forms, there are times where it is easier to opt-in to a two-way bind.

Just because you've used `ng-model` in Angular 1 in the past, don't rush to use `ngModel` right away. There are good reasons to avoid two-way data binding. Of course, `ngModel` can be really handy, but know that we don't necessarily rely on two-way data binding it as much as we did in Angular 1.

Let's change our form a little bit and say we want to input `productName`. We're going to use `ngModel` to keep the component instance variable in sync with the view.

First, here's our component definition class:

**code/forms/app/ts/forms/demo_form_ng_model.ts**

```
39 export class DemoFormNgModel {
40   myForm: FormGroup;
41   productName: string;
42 
43   constructor(fb: FormBuilder) {
44     this.myForm = fb.group({
45       'productName':  ['', Validators.required]
46     });
47   }
48 
49   onSubmit(value: string): void {
50     console.log('you submitted value: ', value);
51   }
52 }
```

Notice that we're simply storing `productName: string` as an instance variable.

Next, let's use `ngModel` on our `input` tag:

**code/forms/app/ts/forms/demo_form_ng_model.ts**

```
23          <label for="productNameInput">Product Name</label>
24          <input type="text"
25                 id="productNameInput"
26                 placeholder="Product Name"
27                 [formControl]="myForm.get('productName')"
28                 [(ngModel)]="productName">
```

Now notice something - the syntax for `ngModel` is funny: we are using both brackets and parenthesis around the `ngModel` attribute! The idea this is intended to invoke is that we're using both the *input* `[]` brackets and the *output* `()` parenthesis. It's an indication of the two-way bind.

Notice something else here: we're still using `formControl` to specify that this `input` should be bound to the `FormControl` on our form. We do this because `ngModel` is only binding the `input` to the instance variable - the `FormControl` is completely separate. But because we still want to validate this value and submit it as part of the form, we keep the `formControl` directive.

Last, let's display our `productName` value in the view:

**code/forms/app/ts/forms/demo_form_ng_model.ts**

```
14      <div class="ui info message">
15        The product name is: {{productName}}
16      </div>
```

Here's what it looks like:

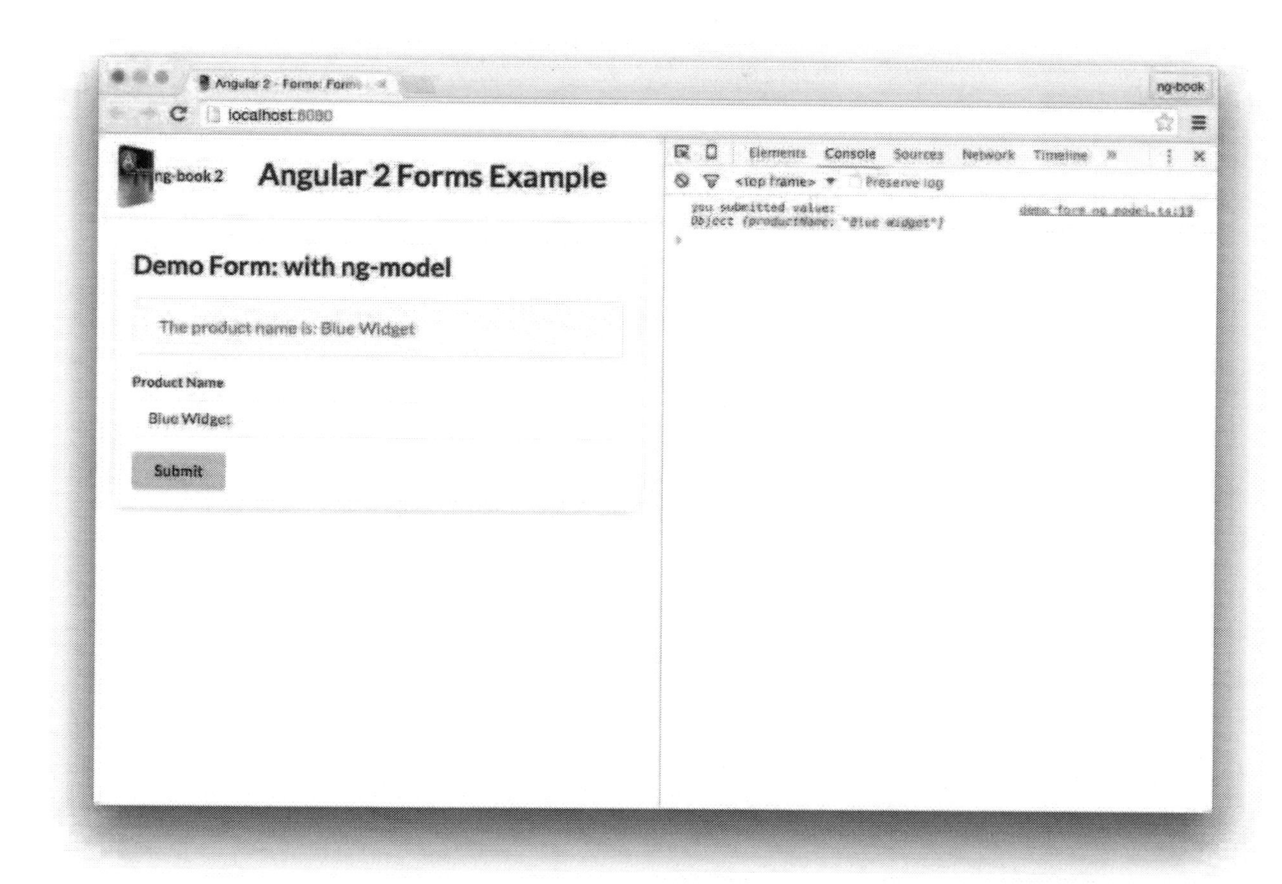


Demo Form with ngModel

Easy!

# Wrapping Up

Forms have a lot of moving pieces, but Angular 2 makes it fairly straightforward. Once you get a handle on how to use `FormGroups`, `FormControls`, and `Validations`, it's pretty easy going from there!

# HTTP

## Introduction

Angular comes with its own HTTP library which we can use to call out to external APIs.

When we make calls to an external server, we want our user to continue to be able to interact with the page. That is, we don't want our page to freeze until the HTTP request returns from the external server. To achieve this effect, our HTTP requests are *asynchronous*.

Dealing with *asynchronous* code is, historically, more tricky than dealing with synchronous code. In Javascript, there are generally three approaches to dealing with async code:

1. Callbacks
2. Promises
3. Observables

In Angular 2, the preferred method of dealing with async code is using Observables, and so that's what we'll cover in this chapter.

**There's a whole chapter on RxJS and Observables**: In this chapter we're going to be using Observables and not explaining them much. If you're just starting to read this book at this chapter, you should know that there's a whole chapter on Observables that goes into RxJS in more detail.

In this chapter we're going to:

1. show a basic example of `Http`
2. create a YouTube search-as-you-type component
3. discuss API details about the `Http` library

**Sample Code** The complete code for the examples in this chapter can be found in the `http` folder of the sample code. That folder contains a `README.md` which gives instructions for building and running the project.

Try running the code while reading the chapter and feel free play around to get a deeper insight about how it all works.

## Using `@angular/http`

HTTP has been split into a separate module in Angular 2. This means that to use it you need to `import` constants from `@angular/http`. For instance, we might import constants from `@angular/http` like this:

```
1 import { Http, Response, RequestOptions, Headers } from '@angular/http';
```

### `import` from `@angular/http`

In our `app.ts` we're going to import `HttpModule` which is a convenience collection of modules.

**code/http/app/ts/app.ts**

```
1  /*
2   * Angular
3   */
4  import {
5    Component
6  } from '@angular/core';
7  import { NgModule } from '@angular/core';
8  import { BrowserModule } from '@angular/platform-browser';
9  import { platformBrowserDynamic } from '@angular/platform-browser-dynamic';
10 import { HttpModule } from '@angular/http';
```

In our `NgModule` we will add `HttpModule` to the list of `imports`. The effect is that we will be able to inject `Http` (and a few other modules) into our components.

**code/http/app/ts/app.ts**

```
48 @NgModule({
49   declarations: [
50     HttpApp,
51     SimpleHTTPComponent,
52     MoreHTTPRequests,
53     YouTubeSearchComponent,
54     SearchBox,
55     SearchResultComponent
56   ],
57   imports: [
58     BrowserModule,
59     HttpModule // <--- right here
```

```
60    ],
61    bootstrap: [ HttpApp ],
62    providers: [
63      youTubeServiceInjectables
64    ]
65  })
66  class HttpAppModule {}
```

Now we can inject the `Http` service into our components (or anywhere we use DI, actually).

```
1  class MyFooComponent {
2    constructor(public http: Http) {
3    }
4
5    makeRequest(): void {
6      // do something with this.http ...
7    }
8  }
```

# A Basic Request

The first thing we're going to do is make a simple `GET` request to the `jsonplaceholder` API[29].

What we're going to do is:

1. Have a `button` that calls `makeRequest`
2. `makeRequest` will call the `http` library to perform a `GET` request on our API
3. When the request returns, we'll update `this.data` with the results of the data, which will be rendered in the view.

Here's a screenshot of our example:

[29]http://jsonplaceholder.typicode.com

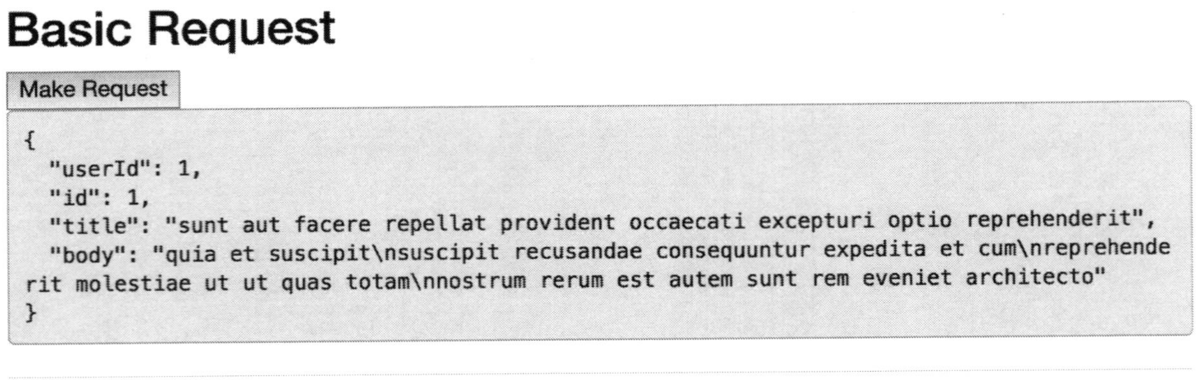

# Basic Request

Make Request


```
{
  "userId": 1,
  "id": 1,
  "title": "sunt aut facere repellat provident occaecati excepturi optio reprehenderit",
  "body": "quia et suscipit\nsuscipit recusandae consequuntur expedita et cum\nreprehenderit molestiae ut ut quas totam\nnostrum rerum est autem sunt rem eveniet architecto"
}
```


**Basic Request**

## Building the `SimpleHTTPComponent` `@Component`

The first thing we're going to do is import a few modules and then specify a `selector` for our `@Component`:

**code/http/app/ts/components/SimpleHTTPComponent.ts**

```
1  /*
2   * Angular
3   */
4  import {Component} from '@angular/core';
5  import {Http, Response} from '@angular/http';
6
7  @Component({
8    selector: 'simple-http',
```

## Building the `SimpleHTTPComponent` `template`

Next we build our view:

**code/http/app/ts/components/SimpleHTTPComponent.ts**

```
 9    template: `
10    <h2>Basic Request</h2>
11    <button type="button" (click)="makeRequest()">Make Request</button>
12    <div *ngIf="loading">loading...</div>
13    <pre>{{data | json}}</pre>
14  `
```

Our template has three interesting parts:

1. The `button`
2. The loading indicator
3. The `data`

On the `button` we bind to `(click)` to call the `makeRequest` function in our controller, which we'll define in a minute.

We want to indicate to the user that our request is loading, so to do that we will show `loading...` if the instance variable `loading` is true, using `ngIf`.

The `data` is an `Object`. A great way to debug objects is to use the `json` pipe as we do here. We've put this in a `pre` tag to give us nice, easy to read formatting.

## Building the `SimpleHTTPComponent` Controller

We start by defining a new `class` for our `SimpleHTTPComponent`:

**code/http/app/ts/components/SimpleHTTPComponent.ts**

```
16  export class SimpleHTTPComponent {
17    data: Object;
18    loading: boolean;
```

We have two instance variables: `data` and `loading`. This will be used for our API return value and loading indicator respectively.

Next we define our `constructor`:

**code/http/app/ts/components/SimpleHTTPComponent.ts**

```
20    constructor(private http: Http) {
21    }
```

The constructor body is empty, but we inject one key module: `Http`.

Remember that when we use the `public` keyword in `public http: Http` TypeScript will assign `http` to `this.http`. It's a shorthand for:

```
1     // other instance variables here
2     http: Http;
3
4     constructor(http: Http) {
5       this.http = http;
6     }
```

Now let's make our first HTTP request by implementing the `makeRequest` function:

**code/http/app/ts/components/SimpleHTTPComponent.ts**

```
23    makeRequest(): void {
24      this.loading = true;
25      this.http.request('http://jsonplaceholder.typicode.com/posts/1')
26        .subscribe((res: Response) => {
27          this.data = res.json();
28          this.loading = false;
29        });
30    }
```

When we call `makeRequest`, the first thing we do is set `this.loading = true`. This will turn on the loading indicator in our view.

To make an HTTP request is straightforward: we call `this.http.request` and pass the URL to which we want to make a `GET` request.

`http.request` returns an `Observable`. We can subscribe to changes (akin to using `then` from a Promise) using `subscribe`.

**code/http/app/ts/components/SimpleHTTPComponent.ts**

```
25      this.http.request('http://jsonplaceholder.typicode.com/posts/1')
26        .subscribe((res: Response) => {
```

When our `http.request` returns (from the server) the stream will emit a `Response` object. We extract the body of the response as an `Object` by using `json` and then we set `this.data` to that `Object`.

Since we have a response, we're not loading anymore so we set `this.loading = false`

`.subscribe` can also handle failures and stream completion by passing a function to the second and third arguments respectively. In a production app it would be a good idea to handle those cases, too. That is, `this.loading` should also be set to `false` if the request fails (i.e. the stream emits an error).

## Full `SimpleHTTPComponent`

Here's what our `SimpleHTTPComponent` looks like altogether:

**code/http/app/ts/components/SimpleHTTPComponent.ts**

```
1   /*
2    * Angular
3    */
4   import {Component} from '@angular/core';
5   import {Http, Response} from '@angular/http';
6
7   @Component({
8     selector: 'simple-http',
9     template: `
10    <h2>Basic Request</h2>
11    <button type="button" (click)="makeRequest()">Make Request</button>
12    <div *ngIf="loading">loading...</div>
13    <pre>{{data | json}}</pre>
14  `
15  })
16  export class SimpleHTTPComponent {
17    data: Object;
18    loading: boolean;
19
20    constructor(private http: Http) {
21    }
```

```
22
23   makeRequest(): void {
24     this.loading = true;
25     this.http.request('http://jsonplaceholder.typicode.com/posts/1')
26       .subscribe((res: Response) => {
27         this.data = res.json();
28         this.loading = false;
29       });
30   }
31 }
```

## Writing a `YouTubeSearchComponent`

The last example was a minimal way to get the data from an API server into your code. Now let's try to build a more involved example.

In this section, we're going to build a way to search YouTube as you type. When the search returns we'll show a list of video thumbnail results, along with a description and link to each video.

Here's a screenshot of what happens when I search for "cats playing ipads":

YouTube Search

cats playing ipads

Funny Cats Playing On iPads Compilation - Funny Videos 2015

You may or may not be surprised, but there are many animals playing on tablet computer. New video funny 2015 Thanks for watching, rating the video and ...

Watch

Animals Playing On iPads Compilation

You may or may not be surprised, but there are many animals playing on tablet computer. Join Us On Facebook http://www.facebook.com/Compilariz No ...

Watch

Cute cats try to catch a mouse from an IPad

Cute cats try to catch a mouse from an IPad.

Watch

Charlie The Cat - Kitten Playing iPad 2 !!! Game For Cats Cute Funny Clever Pets Bloopers

HELLO REDDIT, Thanks for the support! More Charlie the Cat Videos - http://youtu.be/xZHwYNrfWd0 Check My Other Videos Kitten HArlem Shake ...

Watch

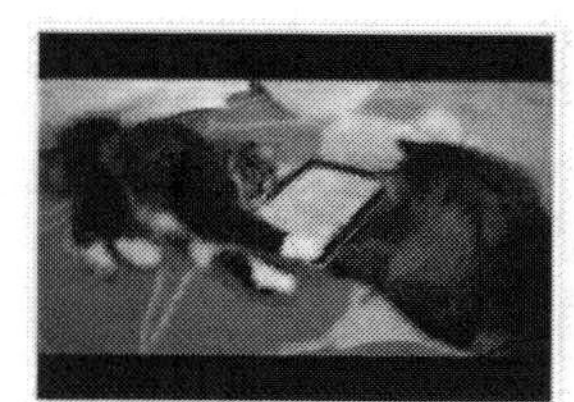

Cats playing "Game for Cats" with Apple iPad

Two Siberian cats like to play "Game for Cats" with Apple iPad :) Note that the iPad has Invisible Shield screen protector. Siperiankissat leikkivät

White Tiger Plays iPad - Game for Cats Gone Wild! Lions, servals, and more!

http://www.ipadgameforcats.com and http://www.conservatorscenter.org/

Cat Plays with iPad - Friskies Games for Cats

Mr. Kitty playing Cat Fishing on my girlfriends 1st gen iPad, via Friskies Games for Cats http://www.gamesforcats.com.

Cute Cat plays on iPad

Cute Cat plays on iPad.

Watch

**Can I get my cat to write Angular 2?**

For this example we're going to write several things:

1. A `SearchResult` object that will hold the data we want from each result
2. A `YouTubeService` which will manage the API request to YouTube and convert the results to a stream of `SearchResult[]`
3. A `SearchBox` component which will call out to the `YouTube` service as the user types
4. A `SearchResultComponent` which will render a specific `SearchResult`
5. A `YouTubeSearchComponent` which will encapsulate our whole YouTube searching app and render the list of results

Let's handle each part one at a time.

Patrick Stapleton has an excellent repository named angular2-webpack-starter[30]. This repo has an RxJS example which autocompletes Github repositories. Some of the ideas in this section are inspired from that example. It's a fantastic project with lots of examples and you should check it out.

## Writing a `SearchResult`

First let's start with writing a basic `SearchResult` class. This class is just a convenient way to store the specific fields we're interested in from our search results.

**code/http/app/ts/components/YouTubeSearchComponent.ts**

```
30 class SearchResult {
31   id: string;
32   title: string;
33   description: string;
34   thumbnailUrl: string;
35   videoUrl: string;
36 
37   constructor(obj?: any) {
38     this.id              = obj && obj.id             || null;
39     this.title           = obj && obj.title          || null;
40     this.description     = obj && obj.description    || null;
41     this.thumbnailUrl    = obj && obj.thumbnailUrl   || null;
42     this.videoUrl        = obj && obj.videoUrl       ||
43                              `https://www.youtube.com/watch?v=${this.id}`;
44   }
45 }
```

This pattern of taking an `obj?: any` lets us simulate keyword arguments. The idea is that we can create a new `SearchResult` and just pass in an object containing the keys we want to specify.

The only thing to point out here is that we're constructing the `videoUrl` using a hard-coded URL format. You're welcome to change this to a function which takes more arguments, or use the video `id` directly in your view to build this URL if you need to.

## Writing the `YouTubeService`

### The API

For this example we're going to be using the YouTube v3 search API[31].

[30]https://github.com/angular-class/angular2-webpack-starter

[31]https://developers.google.com/youtube/v3/docs/search/list

In order to use this API you need to have an API key. I've included an API key in the sample code which you can use. However, by the time you read this, you may find it's over the rate limits. If that happens, you'll need to issue your own key.

To issue your own key see this documentation[32]. For the sake of simplicity, I've registered a server key, but you should probably use a browser key if you're going to put your javascript code online.

We're going to setup two constants for our `YouTubeService` mapping to our API key and the API URL:

```
1 let YOUTUBE_API_KEY: string = "XXX_YOUR_KEY_HERE_XXX";
2 let YOUTUBE_API_URL: string = "https://www.googleapis.com/youtube/v3/search";
```

Eventually we're going to want to test our app. One of the things we find when testing is that we don't always want to test against production - we often want to test against staging or a development API.

To help with this environment configuration, one of the things we can do is **make these constants injectable**.

Why should we inject these constants instead of just using them in the normal way? Because if we make them injectable we can

1. have code that injects the right constants for a given environment at deploy time and
2. replace the injected value easily at test-time

By injecting these values, we have a lot more flexibility about their values down the line.

In order to make these values injectable, we use the `{ provide: ... , useValue: ... }` syntax like this:

**code/http/app/ts/components/YouTubeSearchComponent.ts**

```
82 export var youTubeServiceInjectables: Array<any> = [
83   {provide: YouTubeService, useClass: YouTubeService},
84   {provide: YOUTUBE_API_KEY, useValue: YOUTUBE_API_KEY},
85   {provide: YOUTUBE_API_URL, useValue: YOUTUBE_API_URL}
86 ];
```

Here we're specifying that we want to bind `YOUTUBE_API_KEY` "injectably" to the value of `YOUTUBE_-API_KEY`. (Same for `YOUTUBE_API_URL`, and we'll define `YouTubeService` in a minute.)

If you recall, to make something available to be injected throughout our application, we need to put it in `providers` for our `NgModule`. Since we're exporting `youTubeServiceInjectables` here we can use it in our `app.ts`

[32]https://developers.google.com/youtube/registering_an_application#Create_API_Keys

```
1  // http/app.ts
2  import { HttpModule } from '@angular/http';
3  import { youTubeServiceInjectables } from "components/YouTubeSearchComponent";
4
5  // ...
6  // further down
7  // ...
8
9  @NgModule({
10   declarations: [
11     HttpApp,
12     // others ....
13   ],
14   imports: [ BrowserModule, HttpModule ],
15   bootstrap: [ HttpApp ],
16   providers: [
17     youTubeServiceInjectables // <--- right here
18   ]
19 })
20 class HttpAppModule {}
```

Now we can inject `YOUTUBE_API_KEY` (from the `youTubeServiceInjectables`) instead of using the variable directly.

#### `YouTubeService` constructor

We create our `YouTubeService` by making a `class` and annotating it as `@Injectable`:

**code/http/app/ts/components/YouTubeSearchComponent.ts**

```
47 /**
48  * YouTubeService connects to the YouTube API
49  * See: * https://developers.google.com/youtube/v3/docs/search/list
50  */
51 @Injectable()
52 export class YouTubeService {
53   constructor(private http: Http,
54               @Inject(YOUTUBE_API_KEY) private apiKey: string,
55               @Inject(YOUTUBE_API_URL) private apiUrl: string) {
56   }
```

In the `constructor` we inject three things:

1. `Http`
2. `YOUTUBE_API_KEY`
3. `YOUTUBE_API_URL`

Notice that we make instance variables from all three arguments, meaning we can access them as `this.http`, `this.apiKey`, and `this.apiUrl` respectively.

Notice that we explicitly inject using the `@Inject(YOUTUBE_API_KEY)` notation.

#### `YouTubeService` search

Next let's implement the `search` function. `search` takes a query `string` and returns an `Observable` which will emit a stream of `SearchResult[]`. That is, each item emitted is an *array* of `SearchResults`.

**code/http/app/ts/components/YouTubeSearchComponent.ts**

```
58    search(query: string): Observable<SearchResult[]> {
59      let params: string = [
60        `q=${query}`,
61        `key=${this.apiKey}`,
62        `part=snippet`,
63        `type=video`,
64        `maxResults=10`
65      ].join('&');
66      let queryUrl: string = `${this.apiUrl}?${params}`;
```

We're building the `queryUrl` in a manual way here. We start by simply putting the query params in the `params` variable. (You can find the meaning of each of those values by reading the search API docs[33].)

Then we build the `queryUrl` by concatenating the `apiUrl` and the `params`.

Now that we have a `queryUrl` we can make our request:

[33] https://developers.google.com/youtube/v3/docs/search/list

**code/http/app/ts/components/YouTubeSearchComponent.ts**

```
58    search(query: string): Observable<SearchResult[]> {
59      let params: string = [
60        `q=${query}`,
61        `key=${this.apiKey}`,
62        `part=snippet`,
63        `type=video`,
64        `maxResults=10`
65      ].join('&');
66      let queryUrl: string = `${this.apiUrl}?${params}`;
67      return this.http.get(queryUrl)
68        .map((response: Response) => {
69          return (<any>response.json()).items.map(item => {
70            // console.log("raw item", item); // uncomment if you want to debug
71            return new SearchResult({
72              id: item.id.videoId,
73              title: item.snippet.title,
74              description: item.snippet.description,
75              thumbnailUrl: item.snippet.thumbnails.high.url
76            });
77          });
78        });
79    }
```

Here we take the return value of `http.get` and use `map` to get the `Response` from the request. From that `response` we extract the body as an object using `.json()` and then we iterate over each item and convert it to a `SearchResult`.

If you'd like to see what the raw `item` looks like, just uncomment the `console.log` and inspect it in your browsers developer console.

Notice that we're calling `(<any>response.json()).items`. What's going on here? We're telling TypeScript that we're not interested in doing strict type checking.

When working with a JSON API, we don't generally have typing definitions for the API responses, and so TypeScript won't know that the `Object` returned even has an `items` key, so the compiler will complain.

We could call `response.json()["items"]` and then cast that to an `Array` etc., but in this case (and in creating the `SearchResult`, it's just cleaner to use an `any` type, at the expense of strict type checking

### `YouTubeService` Full Listing

Here's the full listing of our `YouTubeService`:

**code/http/app/ts/components/YouTubeSearchComponent.ts**

```
/**
 * YouTubeSearchComponent is a tiny app that will autocomplete search YouTube.
 */

import {
  Component,
  Injectable,
  OnInit,
  ElementRef,
  EventEmitter,
  Inject
} from '@angular/core';
import { Http, Response } from '@angular/http';
import { Observable } from 'rxjs';

/*
  This API key may or may not work for you. Your best bet is to issue your own
  API key by following these instructions:
  https://developers.google.com/youtube/registering_an_application#Create_API_Ke\
ys

  Here I've used a **server key** and make sure you enable YouTube.

  Note that if you do use this API key, it will only work if the URL in
  your browser is "localhost"
*/
export var YOUTUBE_API_KEY: string = 'AIzaSyDOfT_BO81aEZScosfTYMruJobmpjqNeEk';
export var YOUTUBE_API_URL: string = 'https://www.googleapis.com/youtube/v3/sear\
ch';
let loadingGif: string = ((<any>window).__karma__) ? '' : require('images/loadin\
g.gif');

class SearchResult {
  id: string;
  title: string;
  description: string;
  thumbnailUrl: string;
  videoUrl: string;
```

```
39
40   constructor(obj?: any) {
41     this.id                = obj && obj.id              || null;
42     this.title             = obj && obj.title           || null;
43     this.description       = obj && obj.description     || null;
44     this.thumbnailUrl      = obj && obj.thumbnailUrl    || null;
45     this.videoUrl          = obj && obj.videoUrl        ||
46                                 `https://www.youtube.com/watch?v=${this.id}`;
47   }
48 }
49
50 /**
51  * YouTubeService connects to the YouTube API
52  * See: * https://developers.google.com/youtube/v3/docs/search/list
53  */
54 @Injectable()
55 export class YouTubeService {
56   constructor(private http: Http,
57               @Inject(YOUTUBE_API_KEY) private apiKey: string,
58               @Inject(YOUTUBE_API_URL) private apiUrl: string) {
59   }
60
61   search(query: string): Observable<SearchResult[]> {
62     let params: string = [
63       `q=${query}`,
64       `key=${this.apiKey}`,
65       `part=snippet`,
66       `type=video`,
67       `maxResults=10`
68     ].join('&');
69     let queryUrl: string = `${this.apiUrl}?${params}`;
70     return this.http.get(queryUrl)
71       .map((response: Response) => {
72         return (<any>response.json()).items.map(item => {
73           // console.log("raw item", item); // uncomment if you want to debug
74           return new SearchResult({
75             id: item.id.videoId,
76             title: item.snippet.title,
77             description: item.snippet.description,
78             thumbnailUrl: item.snippet.thumbnails.high.url
79           });
80         });
```

```
81        });
82      }
83    }
84
85    export var youTubeServiceInjectables: Array<any> = [
86      {provide: YouTubeService, useClass: YouTubeService},
87      {provide: YOUTUBE_API_KEY, useValue: YOUTUBE_API_KEY},
88      {provide: YOUTUBE_API_URL, useValue: YOUTUBE_API_URL}
89    ];
90
91    /**
92     * SearchBox displays the search box and emits events based on the results
93     */
94
95    @Component({
96      outputs: ['loading', 'results'],
97      selector: 'search-box',
98      template: `
99        <input type="text" class="form-control" placeholder="Search" autofocus>
100     `
101   })
102   export class SearchBox implements OnInit {
103     loading: EventEmitter<boolean> = new EventEmitter<boolean>();
104     results: EventEmitter<SearchResult[]> = new EventEmitter<SearchResult[]>();
105
106     constructor(private youtube: YouTubeService,
107                 private el: ElementRef) {
108     }
109
110     ngOnInit(): void {
111       // convert the `keyup` event into an observable stream
112       Observable.fromEvent(this.el.nativeElement, 'keyup')
113         .map((e: any) => e.target.value) // extract the value of the input
114         .filter((text: string) => text.length > 1) // filter out if empty
115         .debounceTime(250)                         // only once every 250ms
116         .do(() => this.loading.next(true))         // enable loading
117         // search, discarding old events if new input comes in
118         .map((query: string) => this.youtube.search(query))
119         .switch()
120         // act on the return of the search
121         .subscribe(
122           (results: SearchResult[]) => { // on sucesss
```

```
123             this.loading.next(false);
124             this.results.next(results);
125           },
126           (err: any) => { // on error
127             console.log(err);
128             this.loading.next(false);
129           },
130           () => { // on completion
131             this.loading.next(false);
132           }
133         );
134 
135   }
136 }
137 
138 @Component({
139   inputs: ['result'],
140   selector: 'search-result',
141   template: `
142    <div class="col-sm-6 col-md-3">
143       <div class="thumbnail">
144         <img src="{{result.thumbnailUrl}}">
145         <div class="caption">
146           <h3>{{result.title}}</h3>
147           <p>{{result.description}}</p>
148           <p><a href="{{result.videoUrl}}"
149                 class="btn btn-default" role="button">
150                 Watch</a></p>
151         </div>
152       </div>
153     </div>
154   `
155 })
156 export class SearchResultComponent {
157   result: SearchResult;
158 }
159 
160 @Component({
161   selector: 'youtube-search',
162   template: `
163   <div class='container'>
164       <div class="page-header">
```

```
165         <h1>YouTube Search
166           <img
167             style="float: right;"
168             *ngIf="loading"
169             src='${loadingGif}' />
170         </h1>
171       </div>
172 
173       <div class="row">
174         <div class="input-group input-group-lg col-md-12">
175           <search-box
176              (loading)="loading = $event"
177              (results)="updateResults($event)"
178               ></search-box>
179         </div>
180       </div>
181 
182       <div class="row">
183         <search-result
184           *ngFor="let result of results"
185           [result]="result">
186         </search-result>
187       </div>
188   </div>
189   `
190 })
191 export class YouTubeSearchComponent {
192   results: SearchResult[];
193 
194   updateResults(results: SearchResult[]): void {
195     this.results = results;
196     // console.log("results:", this.results); // uncomment to take a look
197   }
198 }
```

## Writing the `SearchBox`

The `SearchBox` component plays a key role in our app: it is the mediator between our UI and the `YouTubeService`.

The `SearchBox` will:

1. Watch for `keyup` on an `input` and submit a search to the `YouTubeService`

2. Emit a `loading` event when we're loading (or not)
3. Emit a `results` event when we have new results

#### `SearchBox` `@Component` Definition

Let's define our `SearchBox` `@Component`:

**code/http/app/ts/components/YouTubeSearchComponent.ts**

```
88 /**
89  * SearchBox displays the search box and emits events based on the results
90  */
91
92 @Component({
93   outputs: ['loading', 'results'],
94   selector: 'search-box',
```

The `selector` we've seen many times before: this allows us to create a `<search-box>` tag.

The `outputs` key specifies events that will be emitted from this component. That is, we can use the `(output)="callback()"` syntax in our view to listen to events on this component. For example, here's how we will use the `search-box` tag in our view later on:

```
1 <search-box
2  (loading)="loading = $event"
3  (results)="updateResults($event)"
4   ></search-box>
```

In this example, when the `SearchBox` component emits a `loading` event, we will set the variable `loading` in the parent context. Likewise, when the `SearchBox` emits a `results` event, we will call the `updateResults()` function, with the value, in the parent's context.

In the `@Component` configuration we're simply specifying the names of the events with the strings `"loading"` and `"results"`. In this example, each event will have a corresponding `EventEmitter` as an *instance variable of the controller class*. We'll implement that in a few minutes.

For now, remember that `@Component` is like the public API for our component, so here we're just specifying the name of the events, and we'll worry about implementing the `EventEmitters` later.

#### `SearchBox` `template` Definition

Our `template` is straightforward. We have one `input` tag:

code/http/app/ts/components/YouTubeSearchComponent.ts

```
88  /**
89   * SearchBox displays the search box and emits events based on the results
90   */
91
92  @Component({
93    outputs: ['loading', 'results'],
94    selector: 'search-box',
95    template: `
96      <input type="text" class="form-control" placeholder="Search" autofocus>
97    `
98  })
```

#### SearchBox Controller Definition

Our SearchBox controller is a new class:

code/http/app/ts/components/YouTubeSearchComponent.ts

```
99  export class SearchBox implements OnInit {
100   loading: EventEmitter<boolean> = new EventEmitter<boolean>();
101   results: EventEmitter<SearchResult[]> = new EventEmitter<SearchResult[]>();
```

We say that this class `implements OnInit` because we want to use the `ngOnInit` lifecycle callback. If a class `implements OnInit` then the `ngOnInit` function will be called after the first change detection check.

`ngOnInit` is a good place to do initialization (vs. the `constructor`) because inputs set on a component are not available in the `constructor`.

##### SearchBox Controller Definition constructor

Let's talk about the `SearchBox constructor`:

code/http/app/ts/components/YouTubeSearchComponent.ts

```
103   constructor(private youtube: YouTubeService,
104               private el: ElementRef) {
105   }
```

In our `constructor` we inject:

1. Our `YouTubeService` and
2. The element `el` that this component is attached to. `el` is an object of type `ElementRef`, which is an Angular wrapper around a native element.

We set both injections as instance variables.

#### `SearchBox` Controller Definition `ngOnInit`

On this input box we want to watch for `keyup` events. The thing is, if we simply did a search after every `keyup` that wouldn't work very well. There are three things we can do to improve the user experience:

1. Filter out any empty or short queries
2. "debounce" the input, that is, don't search on every character but only after the user has stopped typing after a short amount of time
3. discard any old searches, if the user has made a new search

We could manually bind to `keyup` and call a function on each `keyup` event and then implement filtering and debouncing from there. However, there is a better way: turn the `keyup` events into an observable stream.

RxJS provides a way to listen to events on an element using `Rx.Observable.fromEvent`. We can use it like so:

**code/http/app/ts/components/YouTubeSearchComponent.ts**

```
107   ngOnInit(): void {
108     // convert the `keyup` event into an observable stream
109     Observable.fromEvent(this.el.nativeElement, 'keyup')
```

Notice that in `fromEvent`:

- the first argument is `this.el.nativeElement` (the native DOM element this component is attached to)
- the second argument is the string `'keyup'`, which is the name of the event we want to turn into a stream

We can now perform some RxJS magic over this stream to turn it into `SearchResults`. Let's walk through step by step.

Given the stream of `keyup` events we can chain on more methods. In the next few paragraphs we're going to chain several functions on to our stream which will transform the stream. Then at the end we'll show the whole example together.

First, let's extract the value of the input tag:

```
1 .map((e: any) => e.target.value) // extract the value of the input
```

Above says, map over each `keyup` event, then find the event target (`e.target`, that is, our `input` element) and extract the `value` of that element. This means our stream is now a stream of strings.

Next:

```
.filter((text: string) => text.length > 1)
```

This `filter` means the stream will not emit any search strings for which the length is less than one. You could set this to a higher number if you want to ignore short searches.

```
.debounceTime(250)
```

`debounceTime` means we will throttle requests that come in faster than 250ms. That is, we won't search on every keystroke, but rather after the user has paused a small amount.

```
.do(() => this.loading.next(true))          // enable loading
```

Using `do` on a stream is a way to perform a function mid-stream for each event, but it does not change anything in the stream. The idea here is that we've got our search, it has enough characters, and we've debounced, so now we're about to search, so we turn on `loading`.

`this.loading` is an `EventEmitter`. We "turn on" `loading` by emitting `true` as the next event. We emit something on an `EventEmitter` by calling `next`. Writing `this.loading.next(true)` means, emit a `true` event on the `loading` `EventEmitter`. When we listen to the `loading` event on this component, the `$event` value will now be `true` (we'll look more closely at using `$event` below).

```
.map((query: string) => this.youtube.search(query))
.switch()
```

We use `.map` to call perform a search for each query that is emitted. By using `switch` we're, essentially, saying "ignore all search events but the most recent". That is, if a new search comes in, we want to use the most recent and discard the rest.

> Reactive experts will note that I'm handwaving here. `switch` has a more specific technical definition which you can read about in the RxJS docs here[34].

For each `query` that comes in, we're going to perform a `search` on our `YouTubeService`.

Putting the chain together we have this:

[34] https://github.com/Reactive-Extensions/RxJS/blob/master/doc/api/core/operators/switch.md

**code/http/app/ts/components/YouTubeSearchComponent.ts**

```
107   ngOnInit(): void {
108     // convert the `keyup` event into an observable stream
109     Observable.fromEvent(this.el.nativeElement, 'keyup')
110       .map((e: any) => e.target.value) // extract the value of the input
111       .filter((text: string) => text.length > 1) // filter out if empty
112       .debounceTime(250)                         // only once every 250ms
113       .do(() => this.loading.next(true))         // enable loading
114       // search, discarding old events if new input comes in
115       .map((query: string) => this.youtube.search(query))
116       .switch()
117       // act on the return of the search
118       .subscribe(
```

The API of RxJS can be a little intimidating because the API surface area is large. That said, we've implemented a sophisticated event-handling stream in very few lines of code!

Because we are calling out to our `YouTubeService` our stream is now a stream of `SearchResult[]`. We can `subscribe` to this stream and perform actions accordingly.

`subscribe` takes three arguments: onSuccess, onError, onCompletion.

**code/http/app/ts/components/YouTubeSearchComponent.ts**

```
118       .subscribe(
119         (results: SearchResult[]) => { // on sucesss
120           this.loading.next(false);
121           this.results.next(results);
122         },
123         (err: any) => { // on error
124           console.log(err);
125           this.loading.next(false);
126         },
127         () => { // on completion
128           this.loading.next(false);
129         }
130       );
131
132   }
```

The first argument specifies what we want to do when the stream emits a regular event. Here we emit an event on both of our `EventEmitters`:

1. We call `this.loading.next(false)`, indicating we've stopped loading
2. We call `this.results.next(results)`, which will emit an event containing the list of results

The second argument specifies what should happen when the stream has an error. Here we set `this.loading.next(false)` and log out the error.

The third argument specifies what should happen when the stream completes. Here we also emit that we're done loading.

### `SearchBox` Component: Full Listing

All together, here's the full listing of our `SearchBox` Component:

**code/http/app/ts/components/YouTubeSearchComponent.ts**

```
88  /**
89   * SearchBox displays the search box and emits events based on the results
90   */
91
92  @Component({
93    outputs: ['loading', 'results'],
94    selector: 'search-box',
95    template: `
96      <input type="text" class="form-control" placeholder="Search" autofocus>
97    `
98  })
99  export class SearchBox implements OnInit {
100   loading: EventEmitter<boolean> = new EventEmitter<boolean>();
101   results: EventEmitter<SearchResult[]> = new EventEmitter<SearchResult[]>();
102
103   constructor(private youtube: YouTubeService,
104               private el: ElementRef) {
105   }
106
107   ngOnInit(): void {
108     // convert the `keyup` event into an observable stream
109     Observable.fromEvent(this.el.nativeElement, 'keyup')
110       .map((e: any) => e.target.value) // extract the value of the input
111       .filter((text: string) => text.length > 1) // filter out if empty
112       .debounceTime(250)                         // only once every 250ms
113       .do(() => this.loading.next(true))         // enable loading
114       // search, discarding old events if new input comes in
115       .map((query: string) => this.youtube.search(query))
116       .switch()
```

```
117         // act on the return of the search
118         .subscribe(
119           (results: SearchResult[]) => { // on sucesss
120             this.loading.next(false);
121             this.results.next(results);
122           },
123           (err: any) => { // on error
124             console.log(err);
125             this.loading.next(false);
126           },
127           () => { // on completion
128             this.loading.next(false);
129           }
130         );
131 
132   }
133 }
```

## Writing SearchResultComponent

The SearchBox was pretty complicated. Let's handle a **much** easier component now: the SearchResultComponent. The SearchResultComponent's job is to render a single SearchResult.

There's not really any new ideas here, so let's take it all at once:

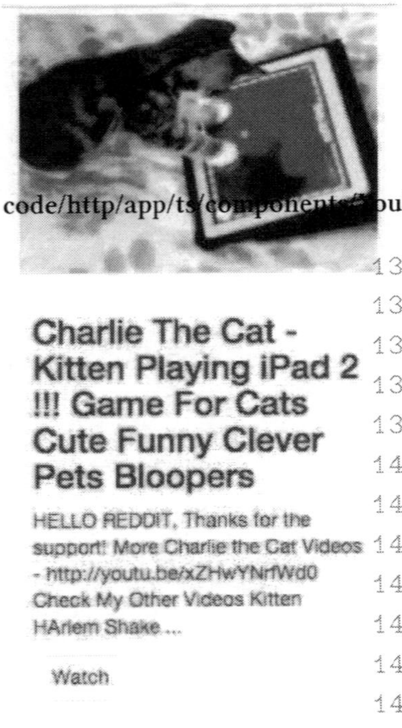

Single Search Result Component

**code/http/app/ts/components/YouTubeSearchComponent.ts**

```
135 @Component({
136   inputs: ['result'],
137   selector: 'search-result',
138   template: `
139    <div class="col-sm-6 col-md-3">
140       <div class="thumbnail">
141         <img src="{{result.thumbnailUrl}}">
142         <div class="caption">
143           <h3>{{result.title}}</h3>
144           <p>{{result.description}}</p>
145           <p><a href="{{result.videoUrl}}"
146                 class="btn btn-default" role="button">
147                 Watch</a></p>
148         </div>
149       </div>
```

```
150         </div>
151     `
152   })
153   export class SearchResultComponent {
154     result: SearchResult;
155   }
```

A few things:

The `@Component` takes a single input `result`, on which we will put the `SearchResult` assigned to this component.

The `template` shows the title, description, and thumbnail of the video and then links to the video via a button.

The `SearchResultComponent` simply stores the `SearchResult` in the instance variable `result`.

## Writing `YouTubeSearchComponent`

The last component we have to implement is the `YouTubeSearch-Component`. This is the component that ties everything together.

### `YouTubeSearchComponent @Component`

**code/http/app/ts/components/YouTubeSearchComponent.ts**

```
157   @Component({
158     selector: 'youtube-search',
```

Our `@Component` annotation is straightforward: use the `selector` `youtube-search`.

### `YouTubeSearchComponent` Controller

Before we look at the `template`, let's take a look at the `YouTube-SearchComponent` controller:

**code/http/app/ts/components/YouTubeSearchComponent.ts**

```
188 export class YouTubeSearchComponent {
189   results: SearchResult[];
190
191   updateResults(results: SearchResult[]): void {
192     this.results = results;
193     // console.log("results:", this.results); // uncomment to take a look
194   }
195 }
```

This component holds one instance variable: `results` which is an array of `SearchResults`.

We also define one function: `updateResults`. `updateResults` simply takes whatever new `SearchResult[]` it's given and sets `this.results` to the new value.

We'll use both `results` and `updateResults` in our `template`.

#### `YouTubeSearchComponent` **template**

Our view needs to do three things:

1. Show the loading indicator, if we're loading
2. Listen to events on the `search-box`
3. Show the search results

Next lets look at our `template`. Let's build some basic structure and show the loading gif next to the header:

**code/http/app/ts/components/YouTubeSearchComponent.ts**

```
159   template: `
160   <div class='container'>
161       <div class="page-header">
162         <h1>YouTube Search
163           <img
164             style="float: right;"
165             *ngIf="loading"
166             src='${loadingGif}' />
167         </h1>
168       </div>
```

Notice that our `img` has a `src` of `${loadingGif}` - that `loadingGif` variable came from a `require` statement earlier in the program. Here we're taking advantage of webpack's image loading feature. If you want to learn more about how this works, take a look at the webpack config in the sample code for this chapter or checkout `image-webpack-loader`[35].

We only want to show this loading image if `loading` is true, so we use `ngIf` to implement that functionality.

Next, let's look at the markup where we use our `search-box`:

**code/http/app/ts/components/YouTubeSearchComponent.ts**

```
169        <div class="row">
170          <div class="input-group input-group-lg col-md-12">
171            <search-box
172               (loading)="loading = $event"
173               (results)="updateResults($event)"
174                ></search-box>
175          </div>
```

The interesting part here is how we bind to the `loading` and `results` outputs. Notice, that we use the `(output)="action()"` syntax here.

For the `loading` output, we run the expression `loading = $event`. `$event` will be substituted with the value of the event that is emitted from the `EventEmitter`. That is, in our `SearchBox` component, when we call `this.loading.next(true)` then `$event` will be `true`.

Similarly, for the `results` output, we call the `updateResults()` function whenever a new set of results are emitted. This has the effect of updating our components `results` instance variable.

Lastly, we want to take the list of `results` in this component and render a `search-result` for each one:

**code/http/app/ts/components/YouTubeSearchComponent.ts**

```
178        <div class="row">
179          <search-result
180            *ngFor="let result of results"
181            [result]="result">
182          </search-result>
183        </div>
184    </div>
```

### `YouTubeSearchComponent` Full Listing

Here's the full listing for the `YouTubeSearchComponent`:

[35] https://github.com/tcoopman/image-webpack-loader

**code/http/app/ts/components/YouTubeSearchComponent.ts**

```
157 @Component({
158   selector: 'youtube-search',
159   template: `
160   <div class='container'>
161       <div class="page-header">
162         <h1>YouTube Search
163           <img
164             style="float: right;"
165             *ngIf="loading"
166             src='${loadingGif}' />
167         </h1>
168       </div>
169 
170       <div class="row">
171         <div class="input-group input-group-lg col-md-12">
172           <search-box
173              (loading)="loading = $event"
174              (results)="updateResults($event)"
175               ></search-box>
176         </div>
177       </div>
178 
179       <div class="row">
180         <search-result
181           *ngFor="let result of results"
182           [result]="result">
183         </search-result>
184       </div>
185   </div>
186   `
187 })
188 export class YouTubeSearchComponent {
189   results: SearchResult[];
190 
191   updateResults(results: SearchResult[]): void {
192     this.results = results;
193     // console.log("results:", this.results); // uncomment to take a look
194   }
195 }
```

There we have it! A functional search-as-you-type implemented for YouTube videos! Try running it from the code examples if you haven't already.

## `@angular/http` API

Of course, all of the HTTP requests we've made so far have simply been `GET` requests. It's important that we know how we can make other requests too.

### Making a `POST` request

Making `POST` request with `@angular/http` is very much like making a `GET` request except that we have one additional parameter: a body.

jsonplaceholder API[36] also provides a convent URL for testing our POST requests, so let's use it for a `POST`:

**code/http/app/ts/components/MoreHTTPRequests.ts**

```
30   makePost(): void {
31     this.loading = true;
32     this.http.post(
33       'http://jsonplaceholder.typicode.com/posts',
34       JSON.stringify({
35         body: 'bar',
36         title: 'foo',
37         userId: 1
38       }))
39       .subscribe((res: Response) => {
40         this.data = res.json();
41         this.loading = false;
42       });
43   }
```

Notice in the second argument we're taking an `Object` and converting it to a JSON string using `JSON.stringify`.

#### PUT / PATCH / DELETE / HEAD

There are a few other fairly common HTTP requests and we call them in much the same way.

- `http.put` and `http.patch` map to `PUT` and `PATCH` respectively and both take a URL and a body
- `http.delete` and `http.head` map to `DELETE` and `HEAD` respectively and both take a URL (no body)

Here's how we might make a `DELETE` request:

[36] http://jsonplaceholder.typicode.com

**code/http/app/ts/components/MoreHTTPRequests.ts**

```
45   makeDelete(): void {
46     this.loading = true;
47     this.http.delete('http://jsonplaceholder.typicode.com/posts/1')
48       .subscribe((res: Response) => {
49         this.data = res.json();
50         this.loading = false;
51       });
52   }
```

### RequestOptions

All of the http methods we've covered so far also take an optional last argument: `RequestOptions`. The `RequestOptions` object encapsulates:

- method
- headers
- body
- mode
- credentials
- cache
- url
- search

Let's say we want to craft a `GET` request that uses a special `X-API-TOKEN` header. We can create a request with this header like so:

**code/http/app/ts/components/MoreHTTPRequests.ts**

```
54   makeHeaders(): void {
55     let headers: Headers = new Headers();
56     headers.append('X-API-TOKEN', 'ng-book');
57 
58     let opts: RequestOptions = new RequestOptions();
59     opts.headers = headers;
60 
61     this.http.get('http://jsonplaceholder.typicode.com/posts/1', opts)
62       .subscribe((res: Response) => {
63         this.data = res.json();
64       });
65   }
```

## Summary

`@angular/http` is flexible and suitable for a wide variety of APIs.

One of the great things about `@angular/http` is that it has support for mocking the backend which is very useful in testing. To learn about testing HTTP, flip on over to the testing chapter.

# Routing

In web development, *routing* means splitting the application into different areas usually based on rules that are derived from the current URL in the browser.

For instance, if we visit the / path of a website, we may be visiting the **home route** of that website. Or if we visit `/about` we want to render the "about page", and so on.

## Why Do We Need Routing?

Defining routes in our application is useful because we can:

- separate different areas of the app;
- maintain the state in the app;
- protect areas of the app based on certain rules;

For example, imagine we are writing an inventory application similar to the one we described in previous chapters.

When we first visit the application, we might see a search form where we can enter a search term and get a list of products that match that term.

After that, we might click a given product to visit that product's details page.

Because our app is client-side, it's not technically required that we change the URL when we change "pages". But it's worth thinking about for a minute: what would be the consequences of using the same URL for all pages?

- You wouldn't be able to refresh the page and keep your location within the app
- You wouldn't be able to bookmark a page and come back to it later
- You wouldn't be able to share the URL of that page with others

Or put in a positive light, routing lets us define a URL string that specifies where within our app a user should be.

In our inventory example we could determine a series of different routes for each activity, for instance:

The initial root URL could be represented by `http://our-app/`. When we visit this page, we could be redirected to our "home" route at `http://our-app/home`.

When accessing the 'About Us' area, the URL could become `http://our-app/about`. This way if we sent the URL `http://our-app/about` to another user they would see same page.

# How client-side routing works

Perhaps you've written server-side routing code before (though, it isn't necessary to complete this chapter). Generally with server-side routing, the HTTP request comes in and the server will render a different controller depending on the incoming URL.

For instance, with Express.js[37] you might write something like this:

```
1 var express = require('express');
2 var router = express.Router();
3
4 // define the about route
5 router.get('/about', function(req, res) {
6   res.send('About us');
7 });
```

Or with Ruby on Rails[38] you might have:

```
1 # routes.rb
2 get '/about', to: 'pages#about'
3
4 # PagesController.rb
5 class PagesController < ActionController::Base
6   def about
7     render
8   end
9 end
```

The pattern varies per framework, but in both of these cases you have a **server** that accepts a request and *routes* to a **controller** and the controller runs a specific **action**, depending on the path and parameters.

Client-side routing is very similar in concept but different in implementation. With client-side routing **we're not necessarily making a request to the server** on every URL change. With our Angular apps, we refer to them as "Single Page Apps" (SPA) because our server only gives us a single page and it's our JavaScript that renders the different pages.

So how can we have different routes in our JavaScript code?

---

[37]http://expressjs.com/guide/routing.html
[38]http://rubyonrails.org/

## The beginning: using anchor tags

Client-side routing started out with a clever hack: Instead of using the page page, instead use the *anchor tag* as the client-side URL.

As you may already know, anchor tags were traditionally used to link directly to a place *within* the webpage and make the browser scroll all the way to where that anchor was defined. For instance, if we define an anchor tag in an HTML page:

```
1  <!-- ... lots of page content here ... -->
2  <a name="about"><h1>About</h1></a>
```

And we visited the URL `http://something/#about`, the browser would jump straight to that H1 tag that identified by the `about` anchor.

The clever move for client-side frameworks used for SPAs was to take the anchor tags and use them represent the routes within the app by formatting them as paths.

For example, the `about` route for an SPA would be something like `http://something/#/about`. This is what is known as **hash-based routing**.

What's neat about this trick is that it looks like a "normal" URL because we're starting our anchor with a slash (`/about`).

## The evolution: HTML5 client-side routing

With the introduction of HTML5, browsers acquired the ability to programmatically create new browser history entries that change the displayed URL *without the need for a new request.*

This is achieved using the `history.pushState` method that exposes the browser's navigational history to JavaScript.

So now, instead of relying on the anchor hack to navigate routes, modern frameworks can rely on `pushState` to perform history manipulation without reloads.

**Angular 1 Note**: This way of routing already works in Angular 1, but it needs to be explicitly enabled using `$locationProvider.html5Mode(true)`.

In Angular 2, however, the HTML5 is the default mode. Later in this chapter we show how to change from HTML5 mode to the old anchor tag mode.

There's two things you need to be aware of when using HTML5 mode routing, though

1. Not all browsers support HTML5 mode routing, so if you need to support older browsers you might be stuck with hash-based routing for a while.
2. **The server has to support HTML5 based routing**.

It may not be immediately clear why the server has to support HTML5 based-routing, we'll talk more about why later in this chapter.

# Writing our first routes

The Angular docs recommends using HTML5 mode routing[39]. But due to the challanges mentioned in the prevoius section we will for simplicity be using hash based routing in our examples.

In Angular we configure routes by mapping *paths* to the component that will handle them.

Let's create a small app that has multiple routes. On this sample application we will have 3 routes:

- A main page route, using the `/#/home` path;
- An about page, using the `/#/about` path;
- A contact us page, using the `/#/contact` path;

And when the user visits the root path (`/#/`), it will redirect to the home path.

# Components of Angular 2 routing

There are three main components that we use to configure routing in Angular:

- `Routes` describes the routes our application supports
- `RouterOutlet` is a "placeholder" component that shows Angular where to put the content of each route
- `RouterLink` directive is used to link to routes

Let's look at each one more closely.

## Imports

In order to use the router in Angular, we import constants from the `@angular/router` package:

[39] https://angular.io/docs/ts/latest/guide/router.html#!#browser-url-styles

**code/routes/basic/app/ts/app.ts**

```
10  import {
11    RouterModule,
12    Routes
13  } from '@angular/router';
```

Now we can define our router configuration.

### Routes

To define routes for our application, create a `Routes` configuration and then use `RouterModule.forRoot(routes)` to provide our application with the dependencies necessary to use the router:

**code/routes/basic/app/ts/app.ts**

```
48  const routes: Routes = [
49    { path: '', redirectTo: 'home', pathMatch: 'full' },
50    { path: 'home', component: HomeComponent },
51    { path: 'about', component: AboutComponent },
52    { path: 'contact', component: ContactComponent },
53    { path: 'contactus', redirectTo: 'contact' },
54  ];
```

Notice a few things about the routes:

- `path` specifies the URL this route will handle
- `component` is what ties a given route path to a component that will handle the route
- the optional `redirectTo` is used to redirect a given path to an existing route

As a summary, the goal of routes is to specify which component will handle a given path.

## Redirections

When we use `redirectTo` on a route definition, it will tell the router that when we visit the `path` of the route, we want the browser to be redirected to another route.

In our sample code above, if we visit the `root` path at http://localhost:8080/#/[40], we'll be redirected to the route `home`.

Another example is the `contactus` route:

[40] http://localhost:8080/#/

**code/routes/basic/app/ts/app.ts**

```
53   { path: 'contactus', redirectTo: 'contact' },
```

In this case, if we visit the URL http://localhost:8080/#/contactus[41], we'll see that the browser redirects to `/contact`.

**Sample Code** The complete code for the examples in this section can be found in the `routes/basic` folder of the sample code. That folder contains a `README.md`, which gives instructions for building and running the project.

There are many different `imports` required for routing and we don't list every single one in every code example below. However we do list the filename and line number from which almost every example is taken from. If you're having trouble figuring out how to `import` a particular class, open up the code using your editor to see the entire code listing.

Try running the code while reading this section and feel free play around to get a deeper insight about how it all works.

## Installing our Routes

Now that we have our `Routes routes`, we need to install it. To use the routes in our app we do two things to our `NgModule`:

1. Import the `RouterModule`
2. Install the routes using `RouterModule.forRoot(routes)` in the `imports` of our `NgModule`

Here's our routes configured into our `NgModule` for this app:

**code/routes/basic/app/ts/app.ts**

```
48 const routes: Routes = [
49   { path: '', redirectTo: 'home', pathMatch: 'full' },
50   { path: 'home', component: HomeComponent },
51   { path: 'about', component: AboutComponent },
52   { path: 'contact', component: ContactComponent },
53   { path: 'contactus', redirectTo: 'contact' },
54 ];
55
56 @NgModule({
57   declarations: [
```

[41] http://localhost:8080/#/contactus

```
58        RoutesDemoApp,
59        HomeComponent,
60        AboutComponent,
61        ContactComponent
62      ],
63      imports: [
64        BrowserModule,
65        RouterModule.forRoot(routes) // <-- routes
66      ],
67      bootstrap: [ RoutesDemoApp ],
68      providers: [
69        { provide: LocationStrategy, useClass: HashLocationStrategy }
70      ]
71    })
72    class RoutesDemoAppModule {}
73
74    platformBrowserDynamic().bootstrapModule(RoutesDemoAppModule)
75      .catch((err: any) => console.error(err));
```

### `RouterOutlet` using `<router-outlet>`

When we change routes, we want to keep our outer "layout" template and only substitute the "inner section" of the page with the route's component.

In order to describe to Angular where in our page we want to render the contents for each route, we use the `RouterOutlet` directive.

Our component `@Component` has a `template` which specifies some `div` structure, a section for `Navigation`, and a directive called `router-outlet`.

**The `router-outlet` element indicates where the contents of each route component will be rendered.**

We are are able to use the `router-outlet` directive in our template because we imported the `RouterModule` in our `NgModule`.

Here's the component and template for the navigation wrapper of our app:

**code/routes/basic/app/ts/app.ts**

```
28 @Component({
29   selector: 'router-app',
30   template: `
31   <div>
32     <nav>
33       <a>Navigation:</a>
34       <ul>
35         <li><a [routerLink]="['home']">Home</a></li>
36         <li><a [routerLink]="['about']">About</a></li>
37         <li><a [routerLink]="['contact']">Contact us</a></li>
38       </ul>
39     </nav>
40 
41     <router-outlet></router-outlet>
42   </div>
43   `
44 })
45 class RoutesDemoApp {
46 }
```

If we look at the `template` contents above, you will note the `router-outlet` element right below the navigation menu. When we visit `/home`, that's where `HomeComponent` template will be rendered. The same happens for the other components.

## `RouterLink` using `[routerLink]`

Now that we know where route templates will be rendered, how do we tell Angular 2 to navigate to a given route?

We might try linking to the routes directly using pure HTML:

```
1 <a href="/#/home">Home</a>
```

But if we do this, we'll notice that clicking the link triggers a page reload and that's definitely not what we want when programming single page apps.

To solve this problem, Angular 2 provides a solution that can be used to link to routes **with no page reload**: the `RouterLink` directive.

This directive allows you to write links using a special syntax:

**code/routes/basic/app/ts/app.ts**

```
33        <a>Navigation:</a>
34        <ul>
35          <li><a [routerLink]="['home']">Home</a></li>
36          <li><a [routerLink]="['about']">About</a></li>
37          <li><a [routerLink]="['contact']">Contact us</a></li>
38        </ul>
```

We can see on the left-hand side the `[routerLink]` that applies the directive to the current element (in our case `a` tags).

Now, on the right-hand side we have an array with the route path as the first element, like `"['home']"` or `"['about']"` that will indicate which route to navigate to when we click the element.

It might seem a little odd that the value of `routerLink` is a string with an array containing a string (`"['home']"`, for example). This is because there are more things you can provide when linking to routes, but we'll look at this into more detail when we talk about child routes and route parameters.

For now, we're only using routes names from the root app component.

# Putting it all together

So now that we have all the basic pieces, let's make them work together to transition from one route to the other.

The first thing we need to write for our application is the `index.html` file.

Here's the full code for that:

**code/routes/basic/app/index.html**

```
1   <!doctype html>
2   <html>
3     <head>
4       <base href="/">
5       <title>ng-book 2: Angular 2 Router</title>
6
7       {% for (var css in o.htmlWebpackPlugin.files.css) { %}
8         <link href="{%=o.htmlWebpackPlugin.files.css[css] %}" rel="stylesheet">
9       {% } %}
10    </head>
11    <body>
12      <router-app></router-app>
13      <script src="/core.js"></script>
```

```
14      <script src="/vendor.js"></script>
15      <script src="/bundle.js"></script>
16    </body>
17  </html>
```

The section describing `htmlWebpackPlugin` comes from the webpack module bundler[42]. We're using `webpack` in this chapter because it's a tool for bundling your assets

The code should be familiar by now, with the exception of this line:

```
1  <base href="/">
```

This line declares the `base` HTML tag. This tag is traditionally used to tell the browser where to look for images and other resources declared using relative paths.

It turns out Angular Router also relies on this tag to determine how to construct its routing information.

For instance, if we have a route with a path of `/hello` and our `base` element declares `href="/app"`, the application will use `/app/#` as the concrete path.

Sometimes though, coders of an Angular application don't have access to the `head` section of the application HTML. This is true for instance, when reusing headers and footers of a larger, pre-existing application.

Fortunately there is a workaround for this case. You can declare the application base path programmatically, when configuring our `NgModule` by using the `APP_BASE_HREF` provider:

```
1  @NgModule({
2    declarations: [ RoutesDemoApp ],
3    imports: [
4      BrowserModule,
5      RouterModule.forRoot(routes) // <-- routes
6    ],
7    bootstrap: [ RoutesDemoApp ],
8    providers: [
9      { provide: LocationStrategy, useClass: HashLocationStrategy },
10     { provide: APP_BASE_HREF, useValue: '/' } // <--- this right here
11   ]
12 })
```

Putting `{ provide: APP_BASE_HREF, useValue: '/' }` in the `providers` is the equivalent of using `<base href="/">` on our application HTML header.

[42]https://webpack.github.io/

## Creating the Components

Before we get to the main app component, let's create 3 simple components, one for each of the routes.

### HomeComponent

The `HomeComponent` will just have an `h1` tag that says "`Welcome!`". Here's the full code for our `HomeComponent`:

**code/routes/basic/app/ts/components/HomeComponent.ts**

```
1  /*
2   * Angular
3   */
4  import {Component} from '@angular/core';
5
6  @Component({
7    selector: 'home',
8    template: `<h1>Welcome!</h1>`
9  })
10 export class HomeComponent {
11 }
```

### AboutComponent

Similarly, the `AboutComponent` will just have a basic `h1`:

**code/routes/basic/app/ts/components/AboutComponent.ts**

```
1  /*
2   * Angular
3   */
4  import {Component} from '@angular/core';
5
6  @Component({
7    selector: 'about',
8    template: `<h1>About</h1>`
9  })
10 export class AboutComponent {
11 }
```

### ContactComponent

And, likewise with `AboutComponent`:

**code/routes/basic/app/ts/components/ContactComponent.ts**

```
/*
 * Angular
 */
import {Component} from '@angular/core';

@Component({
  selector: 'contact',
  template: `<h1>Contact Us</h1>`
})
export class ContactComponent {
}
```

Nothing really very interesting about those components, so let's move on to the main `app.ts` file.

## Application Component

Now we need to create the root-level "application" component that will tie everything together.

We start with the imports we'll need, both from the `core` and `router` bundles:

**code/routes/basic/app/ts/app.ts**

```
/*
 * Angular Imports
 */
import {
  NgModule,
  Component
} from '@angular/core';
import {BrowserModule} from '@angular/platform-browser';
import {platformBrowserDynamic} from '@angular/platform-browser-dynamic';
import {
  RouterModule,
  Routes
} from '@angular/router';
import {LocationStrategy, HashLocationStrategy} from '@angular/common';
```

Next step is to import the three components we created above:

**code/routes/basic/app/ts/app.ts**

```
19 import {HomeComponent} from 'components/HomeComponent';
20 import {AboutComponent} from 'components/AboutComponent';
21 import {ContactComponent} from 'components/ContactComponent';
```

Now let's get to the real component code. We start with the declaration of the component selector and template:

**code/routes/basic/app/ts/app.ts**

```
28 @Component({
29   selector: 'router-app',
30   template: `
31   <div>
32     <nav>
33       <a>Navigation:</a>
34       <ul>
35         <li><a [routerLink]="['home']">Home</a></li>
36         <li><a [routerLink]="['about']">About</a></li>
37         <li><a [routerLink]="['contact']">Contact us</a></li>
38       </ul>
39     </nav>
40 
41     <router-outlet></router-outlet>
42   </div>
43   `
44 })
45 class RoutesDemoApp {
46 }
```

For this component, we're going to use two router directives: `RouterOutlet` and the `RouterLink`. Those directives, along with all other common router directives are imported when we put `RouterModule` in the `imports` section of our `NgModule`.

As a recap, the `RouterOutlet` directive is then used to indicate where in our template the route contents should be rendered. That's represented by the `<router-outlet></router-outlet>` snippet in our `template` code.

The `RouterLink` directive is used to create navigation links to our routes:

**code/routes/basic/app/ts/app.ts**

```
33        <a>Navigation:</a>
34        <ul>
35          <li><a [routerLink]="['home']">Home</a></li>
36          <li><a [routerLink]="['about']">About</a></li>
37          <li><a [routerLink]="['contact']">Contact us</a></li>
38        </ul>
```

Using `[routerLink]` will instruct Angular to take ownership of the `click` event and then initiate a route switch to the right place, based on the route definition.

## Configuring the Routes

Next, we declare the routes creating an array of objects that conform to the `Routes` type:

**code/routes/basic/app/ts/app.ts**

```
48  const routes: Routes = [
49    { path: '', redirectTo: 'home', pathMatch: 'full' },
50    { path: 'home', component: HomeComponent },
51    { path: 'about', component: AboutComponent },
52    { path: 'contact', component: ContactComponent },
53    { path: 'contactus', redirectTo: 'contact' },
54  ];
```

In the last section of the `app.ts` file, we bootstrap the application:

**code/routes/basic/app/ts/app.ts**

```
56  @NgModule({
57    declarations: [
58      RoutesDemoApp,
59      HomeComponent,
60      AboutComponent,
61      ContactComponent
62    ],
63    imports: [
64      BrowserModule,
65      RouterModule.forRoot(routes) // <-- routes
66    ],
67    bootstrap: [ RoutesDemoApp ],
68    providers: [
```

```
69        { provide: LocationStrategy, useClass: HashLocationStrategy }
70      ]
71    })
72    class RoutesDemoAppModule {}
73
74    platformBrowserDynamic().bootstrapModule(RoutesDemoAppModule)
75      .catch((err: any) => console.error(err));
```

Just like we have been doing so far, we are now bootstrapping the app and telling that `RoutesDemoApp` is the root component.

Notice that we put all necessary components in our `declarations`. If we're going to route to a component, then it needs to be declared in *some* `NgModule` (either this module or imported).

In our `imports` we have `RouterModule.forRoot(routes)`. `RouterModule.forRoot(routes)` is a function that will take our routes, configure the router, and return a list of dependencies like `RouteRegistry`, `Location`, and several other classes that are necessary to make routing work.

In our `providers` we have this:

```
1        { provide: LocationStrategy, useClass: HashLocationStrategy }
```

Let's take an in depth look of what we want to achieve with this line.

## Routing Strategies

The way the Angular application parses and creates paths from and to route definitions is called *location strategy*.

In Angular 1 this is called *routing modes* instead

The default strategy is `PathLocationStrategy`, which is what we call HTML5 routing. While using this strategy, routes are represented by regular paths, like `/home` or `/contact`.

We can change the location strategy used for our application by binding the `LocationStrategy` class to a new, concrete strategy class.

Instead of using the default `PathLocationStrategy` we can also use the `HashLocationStrategy`.

The reason we're using the hash strategy as a default is because if we were using HTML5 routing, our URLs would end up being regular paths (that is, not using hash/anchor tags).

This way, the routes would work when you click a link and navigate on the client side, let's say from `/about` to `/contact`.

If we were to refresh the page, instead of asking the server for the root URL, which is what is being served, instead we'd be asking for `/about` or `/contact`. Because there's no known page at `/about` the server would return a 404.

This default strategy works with hash based paths, like `/#/home` or `/#/contact` that the server understands as being the `/` path. (This is also the default mode in Angular 1.)

Let's say you want to use HTML5 mode in production, what can you do?

In order to use HTML5 mode routing, you have to configure your server to redirect every "missing" route to the root URL.

In the `routes/basic` project we've included a script you can use to develop with `webpack-dev-server` and use HTML5 paths at the same time.

To use it `cd routes/basic` and run `node html5-dev-server.js`.

Finally, in order to make our example application work with this new strategy, first we have to import `LocationStrategy` and `HashLocationStrategy`:

**code/routes/basic/app/ts/app.ts**

```
14  import {LocationStrategy, HashLocationStrategy} from '@angular/common';
```

and then just add that location strategy to the providers of our `NgModule`:

**code/routes/basic/app/ts/app.ts**

```
68    providers: [
69      { provide: LocationStrategy, useClass: HashLocationStrategy }
70    ]
```

**You could write your own strategy if you wanted to.** All you need to do is extend the `LocationStrategy` class and implement the methods. A good way to start is reading the Angular 2 source for the `HashLocationStrategy` or `PathLocationStrategy` classes.

## Path location strategy

In our sample application folder, you'll find a file called `app/ts/app.html5.ts`.

If we want to play with the default `PathLocationStrategy`, we just need to copy the contents of that file to `app/ts/app.ts`, then reload the application.

# Running the application

You can now go into the application root folder (`code/routes`) and run `npm run server` to boot the application.

When you type http://localhost:8080/[43] into your browser you should see the home route rendered:

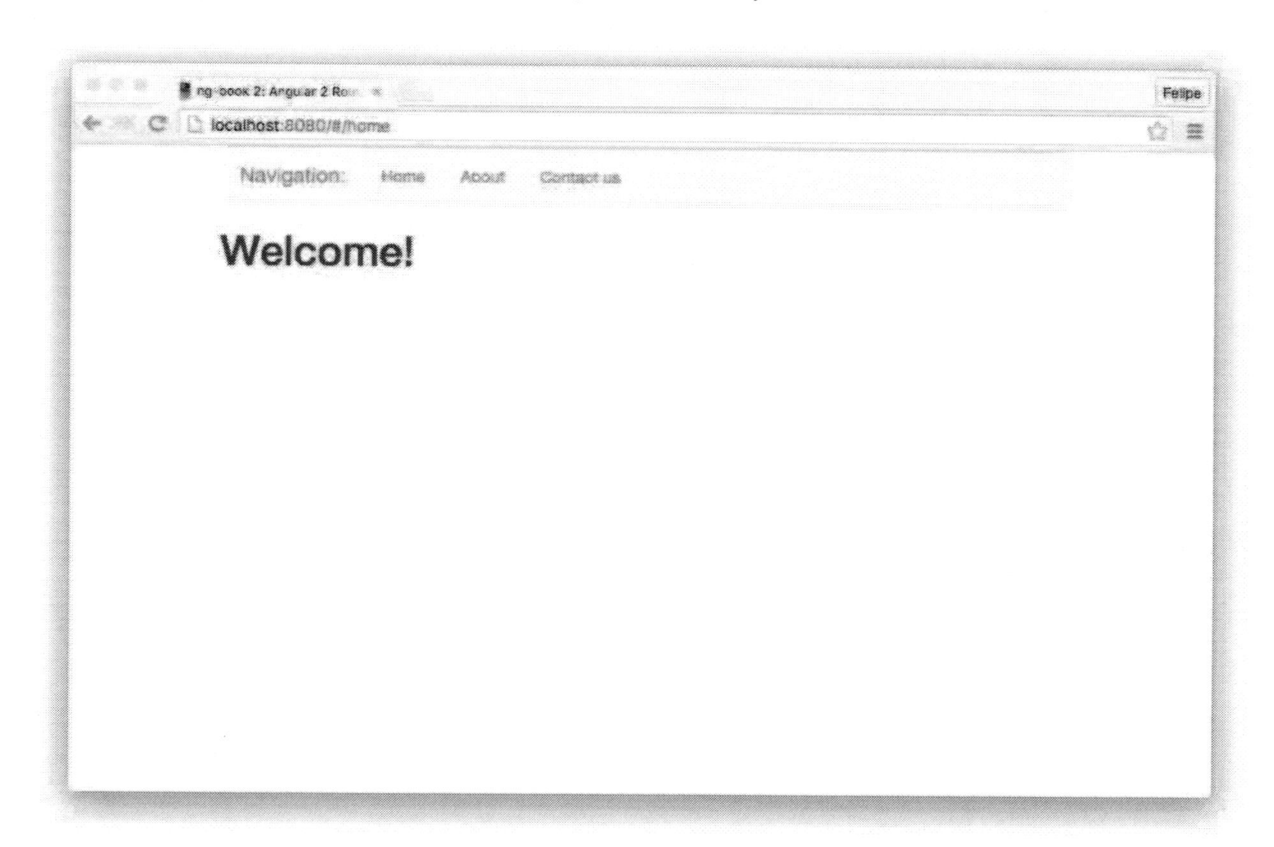


**Home Route**

Notice that the URL in the browser got redirected to http://localhost:8080/#/home[44].

Now clicking each link will render the appropriate routes:

[43]http://localhost:8080/

[44]http://localhost:8080/#/home

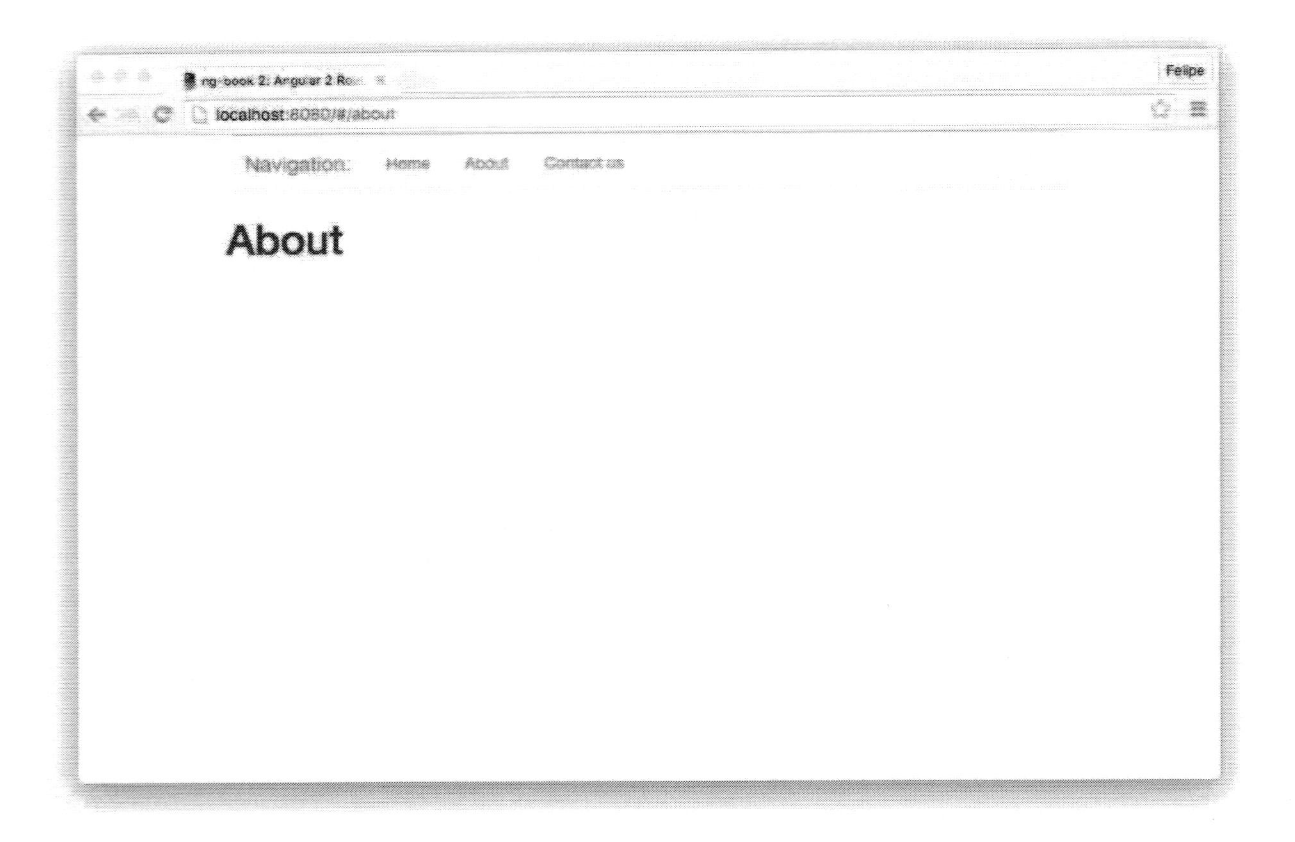
ng-book 2: Angular 2 Rou
Felipe
localhost:8080/#/about
Navigation:
Home
About
Contact us
About

**About Route**

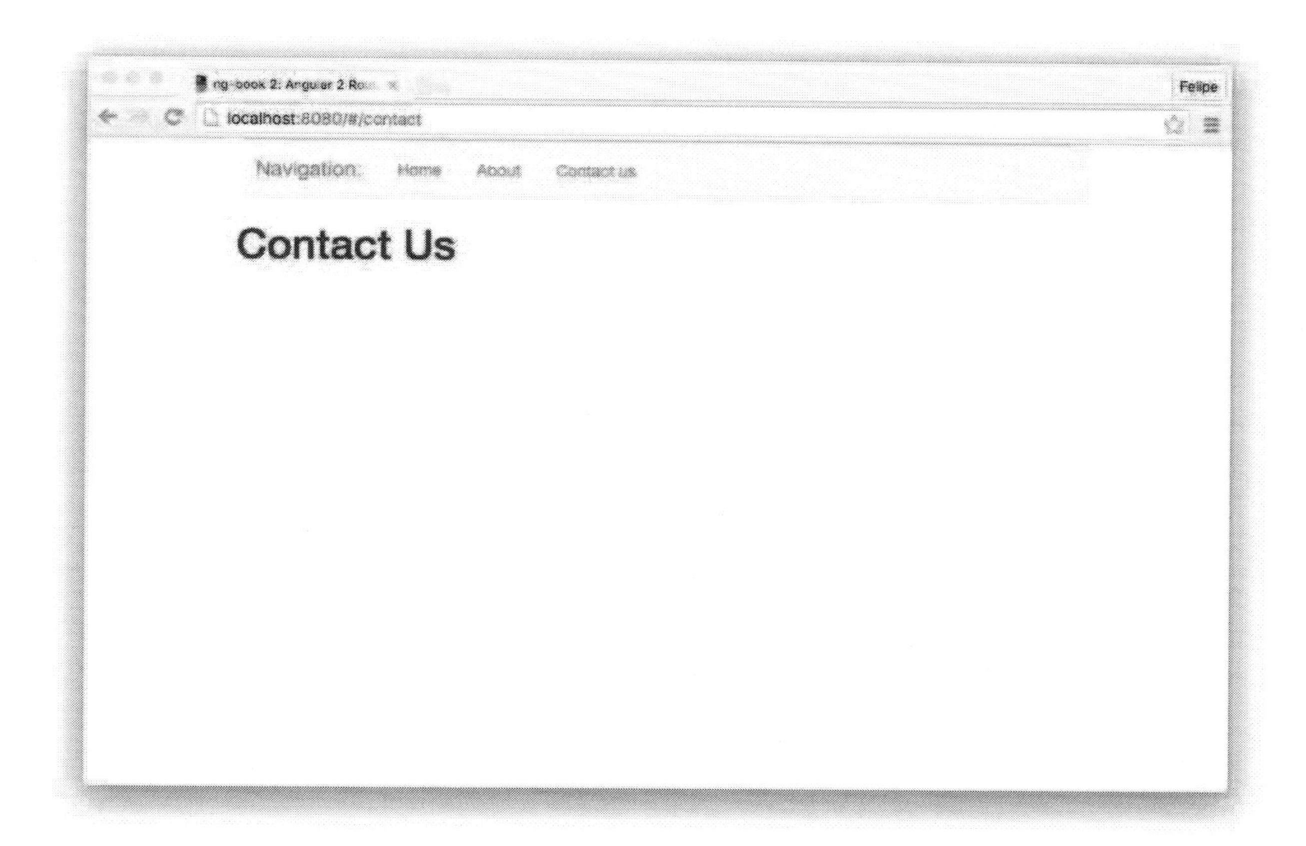


**Contact Us Route**

## Route Parameters

In our apps we often want to navigate to a specific resource. For instance, say we had a news website and we had many articles. Each article may have an ID, and if we had an article with ID `3` then we might navigate to that article by visiting the URL:

`/articles/3`

And if we had an article with an ID of `4` we would access it at

`/articles/4`

and so on.

Obviously we're not going to want to write a route for each article, but instead we want to use a variable, or *route parameter*. We can specify that a route takes a parameter by putting a colon `:` in front of the path segment like this:

`/route/:param`

So in our example news site, we might specify our route as:

`/articles/:id`

To add a parameter to our router configuration, we specify the route path like this:

**code/routes/music/app/ts/app.ts**

```
50 const routes: Routes = [
51   { path: '', redirectTo: 'search', pathMatch: 'full' },
52   { path: 'search', component: SearchComponent },
53   { path: 'artists/:id', component: ArtistComponent },
54   { path: 'tracks/:id', component: TrackComponent },
55   { path: 'albums/:id', component: AlbumComponent },
56 ];
```

When we visit the route `/artist/123`, the `123` part will be passed as the `id` route parameter to our route.

But how can we retrieve the parameter for a given route? That's where we use route parameters.

### ActivatedRoute

In order to use route parameters, we need to first import `ActivatedRoute`:

```
1 import { ActivatedRoute } from '@angular/router';
```

Next, we inject the `ActivatedRoute` into the constructor of our component. For example, let's say we have a `Routes` that specifies the following:

```
1 const routes: Routes = [
2   { path: 'articles/:id', component: ArticlesComponent }
3 ];
```

Then when we write the `ArticleComponent`, we add the `ActivatedRoute` as one of the constructor arguments:

```
1 export class ArticleComponent {
2   id: string;
3
4   constructor(private route: ActivatedRoute) {
5     route.params.subscribe(params => { this.id = params['id']; });
6   }
7 }
```

Notice that `route.params` is an *observable*. We can extract the value of the param into a hard value by using `.subscribe`. In this case, we assign the value of `params['id']` to the `id` instance variable on the component.

Now when we visit `/articles/230`, our component's `id` attribute should receive `230`.

# Music Search App

Let's now work on a more complex application. We will build a music search application that has the following features:

1. **Search for tracks** that match a given term
2. Show **matching tracks** in a grid
3. Show **singer details** when the singer name is clicked
4. Show **album details** and show a list of tracks when the album name is clicked
5. Show **song details** allow the user to **play a preview** when the song name is clicked

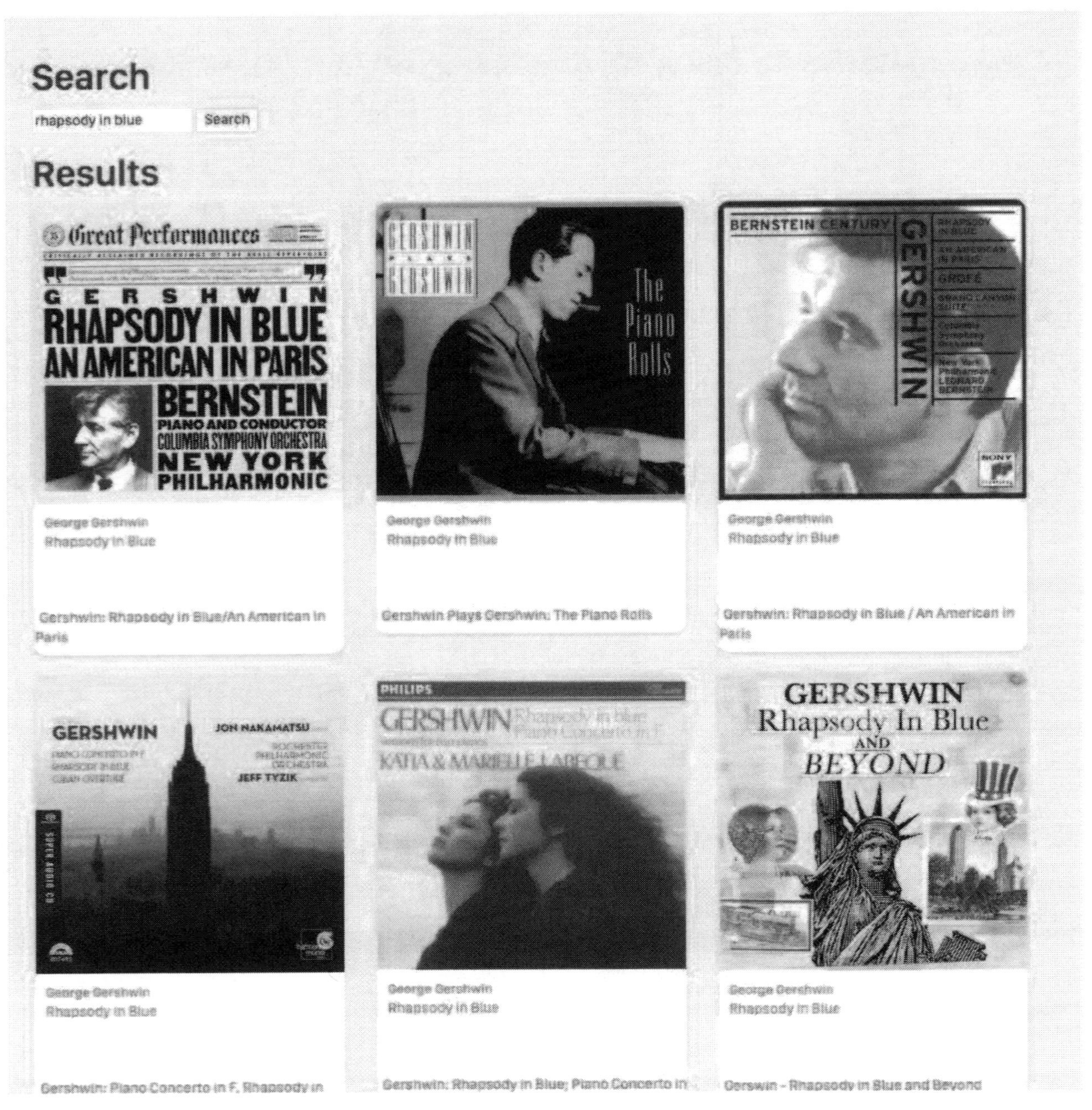


**The Search View of our Music App**

The routes we will need for this application will be:

- `/search` - search form and results
- `/artists/:id` - artist info, represented by a Spotify ID

- `/albums/:id` - album info, with a list of tracks using the Spotify ID
- `/tracks/:id` - track info and preview, also using the Spotify ID

**Sample Code** The complete code for the examples in this section can be found in the `routes/music` folder of the sample code. That folder contains a `README.md`, which gives instructions for building and running the project.

We will use the Spotify API[45] to get information about tracks, artists and albums.

## First Steps

The first file we need work on is `app.ts`. Let's start by importing classes we'll use from Angular:

**code/routes/music/app/ts/app.ts**

```
1  /*
2   * Angular Imports
3   */
4  import {
5    Component
6  } from '@angular/core';
7  import { NgModule } from '@angular/core';
8  import { BrowserModule } from '@angular/platform-browser';
9  import { platformBrowserDynamic } from '@angular/platform-browser-dynamic';
10 import { HttpModule } from '@angular/http';
11 import { FormsModule } from '@angular/forms';
12 import {
13   RouterModule,
14   Routes
15 } from '@angular/router';
16 import {
17   LocationStrategy,
18   HashLocationStrategy,
19   APP_BASE_HREF
20 } from '@angular/common';
21
22 /*
23  * Components
24  */
```

[45] https://developer.spotify.com/web-api

Now that we have the imports there, let's think about the components we'll use for each route.

- For the `Search` route, we'll create a `SearchComponent`. This component will talk to the Spotify API to perform the search and then display the results on a grid.
- For the `Artists` route, we'll create an `ArtistComponent` which will show the artist's information
- For the `Albums` route, we'll create an `AlbumComponent` which will show the list of tracks in the album
- For the `Tracks` route, we'll create a `TrackComponent` which will show the track and let us play a preview of the song

Since this new component will need to interact with the Spotify API, it seems like we need to build a service that uses the `http` module to call out to the API server.

Everything in our app depends on the data, so let's build the `SpotifyService` first.

## The `SpotifyService`

You can find the full code for the `SpotifyService` in the `routes/music/app/ts/services` folder of the sample code.

The first method we'll implement is `searchByTrack` which will search for track, given a search term.

One of the endpoints documented on Spotify API docs is the Search endpoint[46].

This endpoint does exactly what we want: it takes a query (using the `q` parameter) and a `type` parameter.

Query in this case is the search term. And since we're searching for songs, we should use `type=track`.

Here's what a first version of the service could look like:

```
1  class SpotifyService {
2    constructor(public http: Http) {
3    }
4
5    searchByTrack(query: string) {
6      let params: string = [
7        `q=${query}`,
8        `type=track`
9      ].join("&");
```

[46] https://developer.spotify.com/web-api/search-item/

```
10      let queryURL: string = `https://api.spotify.com/v1/search?${params}`;
11      return this.http.request(queryURL).map(res => res.json());
12    }
13  }
```

This code performs an HTTP `GET` request to the URL https://api.spotify.com/v1/search[47], passing our `query` as the search term and `type` hardcoded to `track`.

This `http` call returns an `Observable`. We are going one step further and using the RxJS function `map` to transform the result we would get (which is an `http` module's `Response` object) and parsing it as JSON, resulting on an object.

Any function that calls `searchByQuery` then will have to use the `Observable` API to subscribe to the response like this:

```
1  service
2    .searchTrack('query')
3    .subscribe((res: any) => console.log('Got object', res))
```

## The `SearchComponent`

Now that we have a service that will perform track searches, we can start coding the `SearchComponent`.

Again, we start with an import section:

**code/routes/music/app/ts/components/SearchComponent.ts**

```
1   /*
2    * Angular
3    */
4
5   import {Component, OnInit} from '@angular/core';
6   import {
7     Router,
8     ActivatedRoute,
9   } from '@angular/router';
10
11  /*
12   * Services
13   */
14  import {SpotifyService} from 'services/SpotifyService';
```

[47] https://api.spotify.com/v1/search

Here we're importing, among other things, the SpotifyService class we just created.

The goal here is to render each resulting track side by side on a card like below:

**Music App Card**

We then start coding the component. We're using search as the selector, making a few imports and using the following template. The template is a bit long because we're putting some reasonable styles on it, but it isn't particularly complicated, relative to what we've done so far:

**code/routes/music/app/ts/components/SearchComponent.ts**

```
16 @Component({
17   selector: 'search',
18   template: `
19   <h1>Search</h1>
20 
21   <p>
22     <input type="text" #newquery
23       [value]="query"
24       (keydown.enter)="submit(newquery.value)">
25     <button (click)="submit(newquery.value)">Search</button>
26   </p>
```

```
27
28    <div *ngIf="results">
29      <div *ngIf="!results.length">
30        No tracks were found with the term '{{ query }}'
31      </div>
32
33      <div *ngIf="results.length">
34        <h1>Results</h1>
35
36        <div class="row">
37          <div class="col-sm-6 col-md-4" *ngFor="let t of results">
38            <div class="thumbnail">
39              <div class="content">
40                <img src="{{ t.album.images[0].url }}" class="img-responsive">
41                <div class="caption">
42                  <h3>
43                    <a [routerLink]="['/artists', t.artists[0].id]">
44                      {{ t.artists[0].name }}
45                    </a>
46                  </h3>
47                  <br>
48                  <p>
49                    <a [routerLink]="['/tracks', t.id]">
50                      {{ t.name }}
51                    </a>
52                  </p>
53                </div>
54                <div class="attribution">
55                  <h4>
56                    <a [routerLink]="['/albums', t.album.id]">
57                      {{ t.album.name }}
58                    </a>
59                  </h4>
60                </div>
61              </div>
62            </div>
63          </div>
64        </div>
65      </div>
66    </div>
67    `
68  })
```

### The Search Field

Let's break down the HTML template a bit.

This first section will have the search field:

**code/routes/music/app/ts/components/SearchComponent.ts**

```
21    <p>
22      <input type="text" #newquery
23        [value]="query"
24        (keydown.enter)="submit(newquery.value)">
25      <button (click)="submit(newquery.value)">Search</button>
26    </p>
```

Here we have the input field and we're binding its DOM element `value` property the `query` property of our component.

We also give this element a template variable named `#newquery`. We can now access the value of this input within our template code by using `newquery.value`.

The button will trigger the `submit` method of the component, passing the value of the input field as a parameter.

We also want to trigger `submit` when the user hits "Enter" so we bind to the `keydown.enter` event on the input.

### Search Results and Links

The next section displays the results. We're relying on the `NgFor` directive to iterate through each track from our results object:

**code/routes/music/app/ts/components/SearchComponent.ts**

```
36        <div class="row">
37          <div class="col-sm-6 col-md-4" *ngFor="let t of results">
38            <div class="thumbnail">
```

For each track, we display the artist name:

**code/routes/music/app/ts/components/SearchComponent.ts**

```
42                <h3>
43                  <a [routerLink]="['/artists', t.artists[0].id]">
44                    {{ t.artists[0].name }}
45                  </a>
46                </h3>
```

Notice how we're using the `RouterLink` directive to redirect to `['/artists', t.artists[0].id]`.

This is how we set *route parameters* for a given route. Say we have an artist with an id `abc123`. When this link is clicked, the app would then navigate to `/artist/abc123` (where `abc123` is the `:id` parameter).

Further down we'll show how we can retrieve this value inside the component that handles this route.

Now we display the track:

**code/routes/music/app/ts/components/SearchComponent.ts**

```
48                <p>
49                  <a [routerLink]="['/tracks', t.id]">
50                    {{ t.name }}
51                  </a>
52                </p>
```

And the album:

**code/routes/music/app/ts/components/SearchComponent.ts**

```
55                <h4>
56                  <a [routerLink]="['/albums', t.album.id]">
57                    {{ t.album.name }}
58                  </a>
59                </h4>
```

### `SearchComponent` Class

Let's take a look at the constructor first:

**code/routes/music/app/ts/components/SearchComponent.ts**

```
69 export class SearchComponent implements OnInit {
70   query: string;
71   results: Object;
72 
73   constructor(private spotify: SpotifyService,
74               private router: Router,
75               private route: ActivatedRoute) {
76     this.route
77       .queryParams
78       .subscribe(params => { this.query = params['query'] || ''; });
79   }
```

Here we're declaring two properties:

- `query` for current search term and
- `results` for the search results

On the `constructor` we're injecting the `SpotifyService` (that we created above), `Router`, and the `ActivatedRoute` and making them properties of our class.

In our `constructor` we `subscribe` to the `queryParams` property - this lets us access *query parameters*, such as the search term (`params['query']`).

In a URL like: `http://localhost/#/search?query=cats&order=ascending`, `queryParams` gives us the parameters in an object. This means we could access the `order` with `params['order']` (in this case, `ascending`).

Also note that `queryParams` are different than `route.params`. Whereas `route.params` match parameters in the *route* `queryParams` match parameters in the query string.

In this case, if there is no `query` param, we set `this.query` to the empty string.

#### search

In our `SearchComponent` we will call out to the `SpotifyService` and render the results. There are two cases when we want to run a search:

We want to run a search when the user:

- enters a search query and submits the form
- navigates to this page with a given URL in the query parameters (e.g. someone shared a link or bookmarked the page)

To perform the actual search for both cases, we create the `search` method:

**code/routes/music/app/ts/components/SearchComponent.ts**

```
90    search(): void {
91      console.log('this.query', this.query);
92      if (!this.query) {
93        return;
94      }
95
96      this.spotify
97        .searchTrack(this.query)
98        .subscribe((res: any) => this.renderResults(res));
99    }
```

The `search` function uses the current value of `this.query` to know what to search for. Because we subscribed to the `queryParams` in the constructor, we can be sure that `this.query` will always have the most up-to-date value.

We then subscribe to the `searchTrack` `Observable` and whenever new results are emitted we call `renderResults`.

**code/routes/music/app/ts/components/SearchComponent.ts**

```
101   renderResults(res: any): void {
102     this.results = null;
103     if (res && res.tracks && res.tracks.items) {
104       this.results = res.tracks.items;
105     }
106   }
```

We declared `results` as a component property. Whenever its value is changed, the view will be automatically updated by Angular.

### Searching on Page Load

As we pointed out above, we want to be able to jump straight into the results if the URL includes a search query.

To do that, we are going to implement a hook Angular router provides for us to run whenever our component is initialized.

But isn't that what constructors are for? Well, yes and no. Yes, constructors are used to initialize values, but if you want to write good, testable code, you want to minimize the side effects of *constructing* an object. So keep in mind that you should put your component's initialization login always on a hook like below.

Here's the implementation of the `ngOnInit` method:

**code/routes/music/app/ts/components/SearchComponent.ts**

```
81   ngOnInit(): void {
82     this.search();
83   }
```

> To use `ngOnInit` we imported the `OnInit` class and declared that our component `implements OnInit`.

As you can see, we're just performing the search here. Since the term we're searching for comes from the URL, we're good.

#### `submit`

Now let's see what we do when the user submits the form.

**code/routes/music/app/ts/components/SearchComponent.ts**

```
85   submit(query: string): void {
86     this.router.navigate(['search'], { queryParams: { query: query } })
87       .then(_ => this.search() );
88   }
```

We're manually telling the router to navigate to the search route, and providing a `query` parameter, then performing the actual search.

Doing things this way gives us a great benefit: if we reload the browser, we're going to see the same search result rendered. We can say that we're **persisting the search term on the URL**.

### Putting it all together

Here's the full listing for the `SearchComponent` class:

**code/routes/music/app/ts/components/SearchComponent.ts**

```
1  /*
2   * Angular
3   */
4
5  import {Component, OnInit} from '@angular/core';
6  import {
7    Router,
8    ActivatedRoute,
```

```
9   } from '@angular/router';
10
11  /*
12   * Services
13   */
14  import {SpotifyService} from 'services/SpotifyService';
15
16  @Component({
17    selector: 'search',
18    template: `
19    <h1>Search</h1>
20
21    <p>
22      <input type="text" #newquery
23        [value]="query"
24        (keydown.enter)="submit(newquery.value)">
25      <button (click)="submit(newquery.value)">Search</button>
26    </p>
27
28    <div *ngIf="results">
29      <div *ngIf="!results.length">
30        No tracks were found with the term '{{ query }}'
31      </div>
32
33      <div *ngIf="results.length">
34        <h1>Results</h1>
35
36        <div class="row">
37          <div class="col-sm-6 col-md-4" *ngFor="let t of results">
38            <div class="thumbnail">
39              <div class="content">
40                <img src="{{ t.album.images[0].url }}" class="img-responsive">
41                <div class="caption">
42                  <h3>
43                    <a [routerLink]="['/artists', t.artists[0].id]">
44                      {{ t.artists[0].name }}
45                    </a>
46                  </h3>
47                  <br>
48                  <p>
49                    <a [routerLink]="['/tracks', t.id]">
50                      {{ t.name }}
```

```
51                  </a>
52                </p>
53              </div>
54              <div class="attribution">
55                <h4>
56                  <a [routerLink]="['/albums', t.album.id]">
57                    {{ t.album.name }}
58                  </a>
59                </h4>
60              </div>
61            </div>
62          </div>
63        </div>
64      </div>
65    </div>
66  </div>
67  `
68 })
69 export class SearchComponent implements OnInit {
70   query: string;
71   results: Object;
72 
73   constructor(private spotify: SpotifyService,
74               private router: Router,
75               private route: ActivatedRoute) {
76     this.route
77       .queryParams
78       .subscribe(params => { this.query = params['query'] || ''; });
79   }
80 
81   ngOnInit(): void {
82     this.search();
83   }
84 
85   submit(query: string): void {
86     this.router.navigate(['search'], { queryParams: { query: query } })
87       .then(_ => this.search() );
88   }
89 
90   search(): void {
91     console.log('this.query', this.query);
92     if (!this.query) {
```

```
93          return;
94        }
95
96        this.spotify
97          .searchTrack(this.query)
98          .subscribe((res: any) => this.renderResults(res));
99      }
100
101     renderResults(res: any): void {
102       this.results = null;
103       if (res && res.tracks && res.tracks.items) {
104         this.results = res.tracks.items;
105       }
106     }
107   }
```

## Trying the search

Now that we have completed the code for the search, let's try it out:

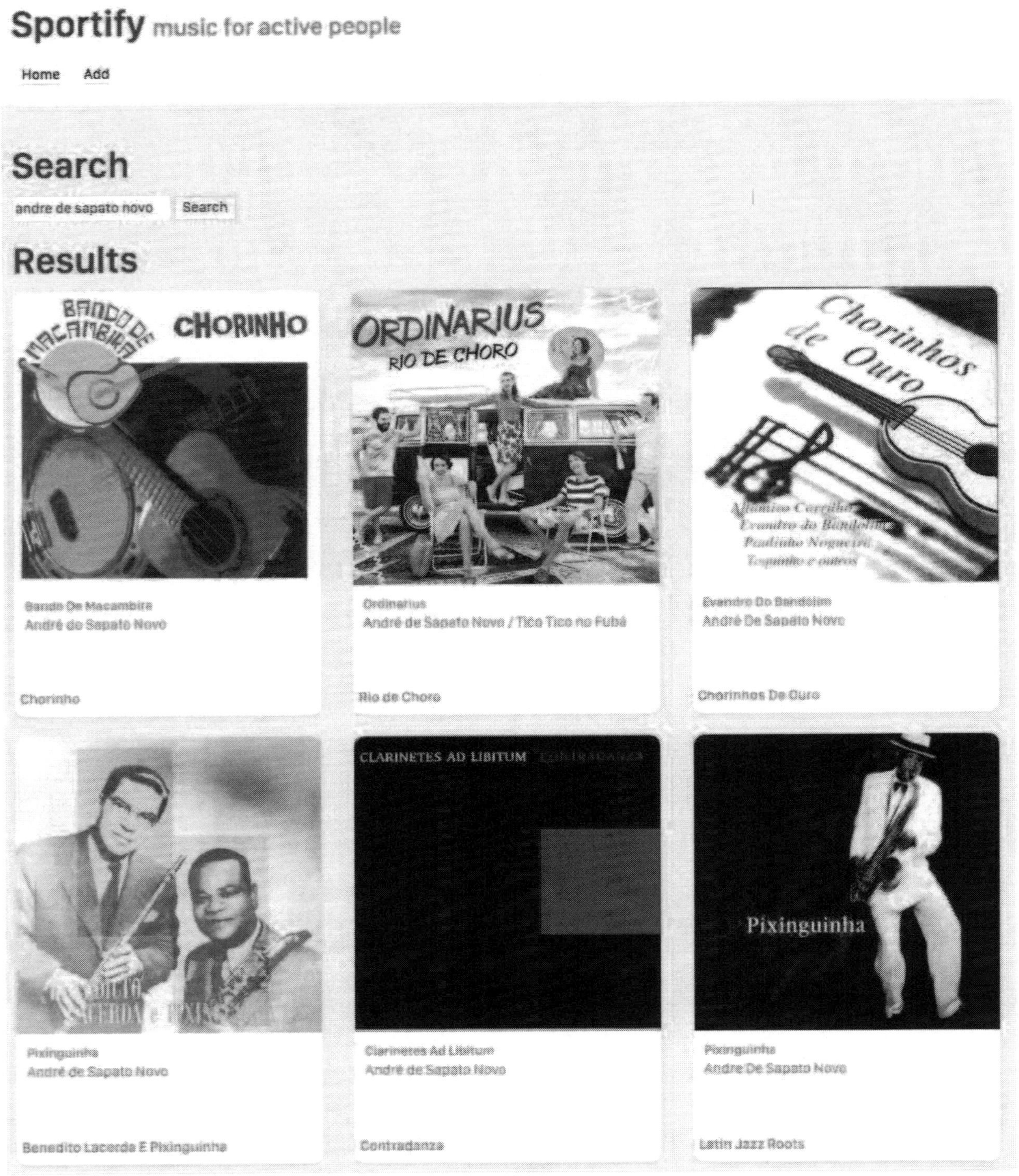

Trying out Search

We can click the artist, track or album links to navigate to the proper route.

### `TrackComponent`

For the track route, we use the `TrackComponent`. It basically displays the track name, the album cover image and allow the user to play a preview using an HTML5 `audio` tag:

**code/routes/music/app/ts/components/TrackComponent.ts**

```
20   template: `
21   <div *ngIf="track">
22     <h1>{{ track.name }}</h1>
23 
24     <p>
25       <img src="{{ track.album.images[1].url }}">
26     </p>
27 
28     <p>
29       <audio controls src="{{ track.preview_url }}"></audio>
30     </p>
31 
32     <p><a href (click)="back()">Back</a></p>
33   </div>
34   `
```

Like we did for the search before, we're going to use the Spotify API. Let's refactor the method `searchTrack` and extract two other useful methods we can reuse:

**code/routes/music/app/ts/services/SpotifyService.ts**

```
13 export class SpotifyService {
14   static BASE_URL: string = 'https://api.spotify.com/v1';
15 
16   constructor(private http: Http) {
17   }
18 
19   query(URL: string, params?: Array<string>): Observable<any[]> {
20     let queryURL: string = `${SpotifyService.BASE_URL}${URL}`;
21     if (params) {
22       queryURL = `${queryURL}?${params.join('&')}`;
23     }
24 
25     return this.http.request(queryURL).map((res: any) => res.json());
26   }
27 
28   search(query: string, type: string): Observable<any[]> {
```

```
29      return this.query(`/search`, [
30        `q=${query}`,
31        `type=${type}`
32      ]);
33    }
```

Now that we've extracted those methods into the `SpotifyService`, notice how much simpler `searchTrack` becomes:

**code/routes/music/app/ts/services/SpotifyService.ts**

```
35    searchTrack(query: string): Observable<any[]> {
36      return this.search(query, 'track');
37    }
```

Now let's create a method to allow the component we're building retrieve track information, based in the track ID:

**code/routes/music/app/ts/services/SpotifyService.ts**

```
39    getTrack(id: string): Observable<any[]> {
40      return this.query(`/tracks/${id}`);
41    }
```

And now we can now use `getTrack` from a new `ngOnInit` method on the `TrackComponent`:

**code/routes/music/app/ts/components/TrackComponent.ts**

```
45    ngOnInit(): void {
46      this.spotify
47        .getTrack(this.id)
48        .subscribe((res: any) => this.renderTrack(res));
49    }
```

The other components work in a similar way and use `get*` methods from the `SpotifyService` to retrieve information about either an Artist or a Track based on their ID.

### Wrapping up music search

Now we have a pretty functional music search and preview app. Try searching for a few of your favorite tunes and try it out!

**It Had to Route You**

## Router Hooks

There are times that we may want to do some action when changing routes. A classical example of that is authentication. Let's say we have a **login** route and a **protected** route.

We want to only allow the app to go to the protected route if the correct username and password were provided on the login page.

In order to do that, we need to hook into the lifecycle of the router and ask to be notified when the protected route is being activated. We then can call an authentication service and ask whether or not the user provided the right credentials.

In order to check if a component can be activated we add a *guard class* to the key `canActivate` in our router configuration.

Let's revisit our initial application, adding login and password input fields and a new protected route that only works if we provide a certain username and password combination.

**Sample Code** The complete code for the examples in this section can be found in the `routes/auth` folder of the sample code. That folder contains a `README.md`, which gives instructions for building and running the project.

### AuthService

Let's create a very simple and minimal implementation of a service, responsible for authentication and authorization of resources:

**code/routes/auth/app/ts/services/AuthService.ts**

```
1  import { Injectable } from '@angular/core';
2
3  @Injectable()
4  export class AuthService {
5    login(user: string, password: string): boolean {
6      if (user === 'user' && password === 'password') {
7        localStorage.setItem('username', user);
8        return true;
9      }
10
11     return false;
12   }
```

The `login` method will return true if the provided user/password pair equals `'user'` and `'password'`, respectively. Also, when it is matched, it's going to use `localStorage` to save the username. This will also serve as a flag to indicate whether or not there is an active logged user.

If you're not familiar, `localStorage` is an HTML5 provided key/value pair that allows you to persist information on the browser. The API is very simple, and basically allows the setting, retrieval and deletion of items. For more information, see the `Storage` interface documents on MDN[48]

The `logout` method just clears the `username` value:

[48] https://developer.mozilla.org/en-US/docs/Web/API/Storage

**code/routes/auth/app/ts/services/AuthService.ts**

```
14    logout(): any {
15      localStorage.removeItem('username');
16    }
```

And the final two methods:

- `getUser` returns the username or null
- `isLoggedIn` uses `getUser()` to return true if we have a user

Here's the code for those methods:

**code/routes/auth/app/ts/services/AuthService.ts**

```
18    getUser(): any {
19      return localStorage.getItem('username');
20    }
21
22    isLoggedIn(): boolean {
23      return this.getUser() !== null;
24    }
```

The last thing we do is export an `AUTH_PROVIDERS`, so it can be injected into our app:

**code/routes/auth/app/ts/services/AuthService.ts**

```
27  export var AUTH_PROVIDERS: Array<any> = [
28    { provide: AuthService, useClass: AuthService }
29  ];
```

Now that we have the `AuthService` we can inject it in our components to log the user in, check for the currently logged in user, log the user out, etc.

In a little bit, we'll also use it in our router to protect the `ProtectedComponent`. But first, let's create the component that we use to log in.

### `LoginComponent`

This component will either show a login form, for the case when there is no logged user, or display a little banner with user information along with a logout link.

The relevant code here is the `login` and `logout` methods:

**code/routes/auth/app/ts/components/LoginComponent.ts**

```
40 export class LoginComponent {
41   message: string;
42 
43   constructor(private authService: AuthService) {
44     this.message = '';
45   }
46 
47   login(username: string, password: string): boolean {
48     this.message = '';
49     if (!this.authService.login(username, password)) {
50       this.message = 'Incorrect credentials.';
51       setTimeout(function() {
52         this.message = '';
53       }.bind(this), 2500);
54     }
55     return false;
56   }
57 
58   logout(): boolean {
59     this.authService.logout();
60     return false;
61   }
```

Once our service validates the credentials, we log the user in.

The component template has two snippets that are displayed based on whether the user is logged in or not.

The first is a login form, protected by `*ngIf="!authService.getUser()"`:

**code/routes/auth/app/ts/components/LoginComponent.ts**

```
18   <form class="form-inline" *ngIf="!authService.getUser()">
19     <div class="form-group">
20       <label for="username">User:</label>
21       <input class="form-control" name="username" #username>
22     </div>
23 
24     <div class="form-group">
25       <label for="password">Password:</label>
26       <input class="form-control" type="password" name="password" #password>
27     </div>
```

```
28
29      <a class="btn btn-default" (click)="login(username.value, password.value)">
30        Submit
31      </a>
32    </form>
```

And the information banner, containing the logout link, protected by the inverse - `*ngIf="authService.getUser()"`:

**code/routes/auth/app/ts/components/LoginComponent.ts**

```
34    <div class="well" *ngIf="authService.getUser()">
35      Logged in as <b>{{ authService.getUser() }}</b>
36      <a href (click)="logout()">Log out</a>
37    </div>
```

There's another snippet of code that is displayed when we have an authentication error:

**code/routes/auth/app/ts/components/LoginComponent.ts**

```
14    <div class="alert alert-danger" role="alert" *ngIf="message">
15      {{ message }}
16    </div>
```

Now that we can handle the user login, let's create a resource that we are going to protect behind a user login.

## `ProtectedComponent` and Route Guards

### The `ProtectedComponent`

Before we can protect the component, it needs to exist. Our `ProtectedComponent` is straightforward:

**code/routes/auth/app/ts/components/ProtectedComponent.ts**

```
/*
 * Angular
 */
import {Component} from '@angular/core';

@Component({
  selector: 'protected',
  template: `<h1>Protected content</h1>`
})
export class ProtectedComponent {
}
```

We want this component to only be accessible to logged in users. But how can we do that?

The answer is to use the router hook `canActivate` with a *guard class* that implements `CanActivate`.

### The `LoggedInGuard`

We create a new folder called `guards` and create `loggedIn.guard.ts`:

**code/routes/auth/app/ts/guards/loggedIn.guard.ts**

```
import { Injectable } from '@angular/core';
import { CanActivate } from '@angular/router';
import { AuthService } from 'services/AuthService';

@Injectable()
export class LoggedInGuard implements CanActivate {
  constructor(private authService: AuthService) {}

  canActivate(): boolean {
    return this.authService.isLoggedIn();
  }
}
```

Our guard states that it implements the `CanActivate` interface. This is satisfied by implementing a method `canActive`.

We inject the `AuthService` into this class in the constructor and save it as a private variable `authService`.

In our `canActivate` function we check `this.authService` to see if the user `isLoggedIn`.

## Configuring the Router

To configure the router to use this guard we need to do the following:

1. `import` the `LoggedInGuard`
2. Use the `LoggedInGuard` in a route configuration
3. Include `LoggedInGuard` in the list of providers (so that it can be injected)

We do all of these steps in our `app.ts`.

We import the `LoggedInGuard`:

**code/routes/auth/app/ts/app.ts**

```
28  import {AUTH_PROVIDERS} from 'services/AuthService';
29  import {LoggedInGuard} from 'guards/loggedIn.guard';
```

We add `canActivate` with our guard to the `protected` route:

**code/routes/auth/app/ts/app.ts**

```
68  const routes: Routes = [
69    { path: '',           redirectTo: 'home', pathMatch: 'full' },
70    { path: 'home',       component: HomeComponent },
71    { path: 'about',      component: AboutComponent },
72    { path: 'contact',    component: ContactComponent },
73    { path: 'protected', component: ProtectedComponent,
74      canActivate: [LoggedInGuard]}
75  ];
```

We add `LoggedInGuard` to our list of providers:

**code/routes/auth/app/ts/app.ts**

```
91    bootstrap: [ RoutesDemoApp ],
```

## Logging in

We import the `LoginComponent`:

**code/routes/auth/app/ts/app.ts**

```
19 import {LoginComponent} from 'components/LoginComponent';
```

And then we have to add:

1. a new link to the protected route
2. a `<login>` tag to the template, to render the new component

Here's what it should look like:

**code/routes/auth/app/ts/app.ts**

```
36 @Component({
37   selector: 'router-app',
38   template: `
39   <div class="page-header">
40     <div class="container">
41       <h1>Router Sample</h1>
42       <div class="navLinks">
43         <a [routerLink]="['/home']">Home</a>
44         <a [routerLink]="['/about']">About</a>
45         <a [routerLink]="['/contact']">Contact us</a>
46         <a [routerLink]="['/protected']">Protected</a>
47       </div>
48     </div>
49   </div>
50 
51   <div id="content">
52     <div class="container">
53 
54       <login></login>
55 
56       <hr>
57 
58       <router-outlet></router-outlet>
59     </div>
60   </div>
61   `
62 })
63 class RoutesDemoApp {
64   constructor(private router: Router) {
65   }
66 }
```

Now when we open the application on the browser, we can see the new login form and the new protected link:

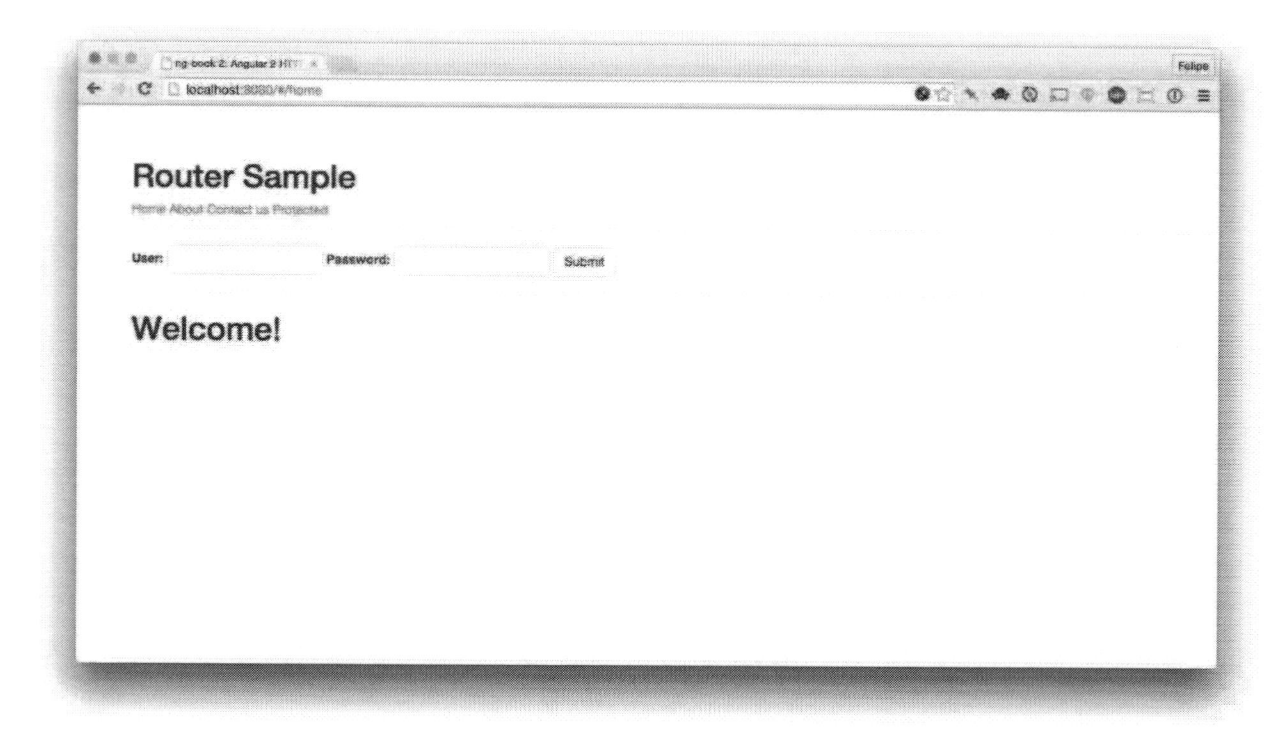


**Auth App - Initial Page**

If you click the Protected link, you'll see nothing happens. The same happens if you try to manually visit http://localhost:8080/#/protected[49].

Now enter *user* and *password* on the form and click **Submit**. You'll see that we now get the current user displayed on a banner:

[49]http://localhost:8080/#/protected

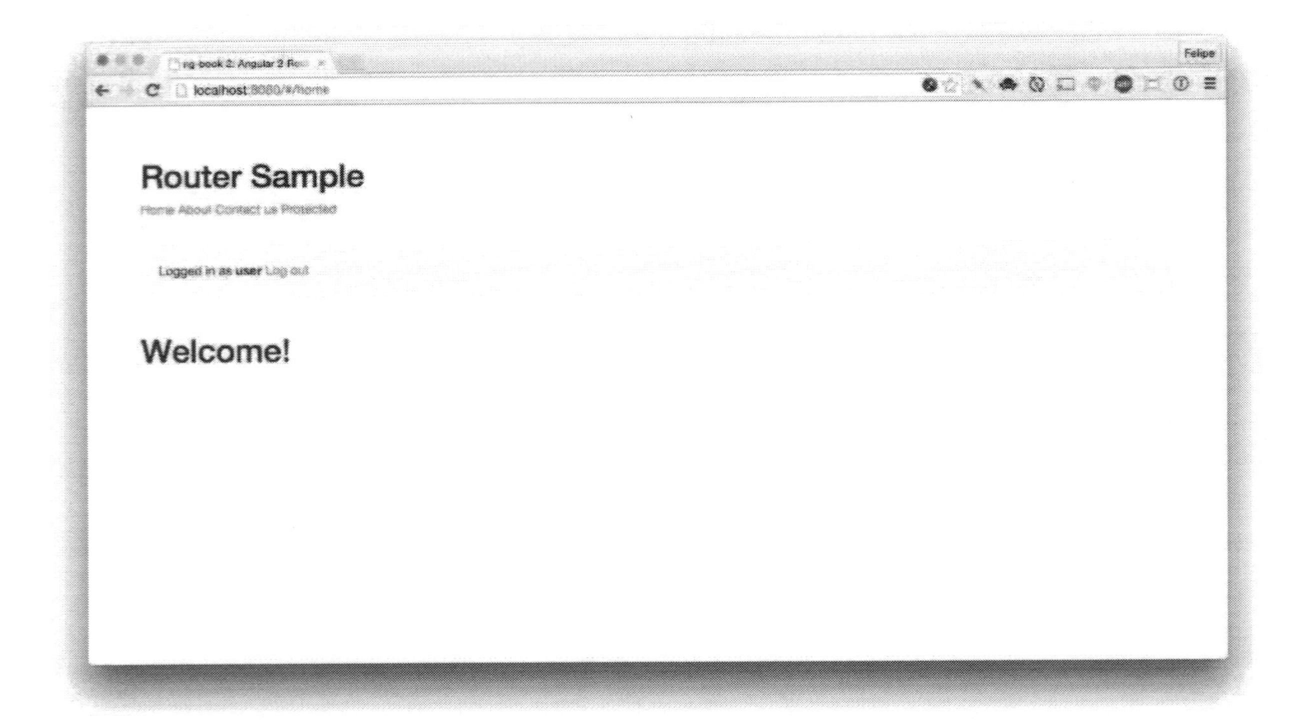


**Auth App - Logged In**

And, sure enough, if we click the Protected link, it gets redirected and the component is rendered:

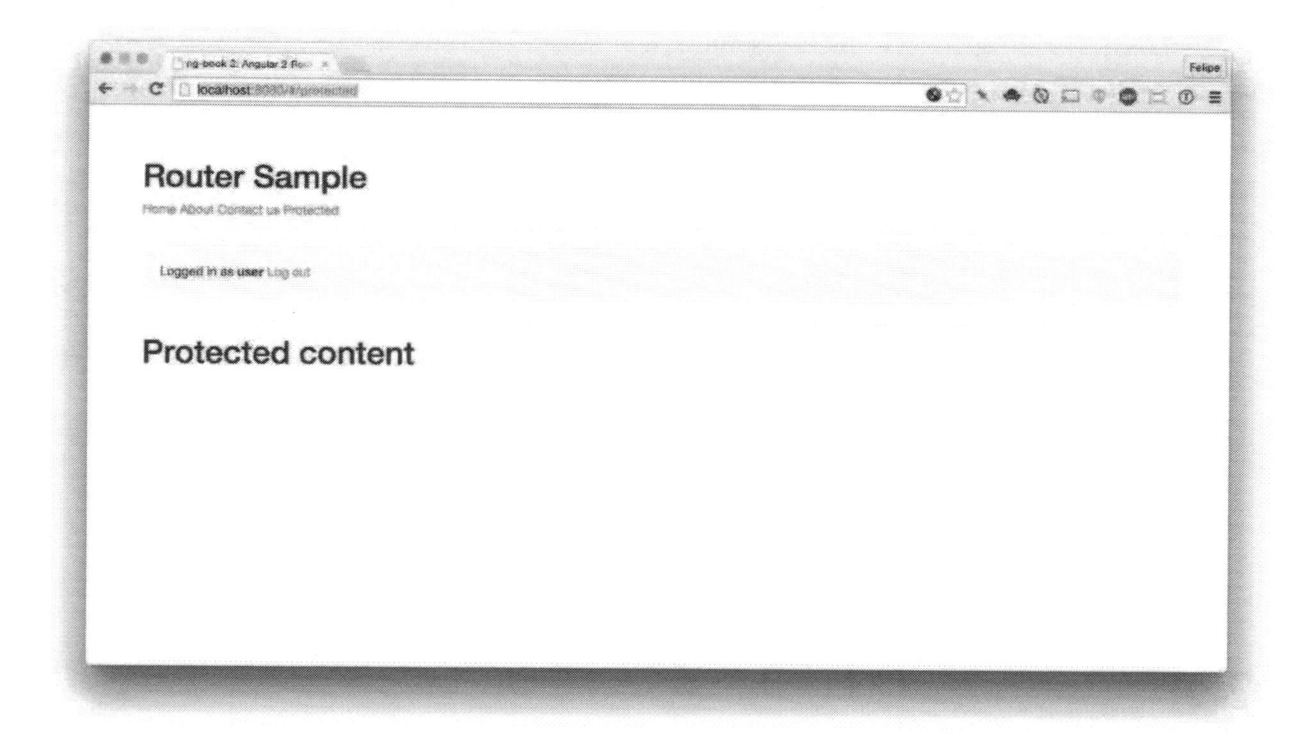


**Auth App - Protected Area**

**A Note on Security**: It's important to know how client-side route protection is working before you rely too heavily on it for security. That is, you should consider client-side route protection a form of *user-experience* and not one of security.

Ultimately all of the javascript in your app that gets served to the client can be inspected, whether the user is logged in or not.

So if you have sensitive data that needs to be protected, you must protect it with **server-side authentication**. That is, require an API key (or auth token) from the user which is validated by the server on every request for data.

Writing a full-stack authentication system is beyond the scope of this book. The important thing to know is that protecting routes on the client-side don't necessarily keep anyone from viewing the javascript pages behind those routes.

# Nested Routes

Nested routes is the concept of containing routes within other routes. With nested routes we're able to encapsulate the functionality of parent routes and have that functionality apply to the child routes.

Let's say we have a website with one area to allow users to know our team, called **Who we are?** and another one for our **Products**.

We could think that the perfect route for **Who we are?** would be `/about` and for products `/products`.

And we're happily displaying all our team and all our products when visiting this areas.

What happens when the website grows and we now need to display individual information about each person in our team and also for each product we sell?

In order to support scenarios like these, the router allows the user to define nested routes.

To do that, you can have multiple, nested `router-outlet`. So each area of our application can have their own child components, that also have their own `router-outlets`.

Let's work on an example to clear things up.

In this example, we'll have a products section where the user will be able to view two highlighted products by visiting a nice URL. For all the other products, the routes will use the product ID.

**Sample Code** The complete code for the examples in this section can be found in the `routes/nested` folder of the sample code. That folder contains a `README.md`, which gives instructions for building and running the project.

## Configuring Routes

We will start by describing two top-level routes on the `app.ts` file:

**code/routes/nested/app/ts/app.ts**

```
58 const routes: Routes = [
59   { path: '', redirectTo: 'home', pathMatch: 'full' },
60   { path: 'home', component: HomeComponent },
61   { path: 'products', component: ProductsComponent, children: childRoutes }
62 ];
```

The `home` route looks very familiar, notice that `products` has a `children` parameter. Where does this come from? We've defined the `childRoutes` alongside the `ProductsComponent`. Let's take a look:

### `ProductsComponent`

This component will have its own route configuration:

**code/routes/nested/app/ts/components/ProductsComponent.ts**

```
51 export const routes: Routes = [
52   { path: '', redirectTo: 'main' },
53   { path: 'main', component: MainComponent },
54   { path: ':id', component: ByIdComponent },
55   { path: 'interest', component: InterestComponent },
56   { path: 'sportify', component: SportifyComponent },
57 ];
```

Notice here that we have an empty `path` on the first object. We do this so that when we visit `/products`, we'll be redirected to the `main` route.

The other route we need to look at is `:id`. In this case, when the user visits something *that doesn't match any other route*, it will fallback to this route. Everything that is passed after / will be extracted to a parameter of the route, called `id`.

Now on the component template, we'll have a link to each of those static child routes:

**code/routes/nested/app/ts/components/ProductsComponent.ts**

```
29     <a [routerLink]="['./main']">Main</a> |
30     <a [routerLink]="['./interest']">Interest</a> |
31     <a [routerLink]="['./sportify']">Sportify</a> |
```

You can see that the route links are all in the format `['./main']`, with a preceding `./`. This indicates that you want to navigate the Main route *relative to the current route context.*

You could also declare the routes with the `['products', 'main']` notation. The downside is that by doing it this way, the child route is aware of the parent route and if you were to move this component around or reuse it, you would have to rewrite your route links.

After the links, we'll add an input where the user will be able to enter a product id, along with a button to navigate to it, and lastly add our `router-outlet`:

**code/routes/nested/app/ts/components/ProductsComponent.ts**

```
25   template: `
26   <h2>Products</h2>
27 
28   <div class="navLinks">
29     <a [routerLink]="['./main']">Main</a> |
30     <a [routerLink]="['./interest']">Interest</a> |
31     <a [routerLink]="['./sportify']">Sportify</a> |
32     Enter id: <input #id size="6">
```

```
33      <button (click)="goToProduct(id.value)">Go</button>
34    </div>
35
36    <div class="products-area">
37      <router-outlet></router-outlet>
38    </div>
39    `
```

Let's look at the ProductsComponent definition:

**code/routes/nested/app/ts/components/ProductsComponent.ts**

```
42  export class ProductsComponent {
43    constructor(private router: Router, private route: ActivatedRoute) {
44    }
45
46    goToProduct(id:string): void {
47      this.router.navigate(['./', id], {relativeTo: this.route});
48    }
49  }
```

First on the constructor we're declaring an instance variable for the `Router`, since we're going to use that instance to navigate to the product by id.

When we want to go to a particular product we use the `goToProduct` method. In `goToProduct` we call the router's `navigate` method and providing the route name and an object with route parameters. In our case we're just passing the `id`.

Notice that we use the relative `./` path in the `navigate` function. In order to use this we also pass the `relativeTo` object to the options, which tells the router what that route is relative to.

Now, if we run the application we will see the main page:

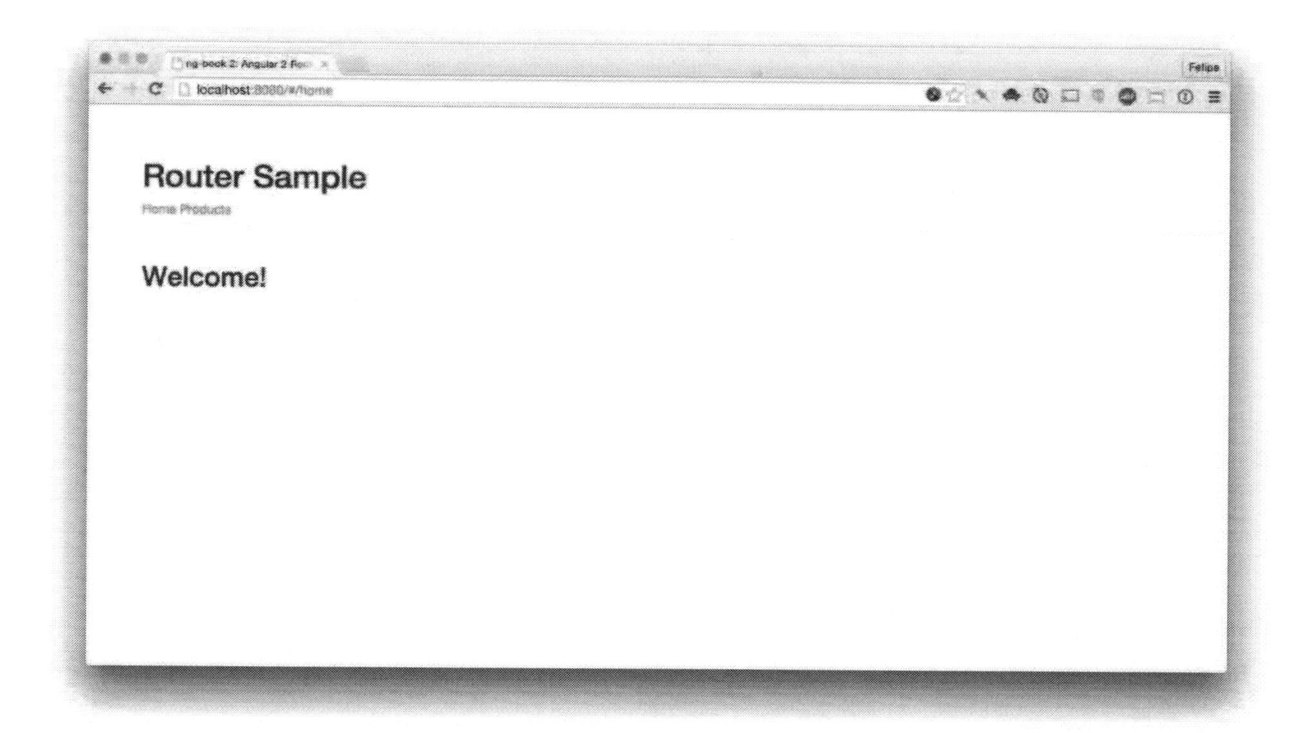


**Nested Routes App**

If you click on the Products link, you'll be redirected to `/products/main` that will render as follows:

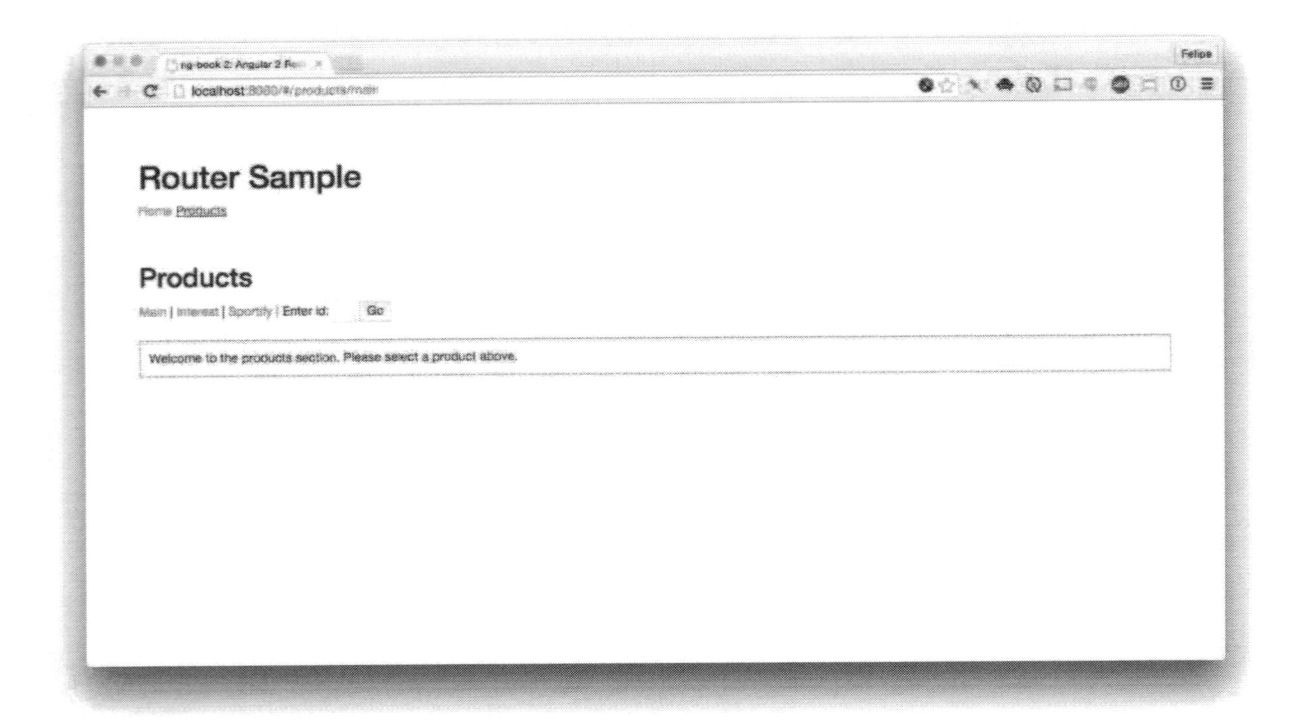


**Nested Routes App - Products Section**

Everything below that thin grey line is being rendered using the main application's `router-outlet`.

And the contents of the dotted red line is being rendered inside the `ProductComponent`'s `router-outlet`. That's how you indicate how the parent and child routes will be rendered.

When we visit one of the product links, or if we an ID on the textbox and click Go, the new content is rendered inside the `ProductComponent`'s outlet:

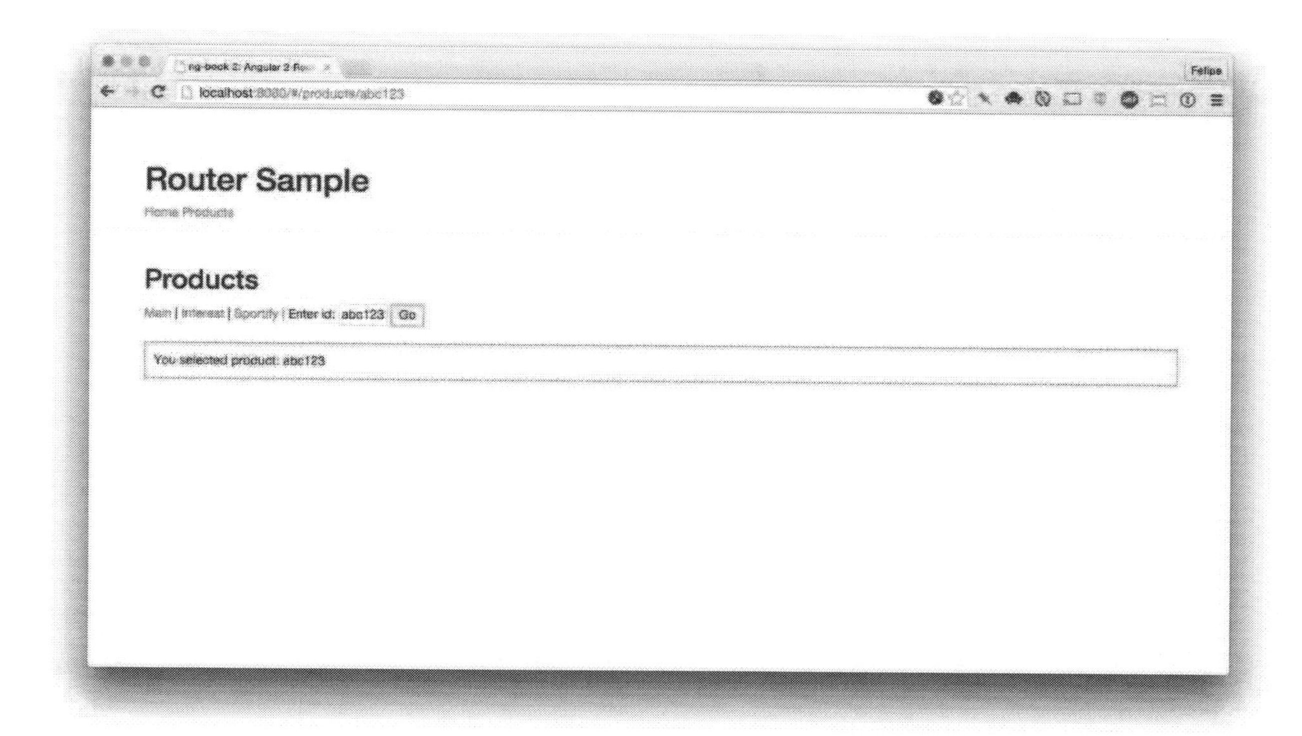


**Nested Routes App - Product By Id**

It's also worth noting that the Angular router is smart enough to prioritize concrete routes first (like `/products/sportify`) over the parameterized ones (like `/products/123`). This way `/products/sportify` will never be handled by the more generic, catch-all route `/products/:id`.

### Redirecting and linking nested routes

Just to recap, if we want to go to a route named `MyRoute` on your top-level routing context, you use `['myRoute']`. This will only work if you're in that same top-level context.

If you are on a child component, and you try to link or redirect to `['myRoute']`, it will try to find a sibling route, and error out. In this case, you need to use `['/myRoute']` with a leading slash.

In a similar way, if we are on the top-level context and we want to link or redirect to a child route, we have to need to use multiple elements on the route definition array.

Let's say we want to visit the `Show` route, which is a child of the `Product` route. In this case, we use `['product', 'show']` as the route definition.

## Summary

As we can see, the new Angular router is very powerful and flexible. Now go out and route your apps!

# Dependency Injection

As our programs grow in size, we often find that different parts of the app need to communicate with other modules. When module `A` requires module `B` to run, we say that `B` is a *dependency* of `A`.

One of the most common ways to get access to dependencies is to simply `import` a file. For instance, in this hypothetical module we might do the following:

```
// in A.ts
import {B} from 'B'; // a dependency!

B.foo(); // using B
```

In many cases, simply `import`ing other code is sufficient. However there are times where we need to provide dependencies in a more sophisticated way. For instance:

- What if we wanted to substitute out the implementation of `B` for `MockB` during testing?
- What if we wanted to share a *single instance* of the `B` class across our whole app (e.g. the *Singleton* pattern)
- What if we wanted to create a *new instance* of the `B` class every time it was used? (e.g. the *Factory* pattern)

Dependency Injection can solve these problems.

Dependency Injection (DI) is a system to make parts of our program accessible to other parts of the program - and we can configure how that happens.

One way to think about an injector is as a replacement for the `new` operator

The term Dependency Injection is used to describe both a design pattern (that used in many different frameworks) and also the specific implementation DI library that is built-in to Angular.

The major benefit of using dependency injection is that the client component doesn't have to be aware of how to create the dependencies, all the component needs to know is how to *interact* with those dependencies.

## Injections Example: `PriceService`

Let's imagine we have a `Product` class. Each product has a base price. In order to calculate the full price for this product, we rely on a service that takes as input

- the **base price** of the product and
- the **state** we're selling it to.

Here's how this would look without dependency injection:

```
1  class Product {
2    constructor(basePrice: number) {
3      this.service = new PriceService();
4      this.basePrice = basePrice;
5    }
6
7    price(state: string) {
8      return this.service.calculate(this.basePrice, state);
9    }
10 }
```

Now let's imagine we need to write a test for this `Product` class. Let's assume the `PriceService` class above uses a database lookup to retrieve taxes for a given state. If we write a test like:

```
1  let product;
2
3  beforeEach(() => {
4    product = new Product(11);
5  });
6
7  describe('price', () => {
8    it('is calculated based on the basePrice and the state', () => {
9      expect(product.price('FL')).toBe(11.66);
10   });
11 })
```

Even though the test may work, there are a few shortcomings to this approach. In order for the test to success a few preconditions have to be met:

1. The database must be running;

2. The tax entry for Florida must be what we're expecting;

Basically we're making our tests more brittle by adding an unexpected dependency between the `Product` class and the `PriceService` that, in turn, depends on a database.

What if we could write the `Product` class a little differently:

```
1  class Product {
2    constructor(service: PriceService, basePrice: number) {
3      this.service = service;
4      this.basePrice = basePrice;
5    }
6
7    price(state: string) {
8      return this.service.calculate(this.basePrice, state);
9    }
10 }
```

Now, when creating a `Product` the client class becomes responsible for deciding which concrete implementation of the `PriceService` is going to be given to the new instance.

With that, we can make our tests a lot simpler by creating a *mock* version of the `PriceService` class:

```
1  class MockPriceService {
2    calculate(basePrice: number, state: string) {
3      if (state === 'FL') {
4        return basePrice * 1.06;
5      }
6
7      return basePrice;
8    }
9  }
```

And with this small change, we can tweak our test slightly and get rid of the database dependency:

```
1  let product;
2
3  beforeEach(() => {
4    const service = new MockPriceService();
5    product = new Product(service, 11);
6  });
7
8  describe('price', () => {
9    it('is calculated based on the basePrice and the state', () => {
10     expect(product.price('FL')).toBe(11.66);
11   });
12 })
```

We also get the bonus of having confidence that we're testing the `Product` class *in isolation*. That is, we're making sure that our class works with a predictable dependency.

## "Don't Call Us..."

This technique of injecting the dependencies relies on a principle called the inversion of control.

The inversion of control (or IoC) principle is also called informally the "Hollywood principle", that is a reference to the Hollywood motto "don't call us, we'll call you".

Over the years it was very common for every component to be aware of the complete application context and be responsible for creating and setting up the dependencies. This can be seen clearly on our example, where the `Product` class had to be aware of the PriceService.

The setback of doing things that way is that once a component becomes aware of the dependency, the component itself becomes more brittle and therefore harder to change. If we make change to a component on which many other components are dependent upon, we end up having to propagate the changes to a lot of different areas of our application and sometimes even outside the boundaries of it. In other words, we're making our components *tightly coupled*.

When we use DI we are moving towards a more *loosely coupled* architecture where changing bits and pieces of a single component affects the other areas of the application less. And, as long as the interface between those components don't change, we can even swap them altogether, without any other components even realizing.

One of the great features that ng2 inherited from ng1 is that both frameworks uses this Inversion of Control pattern. Angular uses *dependency injection* to resolve dependencies out of the box.

Traditionally, if a component A depends on component B, what would happen is that an instance of B would be created inside A. This implies that now A depends on B.

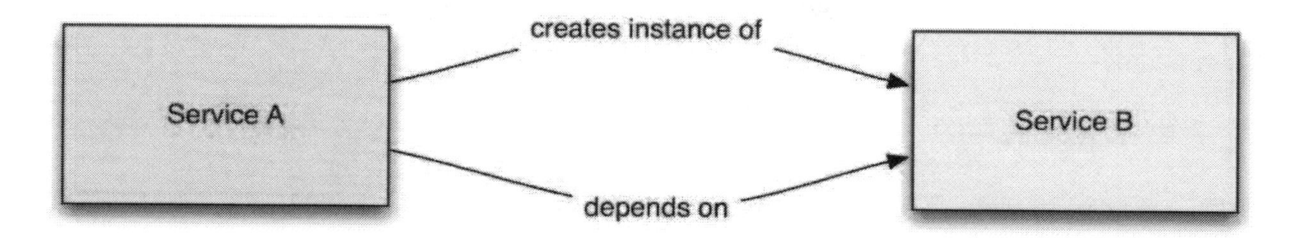


**Without a Dependency Injection Framework**

Angular uses the Dependency Injection to change things around in a way that if we need component B inside component A, we expect that B will be *passed* to A.

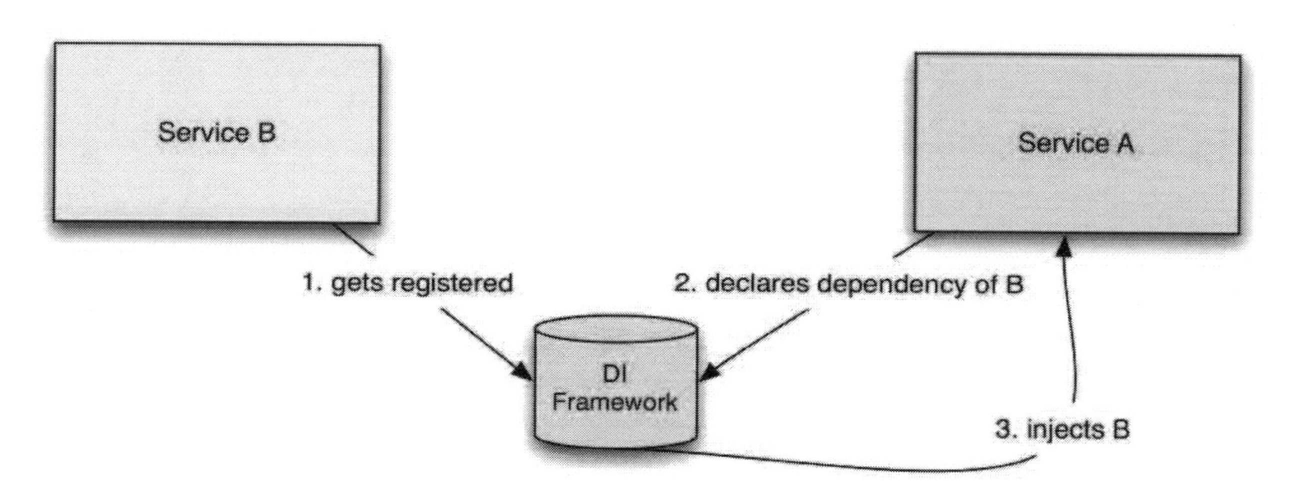


**With a Dependency Injection Framework**

This brings many advantages over the traditional scenario. One example of an advantage is that, if we're testing A in isolation we can easily create a mocked version of B and inject it into A.

We have used services and therefore dependency injection a lot of times earlier in this book. For example, when we created the music application back on the Routing chapter. To interact with the Spotify API, we created the **SpotifyService** that was injected on a number of components as we can see on this snippet from the **AlbumComponent**:

**code/routes/music/app/ts/components/AlbumComponent.ts**

```
37  export class AlbumComponent implements OnInit {
38    id: string;
39    album: Object;
40
41    constructor(private route: ActivatedRoute,
42                private spotify: SpotifyService, // <-- injected
43                private location: Location) {
44      route.params.subscribe(params => { this.id = params['id']; });
45    }
```

Now let's learn how to create our own services and the different forms we can inject them.

## Dependency Injection Parts

To register a dependency we have to bind it to something that will identify that dependency. This identification is called the dependency **token**. For instance, if we want to register the URL of an API, we can use the string `API_URL` as the token. Similarly, if we're registering a class, we can use the class itself as its **token** as we'll see below.

Dependency injection in Angular has three pieces:

- the **Provider** (also often referred to as a binding) maps a *token* (that can be a string or a class) to a list of dependencies. It tells Angular how to create an object, given a token.
- the **Injector** that holds a set of bindings and is responsible for resolving dependencies and injecting them when creating objects
- the **Dependency** that is what's being injected

We can think of the role of each piece as illustrated below:

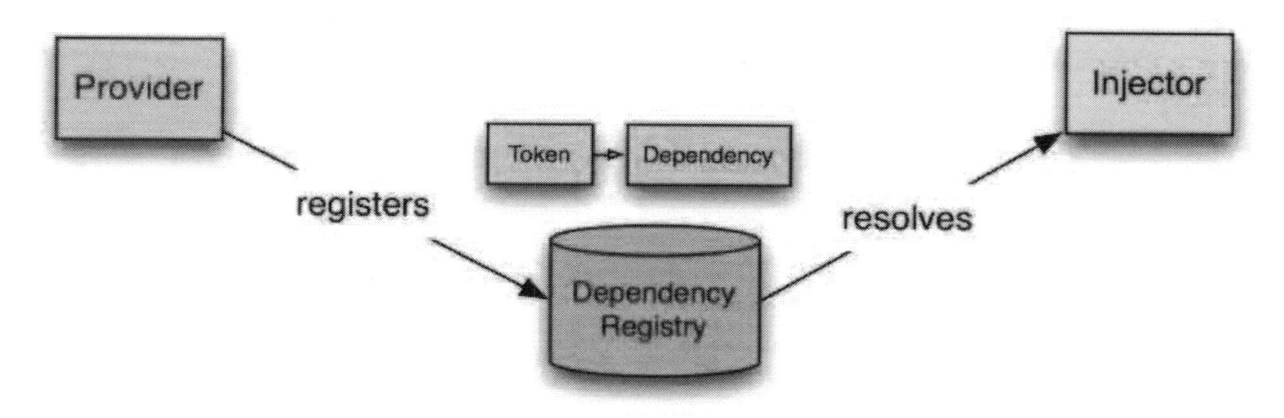


**Dependency Injection**

There are a lot of different options when dealing with DI, so let's see how each of them work.

One of the most common cases is providing a service or value that is the same across our whole application. This scenario would be what we would use 99% of the time in our apps.

If this is all we want to do, we'll cover how to write a basic service in the next section and that is going to be all we need for most of our apps most of the time.

Enough talk, let's code!

## Playing with an Injector

As mentioned above, Angular is going to setup DI for us behind the scenes. But before we deal with annotations and the integrating injections into our components, let's first play with the injector by itself.

Let's create a sample service that only returns a string:

**code/dependency_injection/injector/app/ts/app.ts**

```
17  /*
18   * The injectable service
19   */
20  class MyService {
21    getValue(): string {
22      return 'a value';
23    }
24  }
```

Next, we want to create the app component:

**code/dependency_injection/injector/app/ts/app.ts**

```
26 @Component({
27   selector: 'di-sample-app',
28   template: `
29   <button (click)="invokeService()">Get Value</button>
30   `
31 })
32 class DiSampleApp {
33   myService: MyService;
34 
35   constructor() {
36     let injector: any = ReflectiveInjector.resolveAndCreate([MyService]);
37     this.myService = injector.get(MyService);
38     console.log('Same instance?', this.myService === injector.get(MyService));
39   }
40 
41   invokeService(): void {
42     console.log('MyService returned', this.myService.getValue());
43   }
44 }
```

Let's break things down a bit. We are declaring the `DiSampleApp` component that will render a button. When that button is clicked we call the `invokeService` method.

Focusing on the constructor of the component we can see that we are using a static method from `ReflectiveInjector` called `resolveAndCreate`. That method is responsible for creating a new injector. The parameter we pass in is an array with all the *injectable things* we want this new injector to *know*. In our case, we just wanted it to know about the `MyService` injectable.

The `ReflectiveInjector` is a concrete implementation of `Injector` that uses *reflection* to look up the proper parameter types. While there are other injectors that are possible `ReflectiveInjector` is the "normal" injector we'll be using in most apps.

One important thing to notice is that it will inject a **singleton** instance of the class.

This can be verified by the last two lines of our constructor. We are first asking our newly created injector to give us the instance for the `MyService` class. We then store that into our component's `myService` field. Right after that, we have a `console.log` that asks the injector to give us the instance of `MyService` again. When the result of the comparison of the next line executes:

```
1 console.log('Same instance?', this.myService === injector.get(MyService));
```

We get the confirmation that both instances are actually the exact same object on the console:

```
1 Same instance? true
```

Notice that, since we're using our own injector, we didn't have to add `MyService` to the `NgModule` `providers` list as we're used to during bootstrapping:

**code/dependency_injection/injector/app/ts/app.ts**

```
47 @NgModule({
48   declarations: [ DiSampleApp ],
49   imports: [ BrowserModule ],
50   bootstrap: [ DiSampleApp ]
51 })
52 class DiSampleAppModule {}
53
54 platformBrowserDynamic().bootstrapModule(DiSampleAppModule);
```

## Providing Dependencies with `NgModule`

However, *normally* we **do** need to tell our `NgModule` about the *providers* of things we will inject.

For instance, say we wanted to allow a singleton instance `MyService` to be injected across our app.

In order to be able to inject these things we must *add them to the `providers` key of a `NgModule`*. Here's an example:

```
1 @NgModule({
2   declarations: [
3     MyAppComponent,
4     // other components ...
5   ],
6   providers: [ MyService ]   // <--- here
7 })
8 class MyAppModule {}
```

Now `MyAppComponent` could inject `MyService` in the constructor like so:

```
1 export class MyAppComponent {
2
3   constructor(private myService: MyService /* <-- injected */) {
4     // do something with myService here
5   }
6
7   // ...
8 }
```

When we put the class itself into the list of `providers` like this:

```
1   providers: [ MyService ]
```

That is telling Angular that we want to provide a singleton instance of `MyService` whenever `MyService` is injected. Because this pattern is so common, the class by itself is shorthand notation for the following, equivalent configuration:

```
1   providers: [
2     { provide: MyComponent, useClass: MyComponent }
3   ]
```

There are many different ways of injecting things beyond creating an instance of a class. let's take a look.

# Providers

One of the neat things about Angular's DI system is that there are several ways we can configure the injection. For instance we can:

- Inject a (singleton) instance of a class
- Call any function and inject the return value of that function
- Inject a value
- Create an alias

For instance,

Let's look at how we could create each one:

## Using a Class

Injecting a singleton instance of a class is probably the most common type of injection.

Here's how we configure it:

```
1     { provide: MyComponent, useClass: MyComponent }
```

What's interesting to note is that the `provide` configuration method takes **two** keys. The first `provide` is the *token* that we use to identify the injection and the second `useClass` is how and what to inject.

So here we're mapping the `MyComponent` class to the `MyComponent` token. In this case, the name of the class and the token match. This is the common case, but know that the token and the injected thing *don't have to have the same name.*

As we've seen above, in this case the injector will create a **singleton** behind the scenes and return the same instance every time we inject it .

Of course, the first time it is injected, it hasn't been instantiated yet, so when creating the **MyComponent** instance for the first time, the DI system will trigger the class **constructor** method.

Now what happens if a service's **constructor** requires some parameter? Let's say we have this service:

**code/dependency_injection/misc/app/ts/app.ts**

```
26 class ParamService {
27   constructor(private phrase: string) {
28     console.log('ParamService is being created with phrase', phrase);
29   }
30
31   getValue(): string {
32     return this.phrase;
33   }
34 }
```

Notice how its constructor method takes a phrase as a parameter? If we try to use the regular injection mechanism w:e would see an error on the browser:

```
Cannot resolve all parameters for 'ParameterService'(?). Make sure that all the          lang.js:375
parameters are decorated with Inject or have valid type annotations and that 'ParameterService' is
decorated with Injectable.
```

**Injection error**

This happens because we didn't provide the injector with enough information about the class we're trying to build. In order to resolve this problem, we need to tell the injector which parameter we want it to use when creating the service's instance.

If we need to pass in parameters when creating a service, we would need to use a factory instead.

## Using a Factory

When we use a factory injection, we write a function that can return **any object**.

```
{
  provide: MyComponent,
  useFactory: () => {
    if (loggedIn) {
      return new MyLoggedComponent();
    }
    return new MyComponent();
  }
}
```

Notice in the case above, we inject on the token `MyComponent` but this will check the (out of scope) `loggedIn` variable. If `loggedIn` is truthy then the injection will give an instance of `MyLoggedComponent`, otherwise we will receive `MyComponent`.

Factories can also have dependencies:

```
{
  provide: MyComponent,
  useFactory: (user) => {
    if (user.loggedIn()) {
      return new MyLoggedComponent(user);
    }
    return new MyComponent();
  },
  deps: [ User ]
}
```

So if we wanted to use our `ParamService` from above, we can wrap it with `useFactory` like so:

**code/dependency_injection/misc/app/ts/app.ts**

```
@NgModule({
  declarations: [ DiSampleApp ],
  imports: [ BrowserModule ],
  bootstrap: [ DiSampleApp ],
  providers: [
    SimpleService,
    {
      provide: ParamService,
      useFactory: (): ParamService => new ParamService('YOLO')
    }
  ]
})
```

```
64  class DiSampleAppAppModule {}
65
66  platformBrowserDynamic().bootstrapModule(DiSampleAppAppModule)
67    .catch((err: any) => console.error(err));
```

In the `providers` value we can put `SimpleService` in the list of providers directly because `SimpleService` doesn't need any parameters. It will get translated to:

```
1  { provide: SimpleService, useClass: SimpleService }
```

Using a factory is the most powerful way to create injectables, because we can do whatever we want within the factory function.

### Using a Value

This is useful when we want to register a constant that can be redefined by another part of the application or even by environment (e.g. test or production).

```
1      { provide: 'API_URL', useValue: 'http://my.api.com/v1' }
```

We're going to do a more thorough example that uses values further down on the Substituting Values section.

### Using an alias

We can also make an alias to reference a previously registered token, like so:

```
1      { provide: NewComponent, useClass: MyComponent }
```

## Dependency Injection in Apps

When writing our apps there are three steps we need to take in order to perform an injection:

1. Create the service class
2. Declare the dependencies on the receiving component and
3. Configure the injection (i.e. register the injection with Angular in our `NgModule`)

The first thing we do is create the service class, that is, the class that exposes some behavior we want to use. This will be called the *injectable* because it is the *thing* that our components will receive via the injection.

Here is how we would create a service:

**code/dependency_injection/simple/app/ts/services/ApiService.ts**

```
1 export class ApiService {
2   get(): void {
3     console.log('Getting resource...');
4   }
5 }
```

Now that we have the thing to be injected, we have to take the next step, which is declare the dependencies we want to receive when Angular creates our component.

Earlier we used the `Injector` class directly, but Angular provides two shortcuts for us we can use when writing our components.

The first and typical way of doing it, is by declaring the injectables we want in our component's constructor.

To do that, we require the service:

**code/dependency_injection/simple/app/ts/app.ts**

```
11 /*
12  * Services
13  */
14 import { ApiService } from 'services/ApiService';
```

And then we declare it on the constructor:

**code/dependency_injection/simple/app/ts/app.ts**

```
27 class DiSampleApp {
28   constructor(private apiService: ApiService) {
29   }
```

When we declare the injection in our component constructor, Angular will do some *reflection* to figure out what class to inject. That is, Angular will see that we are looking for an object of type `ApiService` in the constructor and check the DI system for an appropriate injection.

Sometimes we need to give Angular more hints about what we're trying to inject. In that case we use the second method by using the `@Inject` annotation:

```
1  class DiSampleApp {
2    private apiService: ApiService;
3    constructor(@Inject(ApiService) apiService) {
4      this.apiService = apiService;
5    }
```

If you want to play with the equivalent version, use the `app.long.ts` file provided alongside the `app.ts` file, just copy its contents over `app.ts`.

The final step for using dependency injection is to connect the things our components want injected with the injectables. In other words, we are telling Angular which *thing* to inject when a component declares its dependencies.

```
1      { provide: ApiService, useClass: ApiService }
```

In this case, we use the token `ApiService` to expose the singleton of the class `ApiService`.

Finally, we add this `ApiService` to the `providers` key of our `NgModule`:

**code/dependency_injection/simple/app/ts/app.ts**

```
36  @NgModule({
37    declarations: [ DiSampleApp ],
38    imports: [ BrowserModule ],
39    bootstrap: [ DiSampleApp ],
40    providers: [ ApiService ]   // <-- here
41  })
42  class DiSampleAppAppModule {}
43
44  platformBrowserDynamic().bootstrapModule(DiSampleAppAppModule)
45    .catch((err: any) => console.error(err));
```

## Working with Injectors

We've played a little bit with injectors already, so let's talk a little more about when we would need to use them explicitly.

One case would be when we need to control the moment where the singleton instance of our dependency gets created.

To illustrate a scenario where that could happen, let's build another app that uses the **ApiService** we created above, along with a new service.

This service will be used to instantiate two other services, based on the size of the browser viewport. If it's less than 800 pixels, it will return a new instance of a service called **SmallService**. Otherwise, it will return an instance of **LargeService**.

Here's how **SmallService** look like:

**code/dependency_injection/complex/app/ts/services/SmallService.ts**

```
1  export class SmallService {
2    run(): void {
3      console.log('Small service...');
4    }
5  }
```

And **LargeService**:

**code/dependency_injection/complex/app/ts/services/LargeService.ts**

```
1  export class LargeService {
2    run(): void {
3      console.log('Large service...');
4    }
5  }
```

Then we'll write the **ViewPortService** that choses between the two:

**code/dependency_injection/complex/app/ts/services/ViewPortService.ts**

```
1  import {LargeService} from './LargeService';
2  import {SmallService} from './SmallService';
3
4  export class ViewPortService {
5    determineService(): any {
6      let w: number = Math.max(document.documentElement.clientWidth,
7                               window.innerWidth || 0);
8
9      if (w < 800) {
10       return new SmallService();
11     }
12     return new LargeService();
13   }
14 }
```

Now let's create an app that uses our services:

**code/dependency_injection/complex/app/ts/app.ts**

```
31  class DiSampleApp {
32    constructor(private apiService: ApiService,
33                @Inject('ApiServiceAlias') private aliasService: ApiService,
34                @Inject('SizeService') private sizeService: any) {
35    }
```

Here we are getting an instance of **ApiService** the way we saw ealier. But alongside we're also getting an instance of the same service, aliased as `'ApiServiceAlias'`. Finally we're getting an instance of a `'SizeService'` that is not yet defined.

To understand what each service represents, let's look at the `NgModule`:

**code/dependency_injection/complex/app/ts/app.ts**

```
59  @NgModule({
60    declarations: [ DiSampleApp ],
61    imports: [ BrowserModule ],
62    bootstrap: [ DiSampleApp ],
63    providers: [
64      ApiService,
65      ViewPortService,
66      { provide: 'ApiServiceAlias', useExisting: ApiService },
67      {
68        provide: 'SizeService',
69        useFactory: (viewport: any) => {
70          return viewport.determineService();
71        },
72        deps: [ViewPortService]
73      }
74    ]
75  })
76  class DiSampleAppAppModule {}
```

Here we are saying that we want the app injector to be aware of the **ApiService** and **ViewPortService** injectables as they are.

We are then declaring that we want to use the existing **ApiService** with another token: the string `ApiServiceAlias`.

Next, we're declaring another injectable defined by the string token `SizeService`. This factory will receive an instance of the **ViewPortService** that is listed on its `deps` array. Then it will invoke the `determineService()` method on that class and that call will return either an instance of `SmallService` or `LargeService`, depending on our browser's width.

When we click a button in our template, we want to do three calls: one to the **ApiService**, one to its alias **ApiServiceAlias** and finally one to **SizeService**:

**code/dependency_injection/complex/app/ts/app.ts**

```
37    invokeApi(): void {
38      this.apiService.get();
39      this.aliasService.get();
40      this.sizeService.run();
41    }
```

Now if we run the app and click the **Invoke API** button with a small browser window:

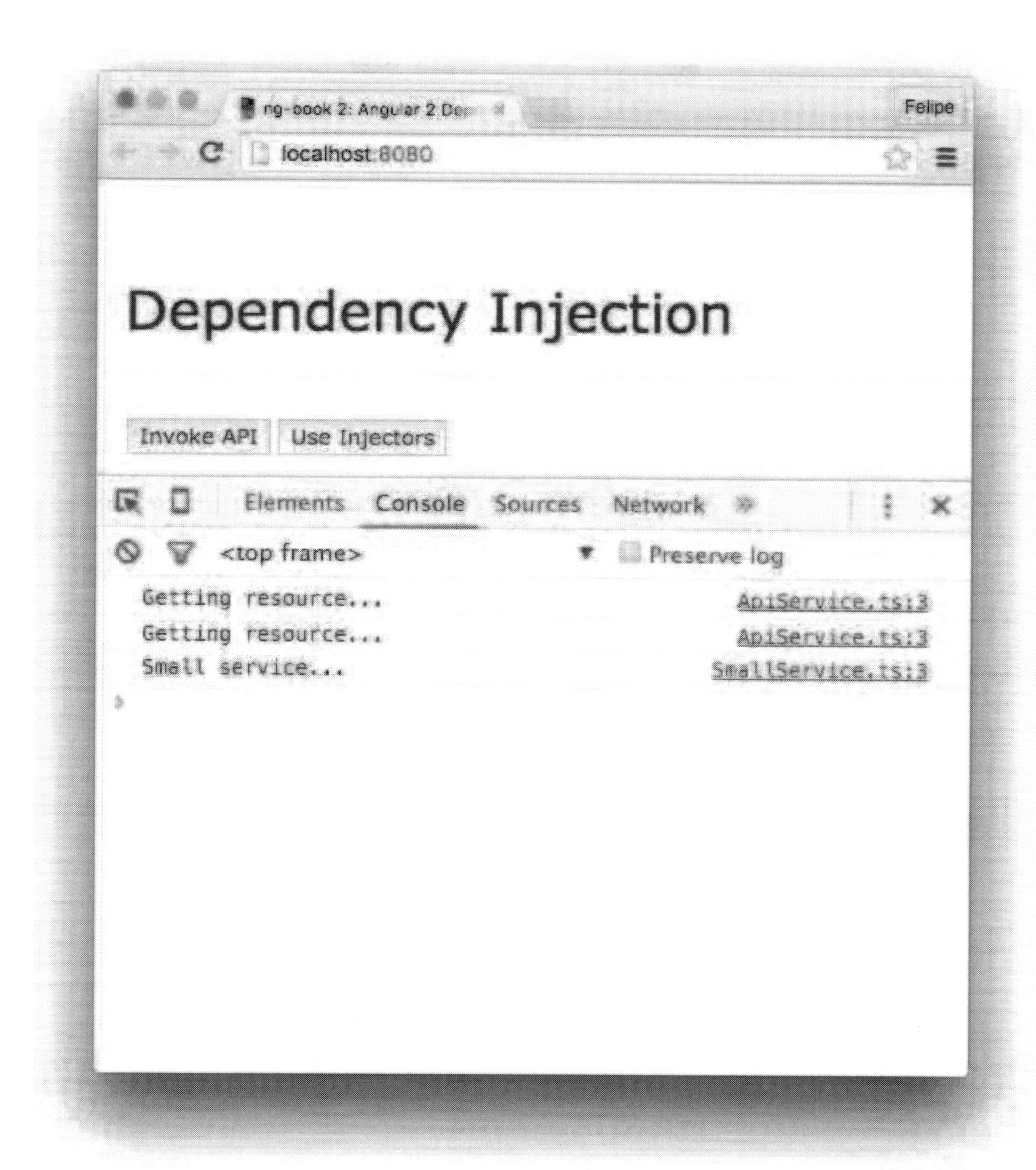


**Small browser window**

We get one log from the **ApiService**, another one from the aliased service and finally we get a log from the **SmallService**.

If we make our browser window larger, reload the page and click the button again:

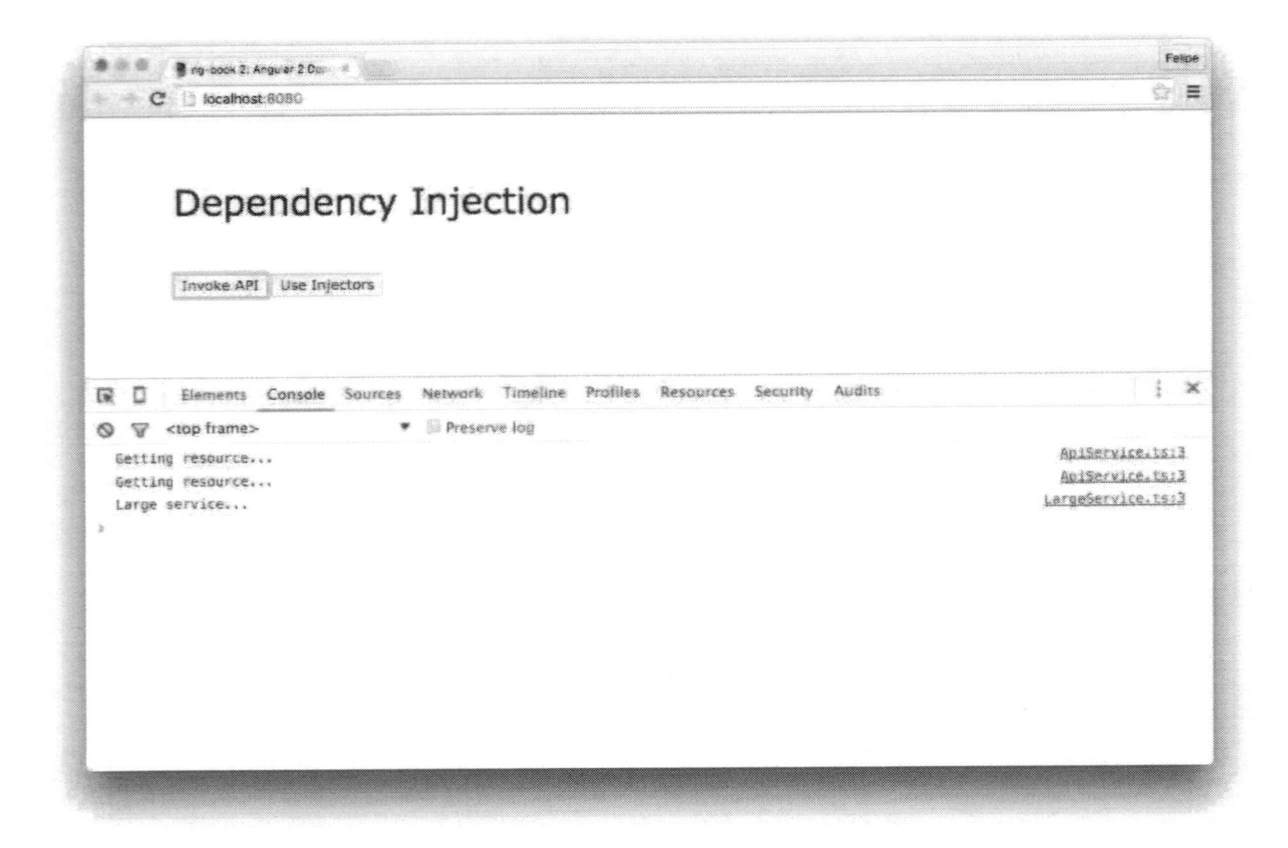


**Large browser window**

We get the **LargeService** log instead. However, if we try to make the browser window smaller and click the button without reloading the page, we still get the **LargeService** log:

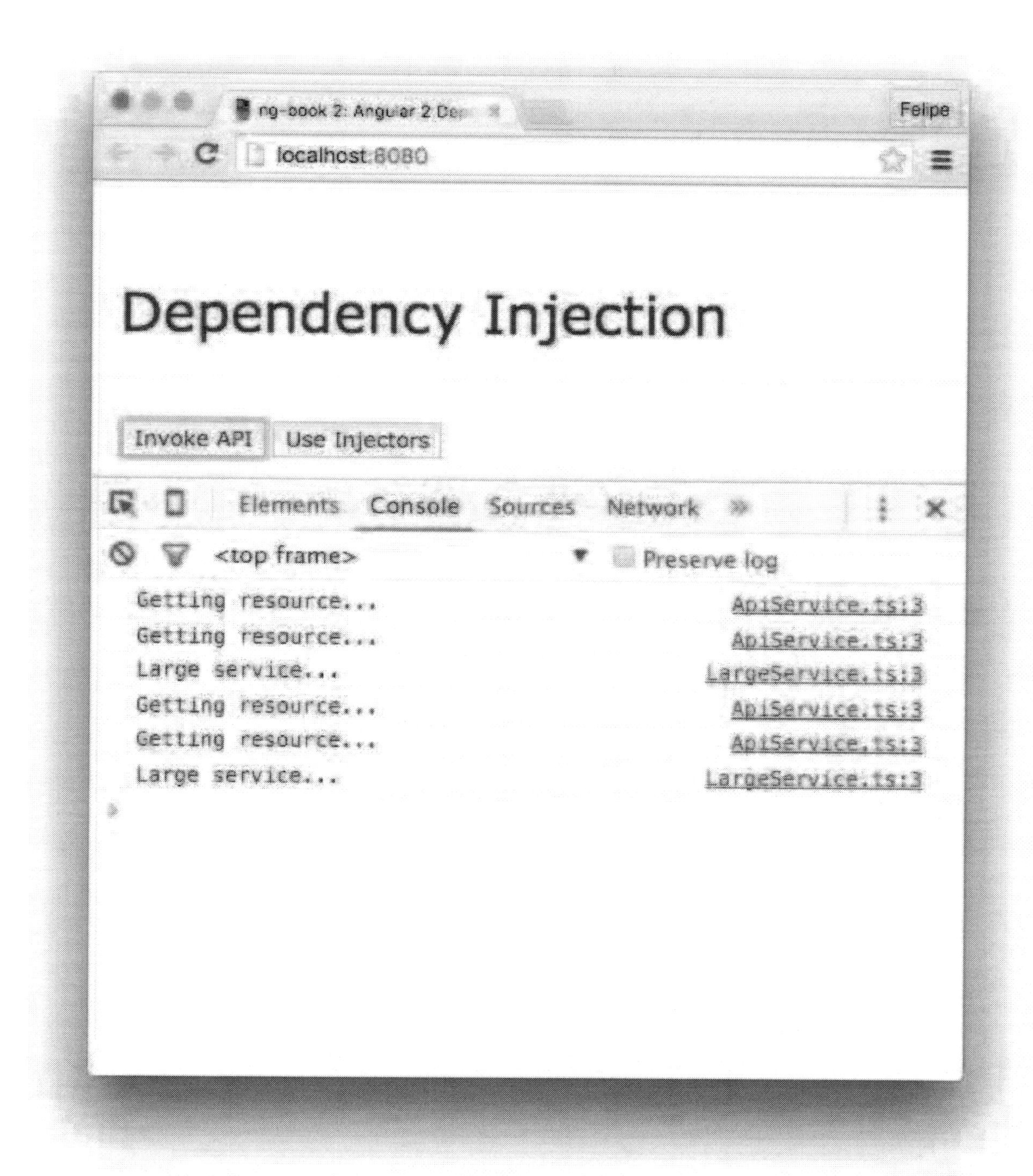


**Small browser window - resized**

That's because the factory was executed once: during the application bootstrap.

To overcome that, we can create our own injectors and get the instance of the proper service by doing the following:

**code/dependency_injection/complex/app/ts/app.ts**

```
43   useInjectors(): void {
44     let injector: any = ReflectiveInjector.resolveAndCreate([
45       ViewPortService,
46       {
47         provide: 'OtherSizeService',
48         useFactory: (viewport: any) => {
49           return viewport.determineService();
50         },
51         deps: [ViewPortService]
52       }
53     ]);
54     let sizeService: any = injector.get('OtherSizeService');
55     sizeService.run();
56   }
```

Here we are creating an injector that knows the **ViewPortService** and another injectable with the string `OtherSizeService` as its token. This injectable uses the same factory as the `SizeService` we used before.

Finally, it uses the injector we created to get an instance of the `OtherSizeService`.line

Now if we run the app with a large browser window and click the **Use Injector** button, we get the large service log. However, if we resize it to a small width, even without reloading we now get the proper log. That's because the factory is being executed every time we click the button, since the injector is being created on demand. Neat!

# Substituting values

Another reason to use DI is to change the hard value of the injection at runtime. That could happen if we have an API service that performs the HTTP requests to our application's API. On the context of our unit or integration tests, we don't want our code to hit the production database. In this case, we could write a Mock API service that seamlessly *replaces* our concrete implementation. Let's take a look at that now.

For instance, if we are running the app in development we might hit a different API server than if we were running the app in production.

This is even more true if we were publishing an open-source or reusable service. In that case we may want the allow the client application define or override an API URL.

Let's write a simple example of an application that injects different values as the API URL depending on whether it's running on production or dev mode. We start with the `ApiService` class:

**code/dependency_injection/value/app/ts/services/ApiService.ts**

```
1  import { Inject } from '@angular/core';
2
3  export const API_URL: string = 'API_URL';
4
5  export class ApiService {
6    constructor(@Inject(API_URL) private apiUrl: string) {
7    }
8
9    get(): void {
10     console.log(`Calling ${this.apiUrl}/endpoint...`);
11   }
12 }
```

We are declaring a constant that will be used as the *token* for our API URL dependency. In other words, Angular will use the string `'API_URL'` to store the information about which URL to call. This way when we `@Inject(API_URL)` the proper value will be injected into the variable.

Notice we are exporting the `API_URL` constant so that client applications can use `API_URL` to inject the correct value from outside the service.

Now that we have the service, let's write the application component that will use the service and provide different values for the URL, depending on the environment the app will be running on.

**code/dependency_injection/value/app/ts/app.ts**

```
21 @Component({
22   selector: 'di-value-app',
23   template: `
24   <button (click)="invokeApi()">Invoke API</button>
25   `
26 })
27 class DiValueApp {
28   constructor(private apiService: ApiService) {
29   }
30
31   invokeApi(): void {
32     this.apiService.get();
33   }
34 }
```

This is the component code. On the constructor we can see that we are declaring a `ApiService` variable called `apiService`. Here Angular will infer that we need to get the `ApiService` dependency and inject it at runtime. If we wanted to be explicit about it we could have used:

```
1 constructor(@Inject(ApiService) private apiService: ApiService) {
2 }
```

The idea behind this component is to have an **Invoke API** button. When we click this button, we'll call the `get()` method of the `ApiService`. This method will then log to the console the `API_URL` we're using.

The next step is to configure the application with the providers:

**code/dependency_injection/value/app/ts/app.ts**

```
36 const isProduction: boolean = false;
37 
38 @NgModule({
39   declarations: [ DiValueApp ],
40   imports: [ BrowserModule ],
41   bootstrap: [ DiValueApp ],
42   providers: [
43     { provide: ApiService, useClass: ApiService },
44     {
45       provide: API_URL,
46       useValue: isProduction ?
47         'https://production-api.sample.com' :
48         'http://dev-api.sample.com'
49     }
50   ]
51 })
52 class DiValueAppAppModule {}
53 
54 platformBrowserDynamic().bootstrapModule(DiValueAppAppModule)
```

First we declare a constant called `isProduction` and set it to `false`. We can pretend that we're doing something here to determine whether or not we are in production mode. This setting could be hardcoded like we're doing, or it could be set using some technique like using WebPack and an `.env` file, for instance.

Finally we bootstrap the application and setup 2 providers: one for `ApiService` using the real class implementation and the other for `API_URL`. If we are in production we're using one value and if not, we're using another value.

To test this we can run the application with `isProduction = true` and when we click the button we'll see the production URL being logged:

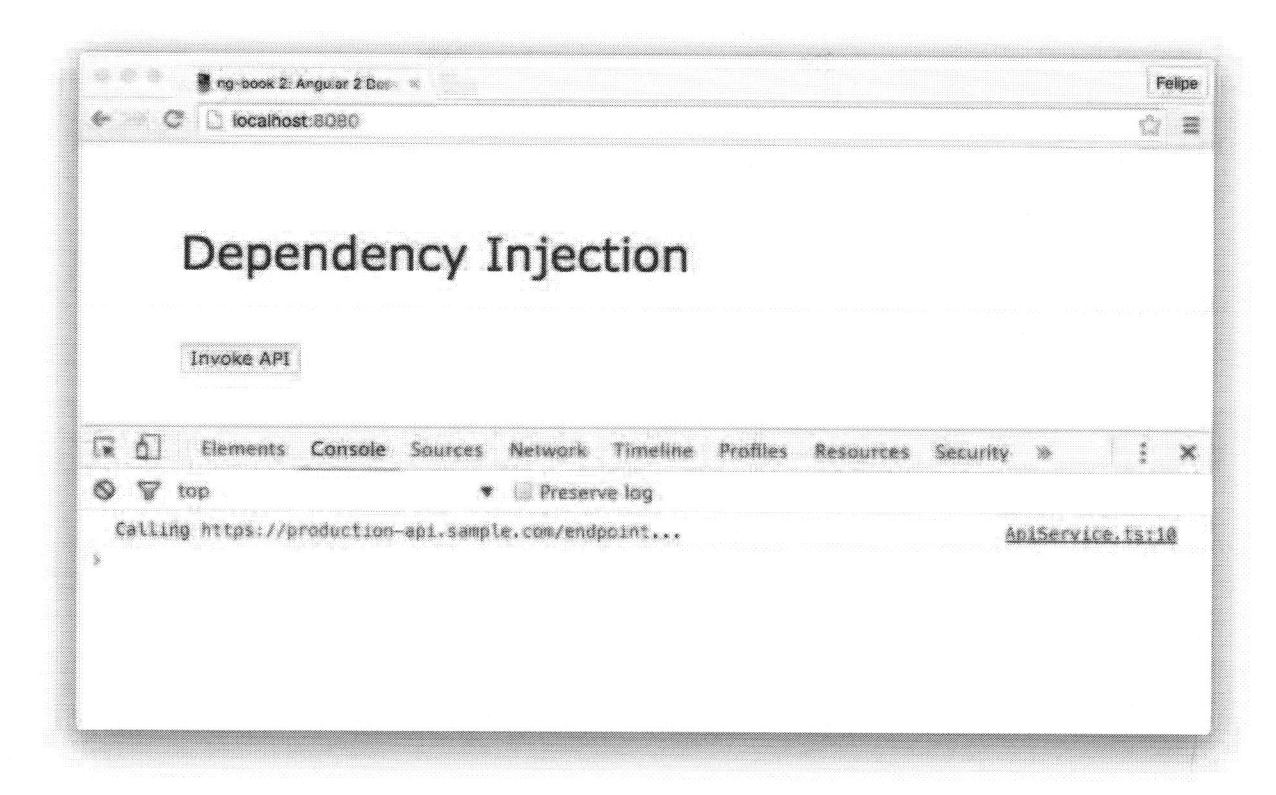


**Production environment**

And if we change it to `isProduction = false`, we see the dev URL instead:

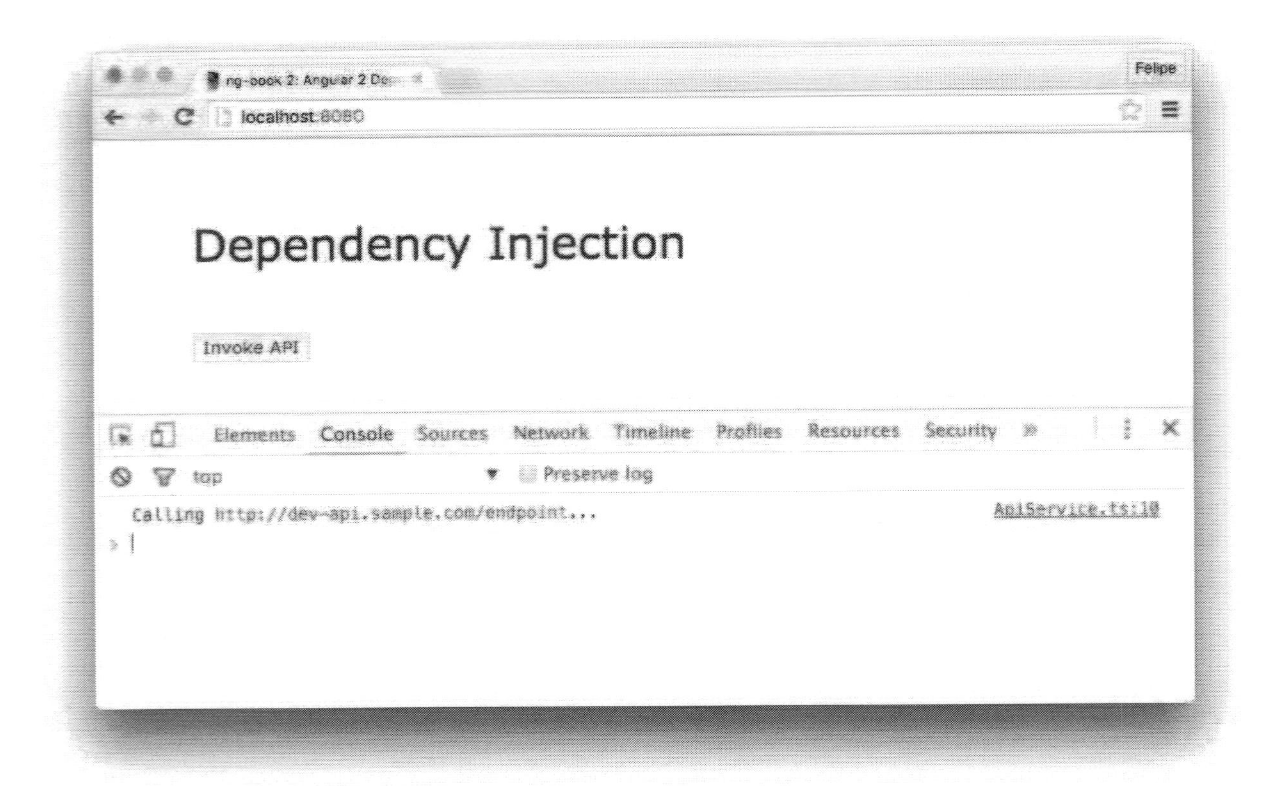


**Dev environment**

# NgModule

`NgModule` is a way to organize your dependencies for 1. the *compiler* and 2. dependency injection. Here we'll explain why we need `NgModules` and how to work with them.

The context here is to think about the two roles of the compiler and dependency injection in Angular. Briefly, Angular needs to know **what components define valid tags** and **where dependencies are coming from**.

## `NgModule` vs. JavaScript Modules

You might be asking, why do we need a new module system at all? Why can't we just use ES6/TypeScript modules?

The reason is, whereas using `import` will load code modules into JavaScript, the `NgModule` system is a way of organizing dependencies *within* the Angular framework. Specifically around **what tags are compiled** and **what dependencies should be injected**.

## The Compiler and Components

For the compiler, when you have an Angular template that has custom tags you have to tell the compiler what tags are valid (and what functionality should be attached to them).

E.g. if we have this component:

```
1  @Component({
2    selector: 'hello-world',
3    template: `<div>Hello world</div>`
4  })
5  class HelloWorld {
6  }
```

We want the compiler to know that the following HTML should use our `hello-world` component (and that `hello-world` isn't some random invalid tag):

```
1  <div>
2    <hello-world></hello-world>
3  </div>
```

In Angular 1, the `hello-world` selector would have been registered *globally* which is convenient until your app grows and you start having naming conflicts. For instance, it's not hard to imagine two open-source projects that might use the same selector.

If you've been using Angular 2 since the earlier versions, you may remember that previous versions required that you specify a `directives` option in your `@Component` annotation. This was good in that it was less "magic" and removed the surface area for conflicts. The problem was it's a bit onerous to specify all directives necessary for all components.

Instead, using `NgModules` we can tell Angular what components are dependencies at a "module" level. More on this in a second.

## Dependency Injection and Providers

Recall that Dependency Injection (DI) is an organized way to make dependencies available across our app. It's an improvement over simply `import`ing code because we have a standardized way to share singletons, create factories, and override dependencies at testing time.

In earlier versions of Angular 2 we had to specify all things-that-would-be-injected (with *providers*) as an argument to the `bootstrap` function.

> Reminder on terminology: a *provider* provides (creates, instantiates, etc.) the *injectable* (the thing you want). In Angular when you want to access an *injectable* you *inject* a dependency into a function and Angular's dependency injection framework will locate it and provide it to you.

Now with `NgModule` each provider is specified as part of a module.

So now that we understand why we need `NgModules` how do we actually use it? Here's the simplest case:

```
1  // app.ts
2
3  @NgModule({
4    imports: [ BrowserModule ],
5    declarations: [ HelloWorld ],
6    bootstrap: [ HelloWorld ]
7  })
8  class HelloWorldAppModule {}
9
10 platformBrowserDynamic().bootstrapModule(HelloWorldAppModule);
```

In this case we're defining a class `HelloWorldAppModule` - this is going to be the entry point of our application. Starting with RC5, instead of bootstrapping our app with a *component*, we bootstrap a module with `bootstrapModule`, as you see here.

`NgModules` can import other modules as dependencies. We're going to be running this app in our browser and so we import `BrowserModule`.

We want to use the `HelloWorld` component in this app. Here's a key thing to keep in mind: **Every component must be declared in *some* `NgModule`**. Here we put `HelloWorld` into the `declarations` of this `NgModule`.

We say the `HelloWorld` component *belongs to* the `HelloWorldAppModule` - every component can belong to **only one** `NgModule`.

You'll often group multiple components together into one `NgModule`, much like you might use a namespace in a language like Java.

If you want to bootstrap this module (that is, use this module as the entry point for an application), then you provide a `bootstrap` key which specifies the component that will be used as the entry-point component for this module.

So in this case we're going to `bootstrap` the `HelloWorld` component as the root component. However, the `bootstrap` key is optional if you're creating a module that doesn't need to be the entry-point of an application.

## Component Visibility

In order to use any component, the current `NgModule` has to know about it. For instance, say we wanted to use a `user-greeting` component in our `hello-world` component like this:

```
<!-- hello-world template -->
<div>
  <user-greeting></user-greeting>
  world
</div>
```

For any component to use another component it must be accessible via the `NgModule` system. There are two ways to make this happen:

1. Either the `user-greeting` component is **part of the same `NgModule`** (e.g. `HelloWorldAppModule`) or
2. The `HelloWorldAppModule` `imports` the module that the `UserGreeting` component is in.

Let's say we want to go the second route. Here's the implementation of our `UserGreeting` component along with the `UserGreetingModule`:

```
@Component({
  selector: 'user-greeting',
  template: `<span>hello</span>`
})
class UserGreeting {
}

@NgModule({
  declarations: [ UserGreeting ],
  exports: [ UserGreeting ]
})
export class UserGreetingModule {}
```

Notice here that we added a new key: `exports`. Think of `exports` as the list of *public* components for this `NgModule`. The implication here is that you can easily have *private* components by simply not listing them in `exports`.

If you forget to put your component in both `declarations` and `exports` (and then try to use it in another module via `imports`) **it won't work**. In order to use a component in another module via `imports` you must put your component in both places.

Now we can use this in our `HelloWorld` component by importing it into the `HelloWorldAppModule` like so:

```
// updated HelloWorldAppModule

@NgModule({
  declarations: [ HelloWorld ],
  imports: [ BrowserModule, UserGreetingModule ], // <-- added
  bootstrap: [ HelloWorld ],
})
class HelloWorldAppModule {}
```

## Specifying Providers

Specifying providers of injectable things is done by adding them to the `providers` key of a `NgModule`.

For instance, say we have this simple service:

```
export class ApiService {
  get(): void {
    console.log('Getting resource...');
  }
}
```

and we want to be able to inject it on a component like this:

```
class ApiDataComponent {
  constructor(private apiService: ApiService) {
  }

  getData(): void {
    this.apiService.get();
  }
}
```

To do this with `NgModule` is easy: we pass `ApiService` to the `providers` key of the module:

```
@NgModule({
  declarations: [ ApiDataComponent ],
  providers: [ ApiService ]   // <-- here
})
class ApiAppModule {}
```

Passing the constant `ApiService` here is the shorthand version of using `provide` like this:

```
@NgModule({
  declarations: [ ApiDataComponent ],
  providers: [
    provide(ApiService, { useClass: ApiService })
  ]
})
class ApiAppModule {}
```

We're telling Angular that when the `ApiService` is to be injected, create and maintain a singleton instance of that class and pass it in the injection.

In order to *use* those providers from another module, you guessed it, you have to `import` that module.

Because the `ApiDataComponent` and `ApiService` are in the same `NgModule` the `ApiDataComponent` is able to inject the `ApiService`. If they were in different modules, then you would need to import the module containing `ApiService` into the `ApiAppModule`.

# Conclusion

As we can see, Dependency Injection and `NgModule` coordinate to provide a powerful way to manage dependencies within our app. Here are a few more resources where you can learn more about it:

- Official Angular DI Docs[50]
- Victor Savkin Comparse DI in Angular 1 vs. Angular 2[51]

[50]https://angular.io/docs/ts/latest/guide/dependency-injection.html
[51]http://victorsavkin.com/post/126514197956/dependency-injection-in-angular-1-and-angular-2

# Data Architecture in Angular 2

## An Overview of Data Architecture

Managing data can be one of the trickiest aspects of writing a maintainable app. There are tons of ways to get data into your application:

- AJAX HTTP Requests
- Websockets
- Indexdb
- LocalStorage
- Service Workers
- etc.

The problem of data architecture addresses questions like:

- How can we aggregate all of these different sources into a coherent system?
- How can we avoid bugs caused by unintended side-effects?
- How can we structure the code sensibly so that it's easier to maintain and on-board new team members?
- How can we make the app run as fast as possible when data changes?

For many years MVC was a standard pattern for architecting data in applications: the Models contained the domain logic, the View displayed the data, and the Controller tied it all together. The problem is, we've learned that MVC doesn't translate directly into client-side web applications very well.

There has been a renaissance in the area of data architectures and many new ideas are being explored. For instance:

- **MVW / Two-way data binding**: *Model-View-Whatever* is a term used[52] to describe Angular 1's default architecture. The `$scope` provides a two-way data-binding - the whole application shares the same data structures and a change in one area propagates to the rest of the app.
- **Flux**[53]: uses a unidirectional data flow. In Flux, Stores hold data, Views render what's in the Store, and Actions change the data in the Store. There is a bit more ceremony to setup Flux, but the idea is that because data only flows in one direction, it's easier to reason about.
- **Observables**: Observables give us streams of data. We subscribe to the streams and then perform operations to react to changes. RxJs[54] is the most popular reactive streams library for

[52]See: Model View Whatever

[53]https://facebook.github.io/flux/

[54]https://github.com/Reactive-Extensions/RxJS

Javascript and it gives us powerful operators for composing operations on streams of data.

There are a lot of variations on these ideas. For instance:

- Flux is a pattern, and not an implementation. There are **many** different implementations of Flux (just like there are many implementations of MVC)
- Immutability is a common variant on all of the above data architectures.
- Falcor[55] is a powerful framework that helps bind your client-side models to the server-side data. Falcor often used with an Observables-type data architecture.

## Data Architecture in Angular 2

Angular 2 is extremely flexible in what it allows for data architecture. A data strategy that works for one project doesn't necessarily work for another. So Angular doesn't prescribe a particular stack, but instead tries to make it easy to use whatever architecture we choose (while still retaining fast performance).

The benefit of this is that you have flexibility to fit Angular into almost any situation. The downside is that you have to make your own decisions about what's right for your project.

Don't worry, we're not going to leave you to make this decision on your own! In the chapters that follow, we're going to cover how to build applications using some of these patterns.

---

[55]http://netflix.github.io/falcor/

# Data Architecture with Observables - Part 1: Services

## Observables and RxJS

In Angular, we can structure our application to use Observables as the backbone of our data architecture. Using Observables to structure our data is called *Reactive Programming*.

But what are Observables, and Reactive Programming anyway? Reactive Programming is a way to work with asynchronous streams of data. Observables are the main data structure we use to implement Reactive Programming. But I'll admit, those terms may not be that clarifying. So we'll look at concrete examples through the rest of this chapter that should be more enlightening.

### Note: Some RxJS Knowledge Required

I want to point out **this book is not primarily about Reactive Programming**. There are several other good resources that can teach you the basics of Reactive Programming and you should read them. We've listed a few below.

**Consider this chapter a tutorial on how to work with RxJS and Angular** rather than an exhaustive introduction to RxJS and Reactive Programming.

In this chapter, I'll **explain in detail the RxJS concepts and APIs that we encounter**. But know that you may need to supplement the content here with other resources if RxJS is still new to you.

**Use of Underscore.js in this chapter**

Underscore.js[56] is a popular library that provides functional operators on Javascript data structures such as `Array` and `Object`. We use it a bunch in this chapter alongside RxJS. If you see the _ in code, such as `_.map` or `_.sortBy` know that we're using the Underscore.js library. You can find the docs for Underscore.js here[57].

### Learning Reactive Programming and RxJS

If you're just learning RxJS I recommend that you read this article first:

[56]http://underscorejs.org/

[57]http://underscorejs.org/

- The introduction to Reactive Programming you've been missing[58] by Andre Staltz

After you've become a bit more familiar with the concepts behind RxJS, here are a few more links that can help you along the way:

- Which static operators to use to create streams?[59]
- Which instance operators to use on streams?[60]
- RxMarbles[61] - Interactive diagrams of the various operations on streams

Throughout this chapter I'll provide links to the API documentation of RxJS. The RxJS docs have tons of great example code that shed light on how the different streams and operators work.

**Do I have to use RxJS to use Angular 2?** - No, you definitely don't. Observables are just one pattern out of many that you can use with Angular 2. We talk more about other data patterns you can use here.

I want to give you fair warning: learning RxJS can be a bit mind-bending at first. But trust me, you'll get the hang of it and it's worth it. Here's a few big ideas about streams that you might find helpful:

1. **Promises emit a single value whereas streams emit many values.** - Streams fulfill the same role in your application as promises. If you've made the jump from callbacks to promises, you know that promises are a big improvement in readability and data maintenance vs. callbacks. In the same way, streams improve upon the promise pattern in that we can continuously respond to data changes on a stream (vs. a one-time resolve from a promise)
2. **Imperative code "pulls" data whereas reactive streams "push" data** - In Reactive Programming our code subscribes to be notified of changes and the streams "push" data to these subscribers
3. **RxJS is *functional*** - If you're a fan of functional operators like `map`, `reduce`, and `filter` then you'll feel right at home with RxJS because streams are, in some sense, lists and so the powerful functional operators all apply
4. **Streams are composable** - Think of streams like a pipeline of operations over your data. You can subscribe to any part of your stream and even combine them to create new streams

[58] https://gist.github.com/staltz/868e7e9bc2a7b8c1f754
[59] https://github.com/Reactive-Extensions/RxJS/blob/master/doc/gettingstarted/which-static.md
[60] https://github.com/Reactive-Extensions/RxJS/blob/master/doc/gettingstarted/which-instance.md
[61] http://staltz.com/rxmarbles

# Chat App Overview

In this chapter, we're going to use RxJS to build a chat app. Here's a screenshot:

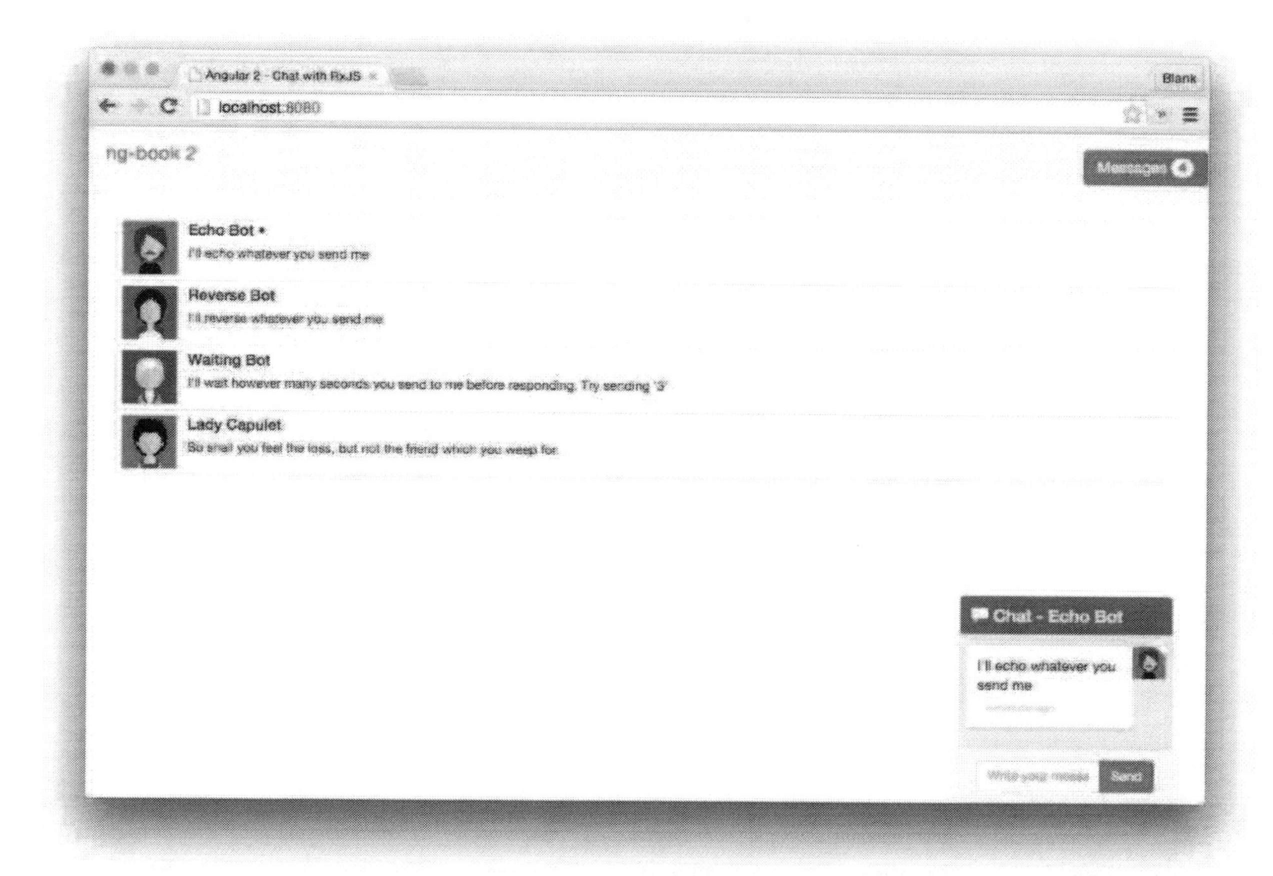


**Completed Chat Application**

Usually we try to show every line of code here in the book text. However, this chat application has a lot of moving parts, so in this chapter we're not going to have every single line of code in the text. You can find the sample code for this chapter in the folder `code/rxjs/chat`. We'll call out each filter where you can view the context, where appropriate.

In this application we've provided a few bots you can chat with. Open up the code and try it out:

```
1   cd code/rxjs/chat
2   npm install
3   npm run go
```

Now open your browser to `http://localhost:8080`.

If the above URL doesn't work, try this URL: `http://localhost:8080/webpack-dev-server/index.html`

Some Windows users may have trouble doing an `npm install` on this repo. If this causes problems for you, make sure you're running these commands inside Cygwin[62].

Notice a few things about this application:

- You can click on the threads to chat with another person
- The bots will send you messages back, depending on their personality
- The unread message count in the top corner stays in sync with the number of unread messages

Let's look at an overview of how this app is constructed. We have

- 3 top-level Angular Components
- 3 models
- and 3 services

Let's look at them one at a time.

## Components

The page is broken down into three top-level components:

[62] https://www.cygwin.com/

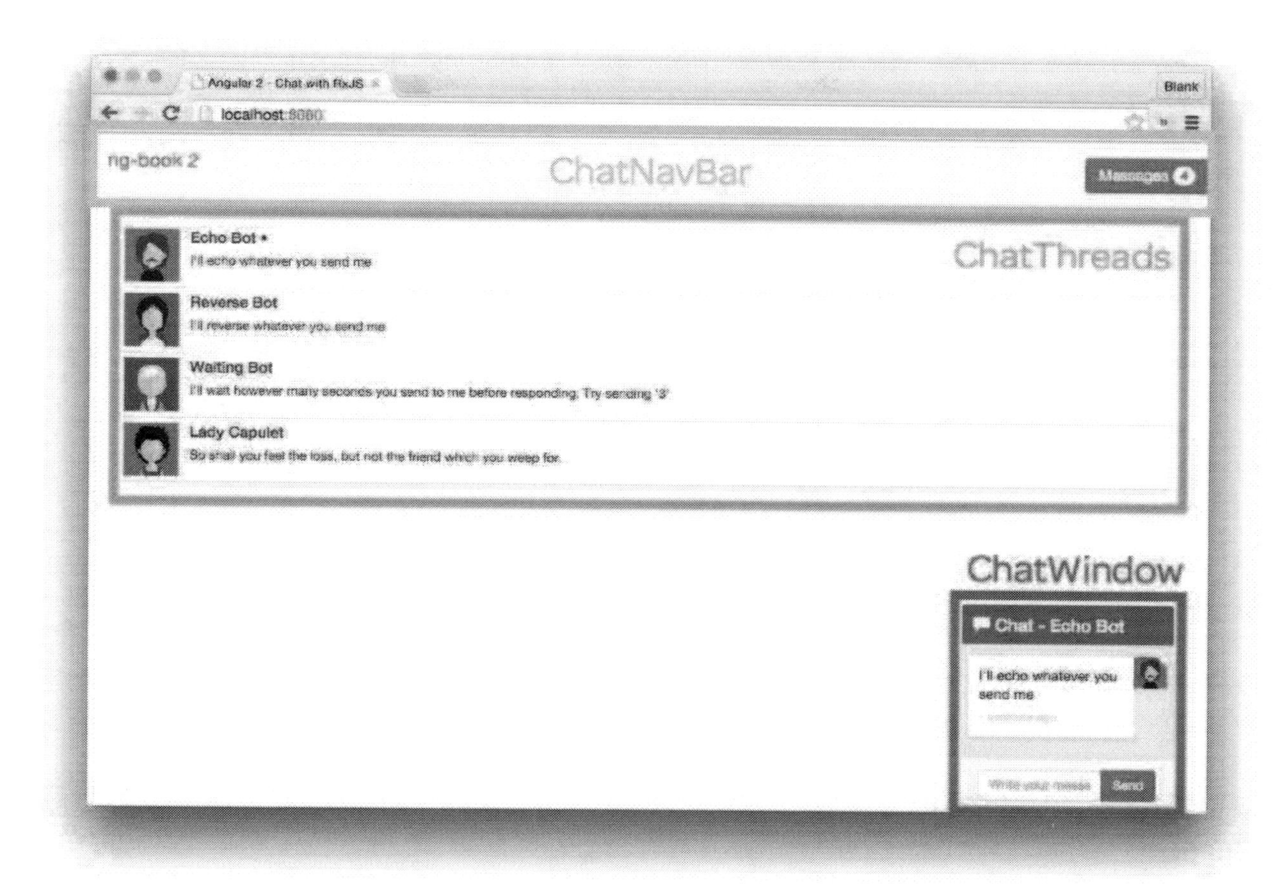


Chat Top-Level Components

- `ChatNavBar` - contains the unread messages count
- `ChatThreads` - shows a clickable list of threads, along with the most recent message and the conversation avatar
- `ChatWindow` - shows the messages in the current thread with an input box to send new messages

## Models

This application also has three models:

**Chat Models**

- `User` - stores information about a chat participant
- `Message` - stores an individual message
- `Thread` - stores a collection of `Messages` as well as some data about the conversation

## Services

In this app, each of our models has a corresponding *service*. The services are singleton objects that play two roles:

1. **Provide streams** of data that our application can subscribe to
2. **Provide operations** to add or modify data

For instance, the `UserService`:

- publishes a stream that emits the current user and
- offers a `setCurrentUser` function which will set the current user (that is, emit the current user from the `currentUser` stream)

## Summary

At a high level, the application data architecture is straightforward:

- The **services** maintain streams which emit models (e.g. `Messages`)
- The **components** subscribe to those streams and render according to the most recent values

For instance, the `ChatThreads` component listens for the most recent list of threads from the `ThreadService` and the `ChatWindow` subscribes for the most recent list of messages.

In the rest of this chapter, we're going to go in-depth on how we implement this using Angular 2 and RxJS. We'll start by implementing our models, then look at how we create Services to manage our streams, and then finally implement the Components.

# Implementing the Models

Let's start with the easy stuff and take a look at the models.

### User

Our `User` class is straightforward. We have an `id`, `name`, and `avatarSrc`.

**code/rxjs/chat/app/ts/models.ts**

```
 3 export class User {
 4   id: string;
 5 
 6   constructor(public name: string,
 7               public avatarSrc: string) {
 8     this.id = uuid();
 9   }
10 }
```

Notice above that we're using a TypeScript shorthand in the constructor. When we say `public name: string` we're telling TypeScript that 1. we want `name` to be a public property on this class and 2. assign the argument value to that property when a new instance is created.

### Thread

Similarly, `Thread` is also a straightforward TypeScript class:

**code/rxjs/chat/app/ts/models.ts**

```
12 export class Thread {
13   id: string;
14   lastMessage: Message;
15   name: string;
16   avatarSrc: string;
17 
18   constructor(id?: string,
19               name?: string,
20               avatarSrc?: string) {
21     this.id = id || uuid();
22     this.name = name;
```

```
23      this.avatarSrc = avatarSrc;
24    }
25  }
```

Note that we store a reference to the `lastMessage` in our `Thread`. This lets us show a preview of the most recent message in the threads list.

#### Message

`Message` is also a simple TypeScript class, however in this case we use a slightly different form of constructor:

**code/rxjs/chat/app/ts/models.ts**

```
14    lastMessage: Message;
```

The pattern you see here in the constructor allows us to simulate using keyword arguments in the constructor. Using this pattern, we can create a new `Message` using whatever data we have available and we don't have to worry about the order of the arguments. For instance we could do this:

```
1    let msg1 = new Message();
2
3    # or this
4
5    let msg2 = new Message({
6      text: "Hello Nate Murray!"
7    })
```

Now that we've looked at our models, let's take a look at our first service: the `UserService`.

## Implementing `UserService`

The point of the `UserService` is to provide a place where our application can learn about the current user and also notify the rest of the application if the current user changes.

The first thing we need to do is create a TypeScript class and make it *injectable* by using the `@Injectable` annotation.

**code/rxjs/chat/app/ts/services/UserService.ts**

```
10  export class UserService {
11    // `currentUser` contains the current user
12    currentUser: Subject<User> = new BehaviorSubject<User>(null);
13
14    public setCurrentUser(newUser: User): void {
15      this.currentUser.next(newUser);
16    }
17  }
```

When we make something *injectable* that means we will be able to use it as a dependency to other components in our application. Briefly, two benefits of dependency-injection are:

1. we let Angular handle the lifecycle of the object and
2. it's easier to test injected components.

We talk more about `@Injectable` in the chapter on dependency injection, but the result is that now we can inject it as a dependency to our components like so:

```
1  class MyComponent {
2    constructor(public userService: UserService) {
3      // do something with `userService` here
4    }
5  }
```

### `currentUser` stream

Next we setup a stream which we will use to manage our current user:

**code/rxjs/chat/app/ts/services/UserService.ts**

```
12    currentUser: Subject<User> = new BehaviorSubject<User>(null);
```

There's a lot going on here, so let's break it down:

- We're defining an instance variable `currentUser` which is a `Subject` stream.
- Concretely, `currentUser` is a `BehaviorSubject` which will contain `User`.

- However, the first value of this stream is `null` (the constructor argument).

If you haven't worked with RxJS much, then you may not know what `Subject` or `BehaviorSubject` are. You can think of a `Subject` as a "read/write" stream.

Technically a `Subject`[63] inherits from both `Observable`[64] and `Observer`[65]

One consequence of streams is that, because messages are published immediately, a new subscriber risks missing the latest value of the stream. BehaviourSubject compensates for this.

**`BehaviourSubject`**[66] has a special property in that it **stores the last value**. Meaning that any subscriber to the stream will receive the latest value. This is great for us because it means that any part of our application can subscribe to the `UserService.currentUser` stream and immediately know who the current user is.

## Setting a new user

We need a way to publish a new user to the stream whenever the current user changes (e.g. logging in).

There's two ways we can expose an API for doing this:

### 1. Add new users to the stream directly:

The most straightforward way to update the current user is to have clients of the `UserService` simply publish a new `User` directly to the stream like this:

```
1  userService.subscribe((newUser) => {
2    console.log('New User is: ', newUser.name);
3  })
4
5  // => New User is: originalUserName
6
7  let u = new User('Nate', 'anImgSrc');
8  userService.currentUser.next(u);
9
10 // => New User is: Nate
```

[63] https://github.com/Reactive-Extensions/RxJS/blob/master/doc/api/subjects/subject.md
[64] https://github.com/Reactive-Extensions/RxJS/blob/master/doc/api/core/observable.md
[65] https://github.com/Reactive-Extensions/RxJS/blob/master/doc/api/core/observer.md
[66] https://github.com/Reactive-Extensions/RxJS/blob/master/doc/api/subjects/behaviorsubject.md

Note here that we use the `next` method on a `Subject` to push a new value to the stream

The pro here is that we're able to reuse the existing API from the stream, so we're not introducing any new code or APIs

## 2. Create a `setCurrentUser(newUser: User)` method

The other way we could update the current user is to create a helper method on the `UserService` like this:

**code/rxjs/chat/app/ts/services/UserService.ts**

```
14    public setCurrentUser(newUser: User): void {
15      this.currentUser.next(newUser);
16    }
```

You'll notice that we're still using the `next` method on the `currentUser` stream, so why bother doing this?

Because there is value in decoupling the implementation of the currentUser from the implementation of the stream. By wrapping the `next` in the `setCurrentUser` call we give ourselves room to change the implementation of the `UserService` without breaking our clients.

In this case, I wouldn't recommend one method very strongly over the other, but it can make a big difference on the maintainability of larger projects.

A third option could be to have the updates expose streams of their own (that is, a stream where we place the action of changing the current user). We explore this pattern in the `MessagesService` below.

### `UserService.ts`

Putting it together, our `UserService` looks like this:

**code/rxjs/chat/app/ts/services/UserService.ts**

```
 1 import {Injectable} from '@angular/core';
 2 import {Subject, BehaviorSubject} from 'rxjs';
 3 import {User} from '../models';
 4 
 5 
 6 /**
 7  * UserService manages our current user
 8  */
 9 @Injectable()
10 export class UserService {
11   // `currentUser` contains the current user
12   currentUser: Subject<User> = new BehaviorSubject<User>(null);
13 
14   public setCurrentUser(newUser: User): void {
15     this.currentUser.next(newUser);
16   }
17 }
18 
19 export var userServiceInjectables: Array<any> = [
20   UserService
21 ];
```

## The `MessagesService`

The `MessagesService` is the backbone of this application. In our app, all messages flow through the `MessagesService`.

Our `MessagesService` has much more sophisticated streams compared to our `UserService`. There are five streams that make up our `MessagesService`: 3 "data management" streams and 2 "action" streams.

The three data management streams are:

- `newMessages` - emits each new `Message` only once
- `messages` - emits **an array** of the current `Messages`
- `updates` - performs operations on `messages`

### the `newMessages` stream

`newMessages` is a `Subject` that will publish each new `Message` only once.

**code/rxjs/chat/app/ts/services/MessagesService.ts**

```
12  export class MessagesService {
13    // a stream that publishes new messages only once
14    newMessages: Subject<Message> = new Subject<Message>();
```

If we want, we can define a helper method to add `Messages` to this stream:

**code/rxjs/chat/app/ts/services/MessagesService.ts**

```
88    addMessage(message: Message): void {
89      this.newMessages.next(message);
90    }
```

It would also be helpful to have a stream that will get all of the messages from a thread that are not from a particular user. For instance, consider the Echo Bot:

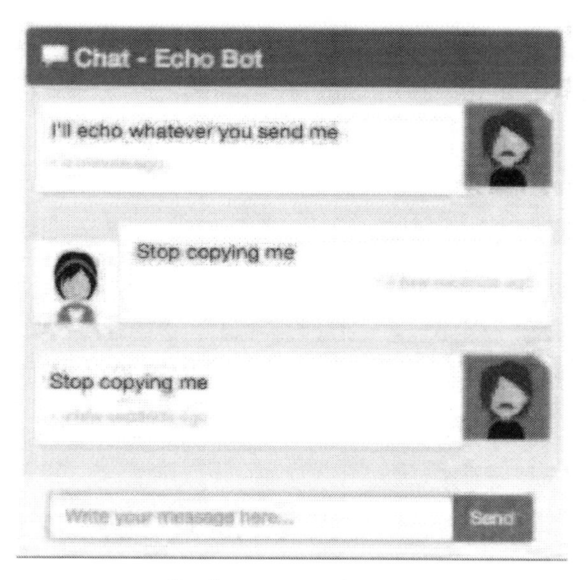


**Real mature, Echo Bot**

When we are implementing the Echo Bot, we don't want to enter an infinite loop and repeat back the bot's messages to itself.

To implement this we can subscribe to the `newMessages` stream and filter out all messages that are

1. part of this thread and
2. not written by the bot.

You can think of this as saying, for a given `Thread` I want a stream of the messages that are "for" this `User`.

**code/rxjs/chat/app/ts/services/MessagesService.ts**

```
92   messagesForThreadUser(thread: Thread, user: User): Observable<Message> {
93     return this.newMessages
94       .filter((message: Message) => {
95               // belongs to this thread
96         return (message.thread.id === thread.id) &&
97               // and isn't authored by this user
98               (message.author.id !== user.id);
99       });
100  }
```

`messagesForThreadUser` takes a `Thread` and a `User` and returns a new stream of `Messages` that are filtered on that `Thread` and not authored by the `User`. That is, it is a stream of "everyone else's" messages in this `Thread`.

### the `messages` stream

Whereas `newMessages` emits individual `Messages`, the `messages` stream emits **an Array of the most recent `Messages`**.

**code/rxjs/chat/app/ts/services/MessagesService.ts**

```
17   messages: Observable<Message[]>;
```

The type `Message[]` is the same as `Array<Message>`. Another way of writing the same thing would be: `Observable<Array<Message>>`. When we define the type of `messages` to be `Observable<Message[]>` we mean that this stream emits an **`Array`** (of `Messages`), not individual `Messages`.

So how does `messages` get populated? For that we need to talk about the `updates` stream and a new pattern: the Operation stream.

## The Operation Stream Pattern

Here's the idea:

- We'll maintain state in `messages` which will hold an `Array` of the most current `Messages`
- We use an `updates` stream which is a **stream of functions** to apply to `messages`

You can think of it this way: any function that is put on the `updates` stream will change the list of the current messages. A function that is put on the `updates` stream should **accept a list of `Messages`** and then **return a list of `Messages`**. Let's formalize this idea by creating an interface in code:

**code/rxjs/chat/app/ts/services/MessagesService.ts**

```
7  interface IMessagesOperation extends Function {
8    (messages: Message[]): Message[];
9  }
```

Let's define our updates stream:

**code/rxjs/chat/app/ts/services/MessagesService.ts**

```
19   // `updates` receives _operations_ to be applied to our `messages`
20   // it's a way we can perform changes on *all* messages (that are currently
21   // stored in `messages`)
22   updates: Subject<any> = new Subject<any>();
```

Remember, updates receives *operations* that will be applied to our list of messages. But how do we make that connection? We do (in the constructor of our MessagesService) like this:

**code/rxjs/chat/app/ts/services/MessagesService.ts**

```
28   constructor() {
29     this.messages = this.updates
30       // watch the updates and accumulate operations on the messages
31       .scan((messages: Message[],
32              operation: IMessagesOperation) => {
33                return operation(messages);
34              },
35             initialMessages)
36       // make sure we can share the most recent list of messages across anyone
```

This code introduces a new stream function: scan[67]. If you're familiar with functional programming, scan is a lot like reduce: it runs the function for each element in the incoming stream and **accumulates a value**. What's special about scan is that it will **emit a value for each intermediate result**. That is, it doesn't wait for the stream to complete before emitting a result, which is exactly what we want.

When we call this.updates.scan, we are creating a new stream that is subscribed to the updates stream. On each pass, we're given:

1. the messages we're accumulating and
2. the new operation to apply.

and then we return the new Message[].

[67]https://github.com/Reactive-Extensions/RxJS/blob/master/doc/api/core/operators/scan.md

## Sharing the Stream

One thing to know about streams is that they aren't shareable by default. That is, if one subscriber reads a value from a stream, it can be gone forever. In the case of our messages, we want to 1. share the same stream among many subscribers and 2. replay the last value for any subscribers who come "late".

To do that, we use two operators: `publishReplay` and `refCount`.

- `publishReplay` let's us share a subscription between multiple subscribers and replay *n* number of values to future subscribers. (see `publish`[68] and `replay`[69])
- `refCount`[70] - makes it easier to use the return value of publish, by managing when the observable will emit values

**Wait, so what does `refCount` do?**

`refCount` can be a little tricky to understand because it relates to how one manages "hot" and "cold" observables. We're not going to dive deep into explaining how this works and we direct the reader to:

- RxJS docs on `refCount`[71]
- Introduction to Rx: Hot and Cold observables[72]
- RefCount Marble Diagram[73]

**code/rxjs/chat/app/ts/services/MessagesService.ts**

```
30      // watch the updates and accumulate operations on the messages
31      .scan((messages: Message[],
32             operation: IMessagesOperation) => {
33               return operation(messages);
34             },
35            initialMessages)
36      // make sure we can share the most recent list of messages across anyone
37      // who's interested in subscribing and cache the last known list of
38      // messages
39      .publishReplay(1)
40      .refCount();
```

[68] https://github.com/Reactive-Extensions/RxJS/blob/master/doc/api/core/operators/publish.md
[69] https://github.com/Reactive-Extensions/RxJS/blob/master/doc/api/core/operators/replay.md
[70] https://github.com/Reactive-Extensions/RxJS/blob/master/doc/api/core/operators/refcount.md
[71] https://github.com/Reactive-Extensions/RxJS/blob/master/doc/api/core/operators/refcount.md
[72] http://www.introtorx.com/Content/v1.0.10621.0/14_HotAndColdObservables.html#RefCount
[73] http://reactivex.io/documentation/operators/refcount.html

## Adding `Message`s to the `messages` Stream

Now we could add a `Message` to the `messages` stream like so:

```
1  var myMessage = new Message(/* params here... */);
2
3  updates.next( (messages: Message[]): Message[] => {
4    return messages.concat(myMessage);
5  })
```

Above, we're adding an operation to the `updates` stream. `messages` is subscribe to that stream and so it will apply that operation which will `concat` our `newMessage` on to the accumulated list of `messages`.

It's okay if this takes a few minutes to mull over. It can feel a little foreign if you're not used to this style of programming.

One problem with the above approach is that it's a bit verbose to use. It would be nice to not have to write that inner function every time. We could do something like this:

```
1  addMessage(newMessage: Message) {
2    updates.next( (messages: Message[]): Message[] => {
3      return messages.concat(newMessage);
4    })
5  }
6
7  // somewhere else
8
9  var myMessage = new Message(/* params here... */);
10 MessagesService.addMessage(myMessage);
```

This is a little bit better, but it's not "the reactive way". In part, because this action of creating a message isn't composable with other streams. (Also this method is circumventing our `newMessages` stream. More on that later.)

A reactive way of creating a new message would be **to have a stream that accepts `Messages` to add to the list**. Again, this can be a bit new if you're not used to thinking this way. Here's how you'd implement it:

First we make an "action stream" called `create`. (The term "action stream" is only meant to describe its role in our service. The stream itself is still a regular `Subject`):

**code/rxjs/chat/app/ts/services/MessagesService.ts**

```
24    // action streams
25    create: Subject<Message> = new Subject<Message>();
```

Next, in our `constructor` we configure the `create` stream:

**code/rxjs/chat/app/ts/services/MessagesService.ts**

```
56      this.create
57        .map( function(message: Message): IMessagesOperation {
58          return (messages: Message[]) => {
59            return messages.concat(message);
60          };
61        })
```

The `map`[74] operator is a lot like the built-in `Array.map` function in Javascript except that it works on streams. That is, it runs the function once for each item in the stream and emits the return value of the function.

In this case, we're saying "for each `Message` we receive as input, return an `IMessagesOperation` that adds this message to the list". Put another way, this stream will emit a **function** which accepts the list of `Messages` and adds this `Message` to our list of messages.

Now that we have the `create` stream, we still have one thing left to do: we need to actually hook it up to the `updates` stream. We do that by using `subscribe`[75].

**code/rxjs/chat/app/ts/services/MessagesService.ts**

```
56      this.create
57        .map( function(message: Message): IMessagesOperation {
58          return (messages: Message[]) => {
59            return messages.concat(message);
60          };
61        })
62        .subscribe(this.updates);
```

What we're doing here is *subscribing* the `updates` stream to listen to the `create` stream. This means that if `create` receives a `Message` it will emit an `IMessagesOperation` that will be received by `updates` and then the `Message` will be added to `messages`.

Here's a diagram that shows our current situation:

[74] https://github.com/Reactive-Extensions/RxJS/blob/master/doc/api/core/operators/select.md
[75] https://github.com/Reactive-Extensions/RxJS/blob/master/doc/api/core/operators/subscribe.md

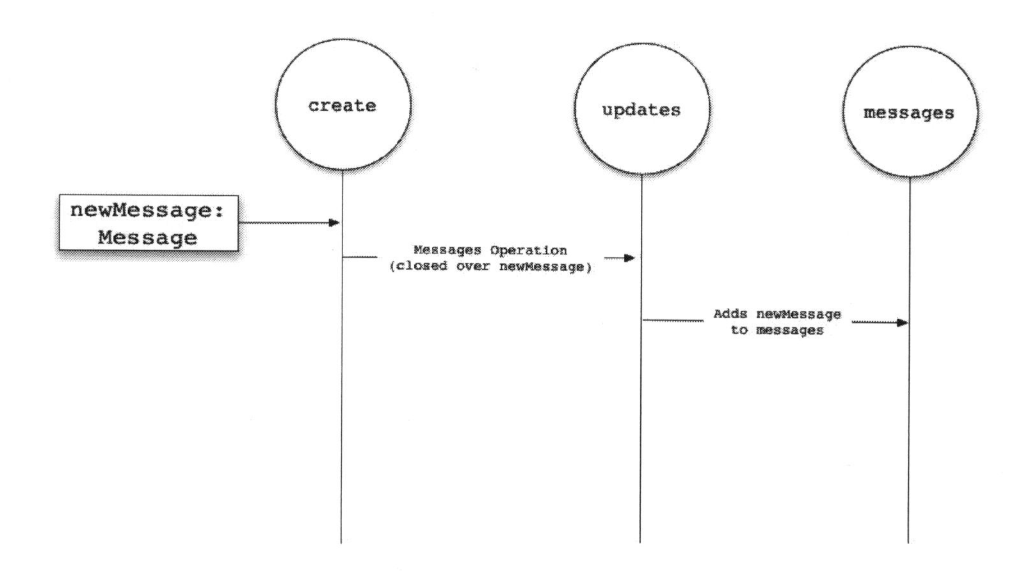


**Creating a new message, starting with the `create` stream**

This is great because it means we get a few things:

1. The current list of messages from `messages`
2. A way to process operations on the current list of messages (via `updates`)
3. An easy-to-use stream to put create operations on our `updates` stream (via `create`)

Anywhere in our code, if we want to get the most current list of messages, we just have to go to the `messages` stream. But we have a problem, **we still haven't connected this flow to the `newMessages` stream.

It would be great if we had a way to easily connect this stream with any `Message` that comes from `newMessages`. It turns out, it's really easy:

**code/rxjs/chat/app/ts/services/MessagesService.ts**

```
64    this.newMessages
65      .subscribe(this.create);
```

Now our diagram looks like this:

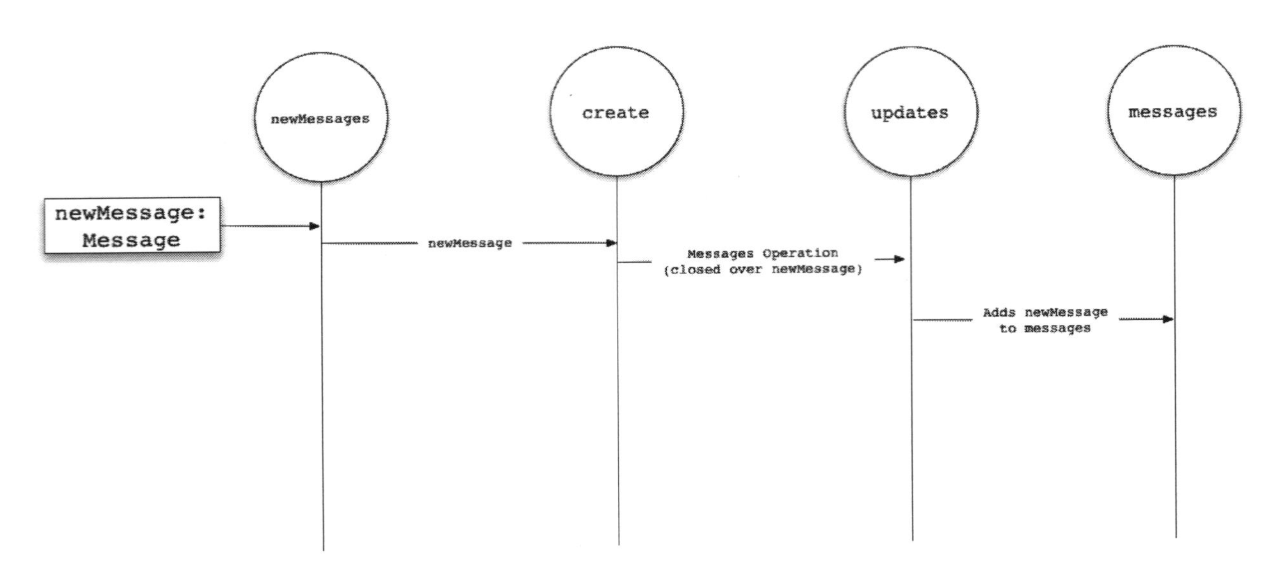


**Creating a new message, starting with the `newMessages` stream**

Now our flow is complete! It's the best of both worlds: we're able to subscribe to the stream of individual messages through `newMessages`, but if we just want the most up-to-date list, we can subscribe to `messages`.

It's worth pointing out some implications of this design: if you subscribe to `newMessages` directly, you have to be careful about changes that may happen downstream. Here are three things to consider:

First, you obviously won't get any downstream updates that are applied to the `Messages`.

Second, in this case, we have **mutable** `Message` objects. So if you subscribe to `newMessages` and store a reference to a `Message`, that `Message`'s attributes may change.

Third, in the case where you want to take advantage of the mutability of our `Messages` you may not be able to. Consider the case where we could put an operation on the `updates` queue that makes a copy of each `Message` and then mutates the copy. (This is probably a better design than what we're doing here.) In this case, you couldn't rely on any `Message` emitted directly from `newMessages` being in its "final" state.

That said, as long as you keep these considerations in mind, you shouldn't have too much trouble.

## Our completed `MessagesService`

Here's what the completed `MessagesService` looks like:

**code/rxjs/chat/app/ts/services/MessagesService.ts**

```
import {Injectable} from '@angular/core';
import {Subject, Observable} from 'rxjs';
import {User, Thread, Message} from '../models';

let initialMessages: Message[] = [];

interface IMessagesOperation extends Function {
  (messages: Message[]): Message[];
}

@Injectable()
export class MessagesService {
  // a stream that publishes new messages only once
  newMessages: Subject<Message> = new Subject<Message>();

  // `messages` is a stream that emits an array of the most up to date messages
  messages: Observable<Message[]>;

  // `updates` receives _operations_ to be applied to our `messages`
  // it's a way we can perform changes on *all* messages (that are currently
  // stored in `messages`)
  updates: Subject<any> = new Subject<any>();

  // action streams
  create: Subject<Message> = new Subject<Message>();
  markThreadAsRead: Subject<any> = new Subject<any>();

  constructor() {
    this.messages = this.updates
      // watch the updates and accumulate operations on the messages
      .scan((messages: Message[],
             operation: IMessagesOperation) => {
               return operation(messages);
             },
            initialMessages)
      // make sure we can share the most recent list of messages across anyone
      // who's interested in subscribing and cache the last known list of
      // messages
      .publishReplay(1)
      .refCount();

```

```
42     // `create` takes a Message and then puts an operation (the inner function)
43     // on the `updates` stream to add the Message to the list of messages.
44     //
45     // That is, for each item that gets added to `create` (by using `next`)
46     // this stream emits a concat operation function.
47     //
48     // Next we subscribe `this.updates` to listen to this stream, which means
49     // that it will receive each operation that is created
50     //
51     // Note that it would be perfectly acceptable to simply modify the
52     // "addMessage" function below to simply add the inner operation function to
53     // the update stream directly and get rid of this extra action stream
54     // entirely. The pros are that it is potentially clearer. The cons are that
55     // the stream is no longer composable.
56     this.create
57       .map( function(message: Message): IMessagesOperation {
58         return (messages: Message[]) => {
59           return messages.concat(message);
60         };
61       })
62       .subscribe(this.updates);
63
64     this.newMessages
65       .subscribe(this.create);
66
67     // similarly, `markThreadAsRead` takes a Thread and then puts an operation
68     // on the `updates` stream to mark the Messages as read
69     this.markThreadAsRead
70       .map( (thread: Thread) => {
71         return (messages: Message[]) => {
72           return messages.map( (message: Message) => {
73             // note that we're manipulating `message` directly here. Mutability
74             // can be confusing and there are lots of reasons why you might want
75             // to, say, copy the Message object or some other 'immutable' here
76             if (message.thread.id === thread.id) {
77               message.isRead = true;
78             }
79             return message;
80           });
81         };
82       })
83       .subscribe(this.updates);
```

```
84
85    }
86
87    // an imperative function call to this action stream
88    addMessage(message: Message): void {
89      this.newMessages.next(message);
90    }
91
92    messagesForThreadUser(thread: Thread, user: User): Observable<Message> {
93      return this.newMessages
94        .filter((message: Message) => {
95                // belongs to this thread
96          return (message.thread.id === thread.id) &&
97                // and isn't authored by this user
98                (message.author.id !== user.id);
99        });
100   }
101 }
102
103 export var messagesServiceInjectables: Array<any> = [
104   MessagesService
105 ];
```

## Trying out `MessagesService`

If you haven't already, this would be a good time to open up the code and play around with the `MessagesService` to get a feel for how it works. We've got an example you can start with in `test/services/MessagesService.spec.ts`.

To run the tests in this project, open up your terminal then:

```
1 cd /path/to/code/rxjs/chat // <-- your path will vary
2 npm install
3 karma start
```

Let's start by creating a few instances of our models to use:

**code/rxjs/chat/test/services/MessagesService.spec.ts**

```
1  import {MessagesService} from '../../app/ts/services/services';
2  import {Message, User, Thread} from '../../app/ts/models';
3
4  describe('MessagesService', () => {
5    it('should test', () => {
6
7      let user: User = new User('Nate', '');
8      let thread: Thread = new Thread('t1', 'Nate', '');
9      let m1: Message = new Message({
10       author: user,
11       text: 'Hi!',
12       thread: thread
13     });
14
15     let m2: Message = new Message({
16       author: user,
17       text: 'Bye!',
18       thread: thread
19     });
```

Next let's subscribe to a couple of our streams:

**code/rxjs/chat/test/services/MessagesService.spec.ts**

```
21     let messagesService: MessagesService = new MessagesService();
22
23     // listen to each message indivdually as it comes in
24     messagesService.newMessages
25       .subscribe( (message: Message) => {
26         console.log('=> newMessages: ' + message.text);
27       });
28
29     // listen to the stream of most current messages
30     messagesService.messages
31       .subscribe( (messages: Message[]) => {
32         console.log('=> messages: ' + messages.length);
33       });
34
35     messagesService.addMessage(m1);
36     messagesService.addMessage(m2);
37
```

```
38         // => messages: 1
39         // => newMessages: Hi!
40         // => messages: 2
41         // => newMessages: Bye!
42     });
43
44
45 });
```

Notice that even though we subscribed to newMessages first and newMessages is called directly by addMessage, our messages subscription is logged first. The reason for this is because messages subscribed to newMessages earlier than our subscription in this test (when MessagesService was instantiated). (You shouldn't be relying on the ordering of independent streams in your code, but why it works this way is worth thinking about.)

Play around with the MessagesService and get a feel for the streams there. We're going to be using them in the next section where we build the ThreadsService.

## The ThreadsService

On our ThreadsService were going to define four streams that emit respectively:

1. A map of the current set of Threads (in threads)
2. A chronological list of Threads, newest-first (in orderedthreads)
3. The currently selected Thread (in currentThread)
4. The list of Messages for the currently selected Thread (in currentThreadMessages)

Let's walk through how to build each of these streams, and we'll learn a little more about RxJS along the way.

### A map of the current set of Threads (in threads)

Let's start by defining our ThreadsService class and the instance variable that will emit the Threads:

**code/rxjs/chat/app/ts/services/ThreadsService.ts**

```
1  import {Injectable} from '@angular/core';
2  import {Subject, BehaviorSubject, Observable} from 'rxjs';
3  import {Thread, Message} from '../models';
4  import {MessagesService} from './MessagesService';
5  import * as _ from 'underscore';
6
7  @Injectable()
8  export class ThreadsService {
9
10   // `threads` is a observable that contains the most up to date list of threads
11   threads: Observable<{ [key: string]: Thread }>;
```

Notice that this stream will emit a map (an object) with the `id` of the `Thread` being the `string` key and the `Thread` itself will be the value.

To create a stream that maintains the current list of threads, we start by attaching to the `messagesService.messages` stream:

**code/rxjs/chat/app/ts/services/ThreadsService.ts**

```
11   threads: Observable<{ [key: string]: Thread }>;
```

Recall that each time a new `Message` is added to the steam, `messages` will emit an array of the current `Messages`. We're going to look at each `Message` and we want to return a unique list of the `Threads`.

**code/rxjs/chat/app/ts/services/ThreadsService.ts**

```
26     this.threads = messagesService.messages
27       .map( (messages: Message[]) => {
28         let threads: {[key: string]: Thread} = {};
29         // Store the message's thread in our accumulator `threads`
30         messages.map((message: Message) => {
31           threads[message.thread.id] = threads[message.thread.id] ||
32             message.thread;
```

Notice above that each time we will create a new list of `threads`. The reason for this is because we might delete some messages down the line (e.g. leave the conversation). Because we're recalculating the list of threads each time, we naturally will "delete" a thread if it has no messages.

In the threads list, we want to show a preview of the chat by using the text of the most recent `Message` in that `Thread`.

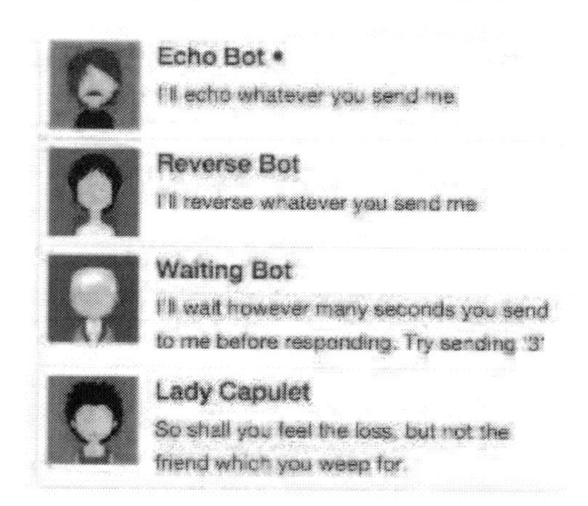


**List of Threads with Chat Preview**

In order to do that, we'll store the most recent `Message` for each `Thread`. We know which `Message` is newest by comparing the `sentAt` times:

**code/rxjs/chat/app/ts/services/ThreadsService.ts**

```
33          // Cache the most recent message for each thread
34          let messagesThread: Thread = threads[message.thread.id];
35          if (!messagesThread.lastMessage ||
36              messagesThread.lastMessage.sentAt < message.sentAt) {
37            messagesThread.lastMessage = message;
38          }
39        });
40        return threads;
41      });
```

Putting it all together, `threads` looks like this:

**code/rxjs/chat/app/ts/services/ThreadsService.ts**

```
26      this.threads = messagesService.messages
27        .map( (messages: Message[]) => {
28          let threads: {[key: string]: Thread} = {};
29          // Store the message's thread in our accumulator `threads`
30          messages.map((message: Message) => {
31            threads[message.thread.id] = threads[message.thread.id] ||
32              message.thread;
33
34            // Cache the most recent message for each thread
```

```
35          let messagesThread: Thread = threads[message.thread.id];
36          if (!messagesThread.lastMessage ||
37              messagesThread.lastMessage.sentAt < message.sentAt) {
38            messagesThread.lastMessage = message;
39          }
40        });
41        return threads;
42      });
```

### Trying out the `ThreadsService`

Let's try out our `ThreadsService`. First we'll create a few models to work with:

**code/rxjs/chat/test/services/ThreadsService.spec.ts**

```
 1  import {MessagesService, ThreadsService} from '../../app/ts/services/services';
 2  import {Message, User, Thread} from '../../app/ts/models';
 3  import * as _ from 'underscore';
 4
 5  describe('ThreadsService', () => {
 6    it('should collect the Threads from Messages', () => {
 7
 8      let nate: User = new User('Nate Murray', '');
 9      let felipe: User = new User('Felipe Coury', '');
10
11      let t1: Thread = new Thread('t1', 'Thread 1', '');
12      let t2: Thread = new Thread('t2', 'Thread 2', '');
13
14      let m1: Message = new Message({
15        author: nate,
16        text: 'Hi!',
17        thread: t1
18      });
19
20      let m2: Message = new Message({
21        author: felipe,
22        text: 'Where did you get that hat?',
23        thread: t1
24      });
25
26      let m3: Message = new Message({
27        author: nate,
28        text: 'Did you bring the briefcase?',
```

```
29        thread: t2
30      });
```

Now let's create an instance of our services:

**code/rxjs/chat/test/services/ThreadsService.spec.ts**

```
32      let messagesService: MessagesService = new MessagesService();
33      let threadsService: ThreadsService = new ThreadsService(messagesService);
```

Notice here that we're passing `messagesService` as an argument to the constructor of our `ThreadsService`. Normally we let the Dependency Injection system handle this for us. But in our test, we can provide the dependencies ourselves.

Let's subscribe to `threads` and log out what comes through:

**code/rxjs/chat/test/services/ThreadsService.spec.ts**

```
33      let threadsService: ThreadsService = new ThreadsService(messagesService);
34
35      threadsService.threads
36        .subscribe( (threadIdx: { [key: string]: Thread }) => {
37          let threads: Thread[] = _.values(threadIdx);
38          let threadNames: string = _.map(threads, (t: Thread) => t.name)
39                                   .join(', ');
40          console.log(`=> threads (${threads.length}): ${threadNames} `);
41        });
42
43      messagesService.addMessage(m1);
44      messagesService.addMessage(m2);
45      messagesService.addMessage(m3);
46
47      // => threads (1): Thread 1
48      // => threads (1): Thread 1
49      // => threads (2): Thread 1, Thread 2
50
51    });
52  });
```

## A chronological list of `Thread`s, newest-first (in `orderedthreads`)

`threads` gives us a map which acts as an "index" of our list of threads. But we want the threads view to be ordered according the most recent message.

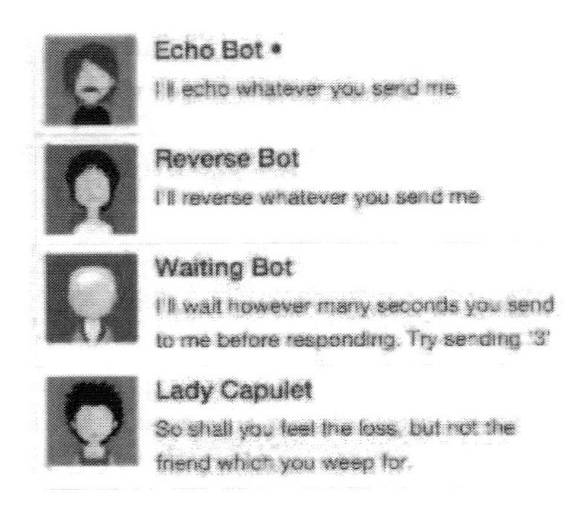


**Time Ordered List of Threads**

Let's create a new stream that returns an `Array` of `Threads` ordered by the most recent `Message` time:

We'll start by defining `orderedThreads` as an instance property:

**code/rxjs/chat/app/ts/services/ThreadsService.ts**

```
13    // `orderedThreads` contains a newest-first chronological list of threads
14    orderedThreads: Observable<Thread[]>;
```

Next, in the `constructor` we'll define `orderedThreads` by subscribing to `threads` and ordered by the most recent message:

**code/rxjs/chat/app/ts/services/ThreadsService.ts**

```
44      this.orderedThreads = this.threads
45        .map((threadGroups: { [key: string]: Thread }) => {
46          let threads: Thread[] = _.values(threadGroups);
47          return _.sortBy(threads, (t: Thread) => t.lastMessage.sentAt).reverse();
48        });
```

## The currently selected `Thread` (in `currentThread`)

Our application needs to know which `Thread` is the currently selected thread. This lets us know:

1. which thread should be shown in the messages window
2. which thread should be marked as the current thread in the list of threads

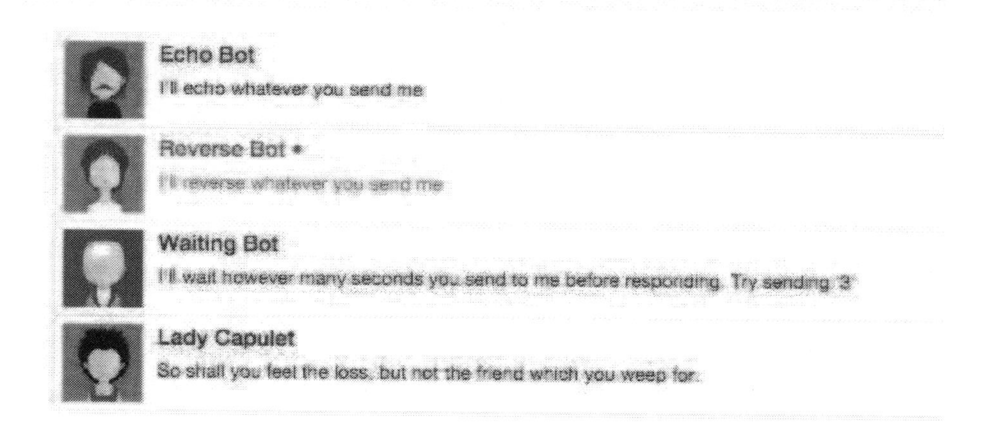


**The current thread is marked by a 'â€¢' symbol**

Let's create a `BehaviorSubject` that will store the `currentThread`:

**code/rxjs/chat/app/ts/services/ThreadsService.ts**

```
16    // `currentThread` contains the currently selected thread
17    currentThread: Subject<Thread> =
18      new BehaviorSubject<Thread>(new Thread());
```

Notice that we're issuing an empty `Thread` as the default value. We don't need to configure the `currentThread` any further.

## Setting the Current `Thread`

To set the current thread we can have clients either

1. submit new threads via `next` directly or
2. add a helper method to do it.

Let's define a helper method `setCurrentThread` that we can use to set the next thread:

code/rxjs/chat/app/ts/services/ThreadsService.ts

```
69    setCurrentThread(newThread: Thread): void {
70      this.currentThread.next(newThread);
71    }
```

### Marking the Current `Thread` as Read

We want to keep track of the number of unread messages. If we switch to a new `Thread` then we want to mark all of the `Messages` in that `Thread` as read. We have the parts we need to do this:

1. The `messagesService.makeThreadAsRead` accepts a `Thread` and then will mark all `Messages` in that `Threaad` as read
2. Our `currentThread` emits a single `Thread` that represents the current `Thread`

So all we need to do is hook them together:

code/rxjs/chat/app/ts/services/ThreadsService.ts

```
66      this.currentThread.subscribe(this.messagesService.markThreadAsRead);
```

## The list of `Message`s for the currently selected `Thread` (in `currentThreadMessages`)

Now that we have the currently selected thread, we need to make sure we can show the list of `Messages` in that `Thread`.

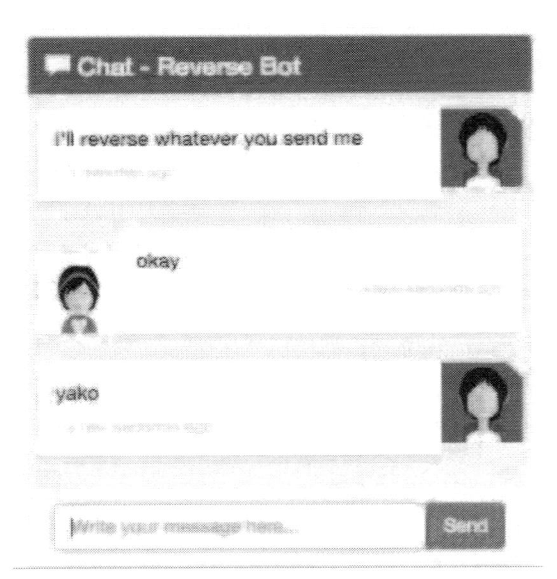


The current list of messages is for the Reverse Bot

Implementing this is a little bit more complicated than it may seem at the surface. Say we implemented it like this:

```
1  var theCurrentThread: Thread;
2
3  this.currentThread.subscribe((thread: Thread) => {
4    theCurrentThread = thread;
5  })
6
7  this.currentThreadMessages.map(
8    (mesages: Message[]) => {
9      return _.filter(messages,
10       (message: Message) => {
11         return message.thread.id == theCurrentThread.id;
12       })
13   })
```

What's wrong with this approach? Well, if the `currentThread` changes, `currentThreadMessages` won't know about it and so we'll have an outdated list of `currentThreadMessages`!

What if we reversed it, and stored the current list of messages in a variable and subscribed to the changing of `currentThread`? We'd have the same problem only this time we would know when the thread changes but not when a new message came in.

How can we solve this problem?

It turns out, RxJS has a set of operators that we can use to **combine multiple streams**. In this case we want to say "if *either* `currentThread` **or** `messagesService.messages` changes, then we want to emit something." For this we use the `combineLatest`[76] operator.

**code/rxjs/chat/app/ts/services/ThreadsService.ts**

```
50      this.currentThreadMessages = this.currentThread
51        .combineLatest(messagesService.messages,
52                       (currentThread: Thread, messages: Message[]) => {
```

When we're combining two streams one or the other will arrive first and there's no guarantee that we'll have a value on both streams, so we need to check to make sure we have what we need otherwise we'll just return an empty list.

Now that we have both the current thread and messages, we can filter out just the messages we're interested in:

[76] https://github.com/Reactive-Extensions/RxJS/blob/master/doc/api/core/operators/combinelatestproto.md

**code/rxjs/chat/app/ts/services/ThreadsService.ts**

```
50      this.currentThreadMessages = this.currentThread
51        .combineLatest(messagesService.messages,
52                       (currentThread: Thread, messages: Message[]) => {
53          if (currentThread && messages.length > 0) {
54            return _.chain(messages)
55              .filter((message: Message) =>
56                      (message.thread.id === currentThread.id))
```

One other detail, since we're already looking that the messages for the current thread, this is a convenient area to mark these messages as read.

**code/rxjs/chat/app/ts/services/ThreadsService.ts**

```
54            return _.chain(messages)
55              .filter((message: Message) =>
56                      (message.thread.id === currentThread.id))
57              .map((message: Message) => {
58                message.isRead = true;
59                return message; })
60              .value();
```

Whether or not we should be marking messages as read here is debatable. The biggest drawback is that we're mutating objects in what is, essentially, a "read" thread. i.e. this is a read operation with a side effect, which is generally a Bad Idea. That said, in this application the `currentThreadMessages` only applies to the `currentThread` and the `currentThread` should always have its messages marked as read. That said, the "read with side-effects" is not a pattern I recommend in general.

Putting it together, here's what `currentThreadMessages` looks like:

**code/rxjs/chat/app/ts/services/ThreadsService.ts**

```
50      this.currentThreadMessages = this.currentThread
51        .combineLatest(messagesService.messages,
52                       (currentThread: Thread, messages: Message[]) => {
53          if (currentThread && messages.length > 0) {
54            return _.chain(messages)
55              .filter((message: Message) =>
56                      (message.thread.id === currentThread.id))
57              .map((message: Message) => {
```

```
58              message.isRead = true;
59              return message; })
60            .value();
61        } else {
62          return [];
63        }
64      });
```

## Our Completed ThreadsService

Here's what our `ThreadService` looks like:

**code/rxjs/chat/app/ts/services/ThreadsService.ts**

```
1   import {Injectable} from '@angular/core';
2   import {Subject, BehaviorSubject, Observable} from 'rxjs';
3   import {Thread, Message} from '../models';
4   import {MessagesService} from './MessagesService';
5   import * as _ from 'underscore';
6
7   @Injectable()
8   export class ThreadsService {
9
10    // `threads` is a observable that contains the most up to date list of threads
11    threads: Observable<{ [key: string]: Thread }>;
12
13    // `orderedThreads` contains a newest-first chronological list of threads
14    orderedThreads: Observable<Thread[]>;
15
16    // `currentThread` contains the currently selected thread
17    currentThread: Subject<Thread> =
18      new BehaviorSubject<Thread>(new Thread());
19
20    // `currentThreadMessages` contains the set of messages for the currently
21    // selected thread
22    currentThreadMessages: Observable<Message[]>;
23
24    constructor(private messagesService: MessagesService) {
25
26      this.threads = messagesService.messages
27        .map( (messages: Message[]) => {
28          let threads: {[key: string]: Thread} = {};
29          // Store the message's thread in our accumulator `threads`
```

```
30          messages.map((message: Message) => {
31            threads[message.thread.id] = threads[message.thread.id] ||
32              message.thread;
33
34            // Cache the most recent message for each thread
35            let messagesThread: Thread = threads[message.thread.id];
36            if (!messagesThread.lastMessage ||
37                messagesThread.lastMessage.sentAt < message.sentAt) {
38              messagesThread.lastMessage = message;
39            }
40          });
41          return threads;
42        });
43
44      this.orderedThreads = this.threads
45        .map((threadGroups: { [key: string]: Thread }) => {
46          let threads: Thread[] = _.values(threadGroups);
47          return _.sortBy(threads, (t: Thread) => t.lastMessage.sentAt).reverse();
48        });
49
50      this.currentThreadMessages = this.currentThread
51        .combineLatest(messagesService.messages,
52                       (currentThread: Thread, messages: Message[]) => {
53          if (currentThread && messages.length > 0) {
54            return _.chain(messages)
55              .filter((message: Message) =>
56                      (message.thread.id === currentThread.id))
57              .map((message: Message) => {
58                message.isRead = true;
59                return message; })
60              .value();
61          } else {
62            return [];
63          }
64        });
65
66      this.currentThread.subscribe(this.messagesService.markThreadAsRead);
67    }
68
69    setCurrentThread(newThread: Thread): void {
70      this.currentThread.next(newThread);
71    }
```

```
72
73 }
74
75 export var threadsServiceInjectables: Array<any> = [
76   ThreadsService
77 ];
```

## Data Model Summary

Our data model and services are complete! Now we have everything we need now to start hooking it up to our view components! In the next chapter we'll build out our 3 major components to render and interact with these streams.

# Data Architecture with Observables - Part 2: View Components

## Building Our Views: The `ChatApp` Top-Level Component

Let's turn our attention to our app and implement our view components.

For the sake of clarity and space, in the following sections I'll be leaving out some `import` statements, CSS, and a few other similar things lines of code. If you're curious about each line of those details, open up the sample code because it contains everything we need to run this app.

The first thing we're going to do is create our top-level component `chat-app`

As we talked about earlier, the page is broken down into three top-level components:

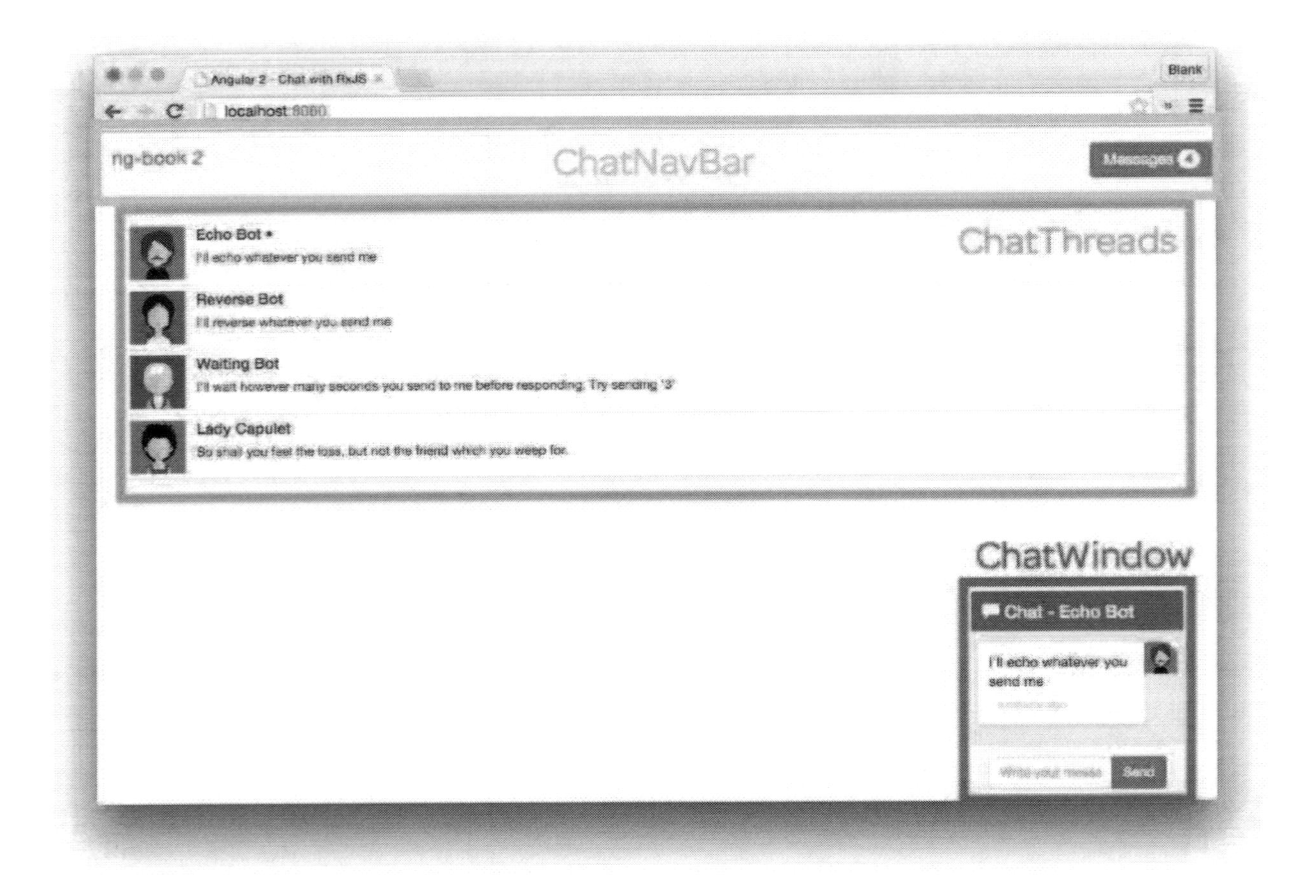


**Chat Top-Level Components**

- `ChatNavBar` - contains the unread messages count
- `ChatThreads` - shows a clickable list of threads, along with the most recent message and the conversation avatar
- `ChatWindow` - shows the messages in the current thread with an input box to send new messages

Here's what our component looks like in code:

**code/rxjs/chat/app/ts/app.ts**

```
54 @Component({
55   selector: 'chat-app',
56   template: `
57   <div>
58     <nav-bar></nav-bar>
59     <div class="container">
60       <chat-threads></chat-threads>
61       <chat-window></chat-window>
62     </div>
63   </div>
64   `
65 })
66 class ChatApp {
67   constructor(private messagesService: MessagesService,
68               private threadsService: ThreadsService,
69               private userService: UserService) {
70     ChatExampleData.init(messagesService, threadsService, userService);
71   }
72 }
73
74 @NgModule({
75   declarations: [
76     ChatApp,
77     ChatNavBar,
78     ChatThreads,
79     ChatThread,
80     ChatWindow,
81     ChatMessage,
82     utilInjectables
83   ],
84   imports: [
85     BrowserModule,
86     FormsModule
87   ],
88   bootstrap: [ ChatApp ],
89   providers: [ servicesInjectables ]
90 })
91 export class ChatAppModule {}
92
93 platformBrowserDynamic().bootstrapModule(ChatAppModule);
```

Take a look at the `constructor`. Here we're injecting our three services: the `MessagesService`, `ThreadsService`, and `UserService`. We're using those services to initialize our example data.

If you're interested in the example data you can find it in `code/rxjs/chat/app/ts/ChatExampleData.ts`.

## The `ChatThreads` Component

Next let's build our thread list in the `ChatThreads` component.

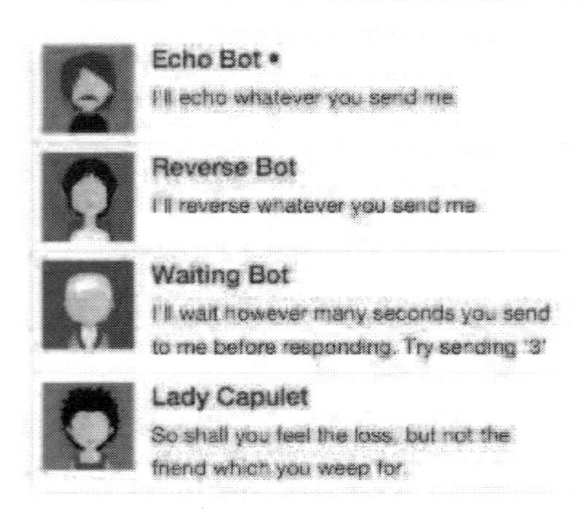


**Time Ordered List of Threads**

Our `selector` is straightforward, we want to match `chat-threads`.

**code/rxjs/chat/app/ts/components/ChatThreads.ts**

```
52 @Component({
53   selector: 'chat-threads',
```

### `ChatThreads` Controller

Take a look at our component controller `ChatThreads`:

**code/rxjs/chat/app/ts/components/ChatThreads.ts**

```
69 export class ChatThreads {
70   threads: Observable<any>;
71 
72   constructor(private threadsService: ThreadsService) {
73     this.threads = threadsService.orderedThreads;
74   }
75 }
```

Here we're injecting `ThreadsService` and then we're keeping a reference to the `orderedThreads`.

### `ChatThreads` template

Lastly, let's look at the `template` and its configuration:

**code/rxjs/chat/app/ts/components/ChatThreads.ts**

```
52 @Component({
53   selector: 'chat-threads',
54   changeDetection: ChangeDetectionStrategy.OnPush,
55   template: `
56     <!-- conversations -->
57     <div class="row">
58       <div class="conversation-wrap">
59 
60         <chat-thread
61              *ngFor="let thread of threads | async"
62              [thread]="thread">
63         </chat-thread>
64 
65       </div>
66     </div>
67   `
```

There's three things to look at here: `NgFor` with the `async` pipe, the `ChangeDetectionStrategy` and `ChatThread`.

The `ChatThread` directive component (which matches `chat-thread` in the markup) will show the view for the `Threads`. We'll define that in a moment.

The `NgFor` iterates over our `threads`, and passes the input `[thread]` to our `ChatThread` directive. But you probably notice something new in our `*ngFor`: the pipe to `async`.

async is implemented by AsyncPipe and it lets us use an RxJS Observable here in our view. What's great about async is that it lets us use our async observable as if it was a sync collection. This is super convenient and really cool.

On this component we specify a custom changeDetection. Angular 2 has a flexible and efficient change detection system. One of the benefits is that if we have a component which has immutable or observable bindings, then we're able to give the change detection system hints that will make our application run very efficiently.

In this case, instead of watching for changes on an array of Threads, Angular will subscribe for changes to the threads observable - and trigger an update when a new event is emitted.

Here's what our total ChatThreads component looks like:

**code/rxjs/chat/app/ts/components/ChatThreads.ts**

```
52  @Component({
53    selector: 'chat-threads',
54    changeDetection: ChangeDetectionStrategy.OnPush,
55    template: `
56      <!-- conversations -->
57      <div class="row">
58        <div class="conversation-wrap">
59
60          <chat-thread
61               *ngFor="let thread of threads | async"
62               [thread]="thread">
63          </chat-thread>
64
65        </div>
66      </div>
67    `
68  })
69  export class ChatThreads {
70    threads: Observable<any>;
71
72    constructor(private threadsService: ThreadsService) {
73      this.threads = threadsService.orderedThreads;
74    }
75  }
```

## The Single ChatThread Component

Let's look at our ChatThread component. This is the component that will be used to display a **single thread**. Starting with the @Component:

**code/rxjs/chat/app/ts/components/ChatThreads.ts**

```
10  @Component({
11    inputs: ['thread'],
12    selector: 'chat-thread',
13    template: `
14    <div class="media conversation">
15      <div class="pull-left">
16        <img class="media-object avatar"
17             src="{{thread.avatarSrc}}">
18      </div>
19      <div class="media-body">
20        <h5 class="media-heading contact-name">{{thread.name}}
21          <span *ngIf="selected">&bull;</span>
22        </h5>
23        <small class="message-preview">{{thread.lastMessage.text}}</small>
24      </div>
25      <a (click)="clicked($event)" class="div-link">Select</a>
26    </div>
27    `
28  })
```

We'll come back and look at the `template` in a minute, but first let's look at the component definition controller.

## `ChatThread` Controller and `ngOnInit`

**code/rxjs/chat/app/ts/components/ChatThreads.ts**

```
29  export class ChatThread implements OnInit {
30    thread: Thread;
31    selected: boolean = false;
32
33    constructor(private threadsService: ThreadsService) {
34    }
35
36    ngOnInit(): void {
37      this.threadsService.currentThread
38        .subscribe( (currentThread: Thread) => {
39          this.selected = currentThread &&
40            this.thread &&
41            (currentThread.id === this.thread.id);
```

```
42       });
43   }
44
45   clicked(event: any): void {
46     this.threadsService.setCurrentThread(this.thread);
47     event.preventDefault();
48   }
49 }
```

Notice that we're implementing a new interface here: `OnInit`. Angular components can declare that they listen for certain lifecycle events. We talk more about lifecycle events here in the Advanced Components chapter.

In this case, because we declared that we implement `OnInit`, the method `ngOnInit` will be called on our component after the component has been checked for changes the first time.

A key reason we will use `ngOnInit` is because **our `thread` property won't be available in the `constructor`**.

Above you can see that in `ngOnInit` we subscribe to `threadsService.currentThread` and if the `currentThread` matches the `thread` property of this component, we set `selected` to `true` (conversely, if the `Thread` doesn't match, we set `selected` to false).

We also setup an event handler `clicked`. This is how we handle selecting the current thread. In our `template` (below), we will bind `clicked()` to clicking on the thread view. If we receive `clicked()` then we tell the `threadsService` we want to set the current thread to the `Thread` of this component.

### `ChatThread` template

Here's the code for our `template`:

**code/rxjs/chat/app/ts/components/ChatThreads.ts**

```
13   template: `
14   <div class="media conversation">
15     <div class="pull-left">
16       <img class="media-object avatar"
17            src="{{thread.avatarSrc}}">
18     </div>
19     <div class="media-body">
20       <h5 class="media-heading contact-name">{{thread.name}}
21         <span *ngIf="selected">&bull;</span>
22       </h5>
23       <small class="message-preview">{{thread.lastMessage.text}}</small>
24     </div>
```

```
25      <a (click)="clicked($event)" class="div-link">Select</a>
26    </div>
27    `
```

Notice we've got some straight-forward bindings like `{{thread.avatarSrc}}`, `{{thread.name}}`, and `{{thread.lastMessage.text}}`.

We've got an `*ngIf` which will show the `•` symbol only if this is the selected thread.

Lastly, we're binding to the `(click)` event to call our `clicked()` handler. Notice that when we call `clicked` we're passing the argument `$event`. This is a special variable provided by Angular that describes the event. We use that in our `clicked` handler by calling `event.preventDefault();`. This makes sure that we don't navigate to a different page.

## `ChatThread` Complete Code

Here's the whole of the `ChatThread` component:

**code/rxjs/chat/app/ts/components/ChatThreads.ts**

```
10  @Component({
11    inputs: ['thread'],
12    selector: 'chat-thread',
13    template: `
14    <div class="media conversation">
15      <div class="pull-left">
16        <img class="media-object avatar"
17             src="{{thread.avatarSrc}}">
18      </div>
19      <div class="media-body">
20        <h5 class="media-heading contact-name">{{thread.name}}
21          <span *ngIf="selected">&bull;</span>
22        </h5>
23        <small class="message-preview">{{thread.lastMessage.text}}</small>
24      </div>
25      <a (click)="clicked($event)" class="div-link">Select</a>
26    </div>
27    `
28  })
29  export class ChatThread implements OnInit {
30    thread: Thread;
31    selected: boolean = false;
32
33    constructor(private threadsService: ThreadsService) {
```

```
  }

  ngOnInit(): void {
    this.threadsService.currentThread
      .subscribe( (currentThread: Thread) => {
        this.selected = currentThread &&
          this.thread &&
          (currentThread.id === this.thread.id);
      });
  }

  clicked(event: any): void {
    this.threadsService.setCurrentThread(this.thread);
    event.preventDefault();
  }
}
```

## The `ChatWindow` Component

The `ChatWindow` is the most complicated component in our app. Let's take it one section at a time:

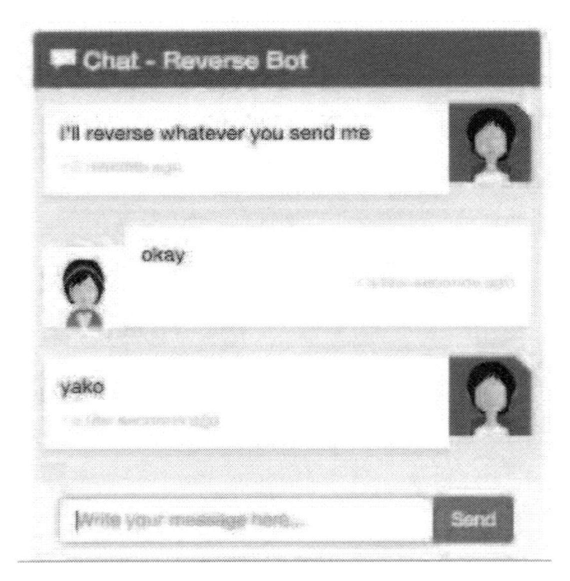


**The Chat Window**

We start by defining our `@Component`:

**code/rxjs/chat/app/ts/components/ChatWindow.ts**

```
61  @Component({
62    selector: 'chat-window',
63    changeDetection: ChangeDetectionStrategy.OnPush,
```

### `ChatWindow` Component Class Properties

Our `ChatWindow` class has four properties:

**code/rxjs/chat/app/ts/components/ChatWindow.ts**

```
110 export class ChatWindow implements OnInit {
111   messages: Observable<any>;
112   currentThread: Thread;
113   draftMessage: Message;
114   currentUser: User;
```

Here's a diagram of where each one is used:

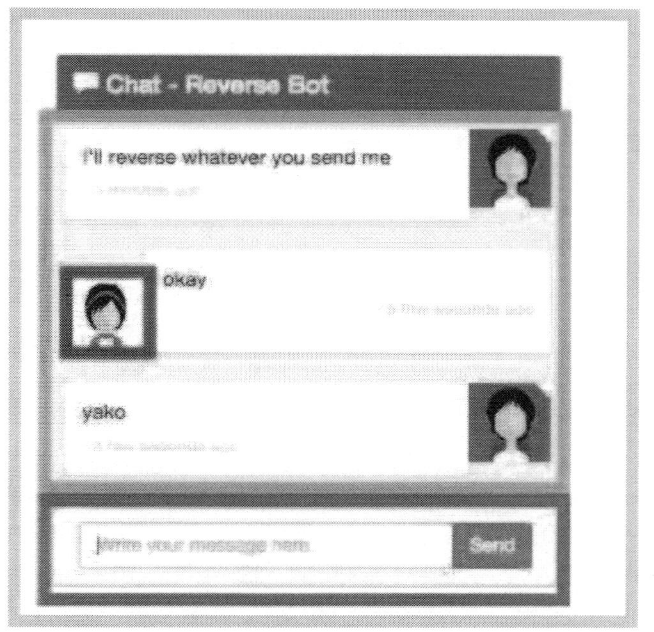


**Chat Window Properties**

In our `constructor` we're going to inject four things:

**code/rxjs/chat/app/ts/components/ChatWindow.ts**

```
116   constructor(private messagesService: MessagesService,
117               private threadsService: ThreadsService,
118               private userService: UserService,
119               private el: ElementRef) {
120   }
```

The first three are our services. The fourth, `el` is an `ElementRef` which we can use to get access to the host DOM element. We'll use that when we scroll to the bottom of the chat window when we create and receive new messages.

Remember: by using `public messagesService: MessagesService` in the constructor, we are not only injecting the `MessagesService` but setting up an instance variable that we can use later in our class via `this.messagesService`

#### `ChatWindow ngOnInit`

We're going to put the initialization of this component in `ngOnInit`. The main thing we're going to be doing here is setting up the subscriptions on our observables which will then change our component properties.

**code/rxjs/chat/app/ts/components/ChatWindow.ts**

```
122   ngOnInit(): void {
123     this.messages = this.threadsService.currentThreadMessages;
124 
125     this.draftMessage = new Message();
```

First, we'll save the `currentThreadMessages` into `messages`. Next we create an empty `Message` for the default `draftMessage`.

When we send a new message we need to make sure that `Message` stores a reference to the sending `Thread`. The sending thread is always going to be the current thread, so let's store a reference to the currently selected thread:

**code/rxjs/chat/app/ts/components/ChatWindow.ts**

```
127      this.threadsService.currentThread.subscribe(
128        (thread: Thread) => {
129          this.currentThread = thread;
130        });
```

We also want new messages to be sent from the current user, so let's do the same with `currentUser`:

**code/rxjs/chat/app/ts/components/ChatWindow.ts**

```
132      this.userService.currentUser
133        .subscribe(
134          (user: User) => {
135            this.currentUser = user;
136          });
```

#### ChatWindow sendMessage

Since we're talking about it, let's implement a `sendMessage` function that will send a new message:

**code/rxjs/chat/app/ts/components/ChatWindow.ts**

```
152    sendMessage(): void {
153      let m: Message = this.draftMessage;
154      m.author = this.currentUser;
155      m.thread = this.currentThread;
156      m.isRead = true;
157      this.messagesService.addMessage(m);
158      this.draftMessage = new Message();
159    }
```

The `sendMessage` function above takes the `draftMessage`, sets the `author` and `thread` using our component properties. Every message we send has "been read" already (we wrote it) so we mark it as read.

Notice here that we're not updating the `draftMessage` text. That's because we're going to bind the value of the messages text in the view in a few minutes.

After we've updated the `draftMessage` properties we send it off to the `messagesService` and then **create a new `Message`** and set that new `Message` to `this.draftMessage`. We do this to make sure we don't mutate an already sent message.

#### ChatWindow onEnter

In our view, we want to send the message in two scenarios

1. the user hits the "Send" button or
2. the user hits the Enter (or Return) key.

Let's define a function that will handle that event:

**code/rxjs/chat/app/ts/components/ChatWindow.ts**

```
147   onEnter(event: any): void {
148     this.sendMessage();
149     event.preventDefault();
150   }
```

#### ChatWindow scrollToBottom

When we send a message, or when a new message comes in, we want to scroll to the bottom of the chat window. To do that, we're going to set the `scrollTop` property of our host element:

**code/rxjs/chat/app/ts/components/ChatWindow.ts**

```
161   scrollToBottom(): void {
162     let scrollPane: any = this.el
163       .nativeElement.querySelector('.msg-container-base');
164     scrollPane.scrollTop = scrollPane.scrollHeight;
165   }
```

Now that we have a function that will scroll to the bottom, we have to make sure that we call it at the right time. Back in `ngOnInit` let's subscribe to the list of `currentThreadMessages` and scroll to the bottom any time we get a new message:

**code/rxjs/chat/app/ts/components/ChatWindow.ts**

```
137     this.messages
138       .subscribe(
139         (messages: Array<Message>) => {
140           setTimeout(() => {
141             this.scrollToBottom();
142           });
143         });
144   }
```

**Why do we have the `setTimeout`?**

If we call `scrollToBottom` immediately when we get a new message then what happens is we scroll to the bottom before the new message is rendered. By using a `setTimeout` we're telling Javascript that we want to run this function when it is finished with the current execution queue. This happens **after** the component is rendered, so it does what we want.

#### ChatWindow template

The opening of our `template` should look familiar, we start by defining some markup and the panel header:

**code/rxjs/chat/app/ts/components/ChatWindow.ts**

```
61  @Component({
62    selector: 'chat-window',
63    changeDetection: ChangeDetectionStrategy.OnPush,
64    template: `
65      <div class="chat-window-container">
66        <div class="chat-window">
67          <div class="panel-container">
68            <div class="panel panel-default">
69
70              <div class="panel-heading top-bar">
71                <div class="panel-title-container">
72                  <h3 class="panel-title">
73                    <span class="glyphicon glyphicon-comment"></span>
74                    Chat - {{currentThread.name}}
75                  </h3>
76                </div>
77                <div class="panel-buttons-container">
78                  <!-- you could put minimize or close buttons here -->
79                </div>
80              </div>
```

Next we show the list of messages. Here we use `ngFor` along with the `async` pipe to iterate over our list of messages. We'll describe the individual `chat-message` component in a minute.

**code/rxjs/chat/app/ts/components/ChatWindow.ts**

```
82            <div class="panel-body msg-container-base">
83              <chat-message
84                   *ngFor="let message of messages | async"
85                   [message]="message">
86              </chat-message>
87            </div>
```

Lastly we have the message input box and closing tags :

**code/rxjs/chat/app/ts/components/ChatWindow.ts**

```
88            <div class="panel-footer">
89              <div class="input-group">
90                <input type="text"
91                       class="chat-input"
92                       placeholder="Write your message here..."
93                       (keydown.enter)="onEnter($event)"
94                       [(ngModel)]="draftMessage.text" />
95                <span class="input-group-btn">
96                  <button class="btn-chat"
97                     (click)="onEnter($event)"
98                     >Send</button>
99                </span>
100             </div>
101           </div>
102
103         </div>
104       </div>
105     </div>
106   </div>
107   `
```

The message `input` box is the most interesting part of this view, so let's talk about two interesting properties: 1. `(keydown.enter)` and 2. `[(ngModel)]`.

## Handling keystrokes

Angular provides a straightforward way to handle keyboard actions: we bind to the event on an element. In this case, we're binding to `keydown.enter` which says if "Enter" is pressed, call the function in the expression, which in this case is `onEnter($event)`.

**code/rxjs/chat/app/ts/components/ChatWindow.ts**

```
91                <input type="text"
92                       class="chat-input"
93                       placeholder="Write your message here..."
94                       (keydown.enter)="onEnter($event)"
95                       [(ngModel)]="draftMessage.text" />
```

### Using ngModel

As we've talked about before, Angular doesn't have a general model for two-way binding. However it can be very useful to have a two-way binding between a component and its view. As long as the side-effects are kept local to the component, it can be a very convenient way to keep a component property in sync with the view.

In this case, we're establishing a two-way bind **between the value of the input tag and `draftMessage.text`**. That is, if we type into the `input` tag, `draftMessage.text` will automatically be set to the value of that `input`. Likewise, if we were to update `draftMessage.text` in our code, the value in the `input` tag would change in the view.

**code/rxjs/chat/app/ts/components/ChatWindow.ts**


```
91                <input type="text"
92                       class="chat-input"
93                       placeholder="Write your message here..."
94                       (keydown.enter)="onEnter($event)"
95                       [(ngModel)]="draftMessage.text" />
```


### Clicking "Send"

On our "Send" button we bind the `(click)` property to the `onEnter` function of our component:

**code/rxjs/chat/app/ts/components/ChatWindow.ts**

```
96                <span class="input-group-btn">
97                  <button class="btn-chat"
98                     (click)="onEnter($event)"
99                     >Send</button>
100               </span>
```

### The Entire `ChatWindow` Component

Here's the code listing for the entire `ChatWindow` Component:

**code/rxjs/chat/app/ts/components/ChatWindow.ts**

```
61 @Component({
62   selector: 'chat-window',
63   changeDetection: ChangeDetectionStrategy.OnPush,
64   template: `
65     <div class="chat-window-container">
66       <div class="chat-window">
67         <div class="panel-container">
68           <div class="panel panel-default">
69
70             <div class="panel-heading top-bar">
71               <div class="panel-title-container">
72                 <h3 class="panel-title">
73                   <span class="glyphicon glyphicon-comment"></span>
74                   Chat - {{currentThread.name}}
75                 </h3>
76               </div>
77               <div class="panel-buttons-container">
78                 <!-- you could put minimize or close buttons here -->
79               </div>
80             </div>
81
82             <div class="panel-body msg-container-base">
83               <chat-message
84                    *ngFor="let message of messages | async"
85                    [message]="message">
86               </chat-message>
87             </div>
88
89             <div class="panel-footer">
90               <div class="input-group">
91                 <input type="text"
92                        class="chat-input"
93                        placeholder="Write your message here..."
94                        (keydown.enter)="onEnter($event)"
95                        [(ngModel)]="draftMessage.text" />
96                 <span class="input-group-btn">
97                   <button class="btn-chat"
98                      (click)="onEnter($event)"
99                      >Send</button>
100                </span>
101              </div>
```

```
102             </div>
103
104           </div>
105         </div>
106       </div>
107     </div>
108   `
109 })
110 export class ChatWindow implements OnInit {
111   messages: Observable<any>;
112   currentThread: Thread;
113   draftMessage: Message;
114   currentUser: User;
115
116   constructor(private messagesService: MessagesService,
117               private threadsService: ThreadsService,
118               private userService: UserService,
119               private el: ElementRef) {
120   }
121
122   ngOnInit(): void {
123     this.messages = this.threadsService.currentThreadMessages;
124
125     this.draftMessage = new Message();
126
127     this.threadsService.currentThread.subscribe(
128       (thread: Thread) => {
129         this.currentThread = thread;
130       });
131
132     this.userService.currentUser
133       .subscribe(
134         (user: User) => {
135           this.currentUser = user;
136         });
137
138     this.messages
139       .subscribe(
140         (messages: Array<Message>) => {
141           setTimeout(() => {
142             this.scrollToBottom();
143           });
```

```
144         });
145     }
146 
147     onEnter(event: any): void {
148       this.sendMessage();
149       event.preventDefault();
150     }
151 
152     sendMessage(): void {
153       let m: Message = this.draftMessage;
154       m.author = this.currentUser;
155       m.thread = this.currentThread;
156       m.isRead = true;
157       this.messagesService.addMessage(m);
158       this.draftMessage = new Message();
159     }
160 
161     scrollToBottom(): void {
162       let scrollPane: any = this.el
163         .nativeElement.querySelector('.msg-container-base');
164       scrollPane.scrollTop = scrollPane.scrollHeight;
165     }
166 
167 }
```

## The `ChatMessage` Component

Each `Message` is rendered by the `ChatMessage` component.

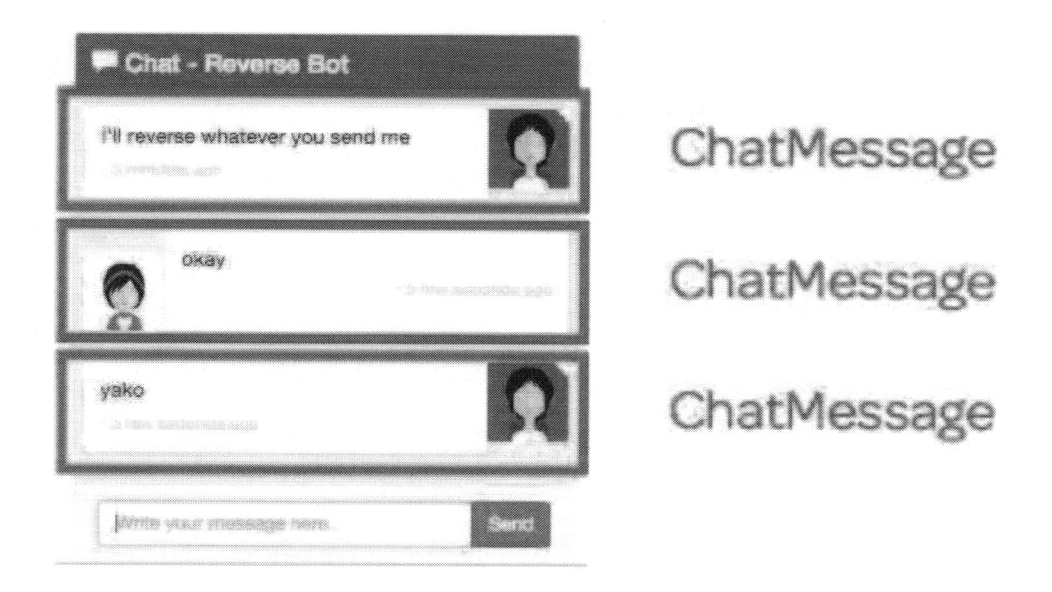


The `ChatMessage` Component

This component is relatively straightforward. The main logic here is rendering a slightly different view depending on if the message was authored by the current user. If the `Message` was **not** written by the current user, then we consider the message `incoming`.

We start by defining the `@Component`:

**code/rxjs/chat/app/ts/components/ChatWindow.ts**

```
15  @Component({
16    inputs: ['message'],
17    selector: 'chat-message',
```

## Setting `incoming`

Remember that each `ChatMessage` belongs to one `Message`. So in `ngOnInit` we will subscribe to the `currentUser` stream and set `incoming` depending on if this `Message` was written by the current user:

**code/rxjs/chat/app/ts/components/ChatWindow.ts**

```
40 export class ChatMessage implements OnInit {
41   message: Message;
42   currentUser: User;
43   incoming: boolean;
44 
45   constructor(private userService: UserService) {
46   }
47 
48   ngOnInit(): void {
49     this.userService.currentUser
50       .subscribe(
51         (user: User) => {
52           this.currentUser = user;
53           if (this.message.author && user) {
54             this.incoming = this.message.author.id !== user.id;
55           }
56         });
57   }
58 
59 }
```

### The `ChatMessage` template

In our `template` we have two interesting ideas:

1. the `FromNowPipe`
2. `[ngClass]`

First, here's the code:

**code/rxjs/chat/app/ts/components/ChatWindow.ts**

```
15 @Component({
16   inputs: ['message'],
17   selector: 'chat-message',
18   template: `
19   <div class="msg-container"
20        [ngClass]="{'base-sent': !incoming, 'base-receive': incoming}">
21 
22     <div class="avatar"
```

```
23          *ngIf="!incoming">
24        <img src="{{message.author.avatarSrc}}">
25      </div>
26
27      <div class="messages"
28        [ngClass]="{'msg-sent': !incoming, 'msg-receive': incoming}">
29        <p>{{message.text}}</p>
30        <p class="time">{{message.author.name}} • {{message.sentAt | fromNow}}</p>
31      </div>
32
33      <div class="avatar"
34          *ngIf="incoming">
35        <img src="{{message.author.avatarSrc}}">
36      </div>
37    </div>
38    `
39  })
```

The `FromNowPipe` is a pipe that casts our `Messages` sent-at time to a human-readable "x seconds ago" message. You can see that we use it by: `{{message.sentAt | fromNow}}`

`FromNowPipe` uses the excellent `moment.js`[77] library. If you'd like to learn about creating your own custom pipes read the source of the `FromNowPipe` in `code/rxjs/chat/app/ts/util/FromNowPipe.ts`

We also make extensive use of `ngClass` in this view. The idea is, when we say:

```
1     [ngClass]="{'msg-sent': !incoming, 'msg-receive': incoming}"
```

We're asking Angular to apply the `msg-receive` class if `incoming` is truthy (and apply `msg-sent` if `incoming` is falsey).

By using the `incoming` property, we're able to display incoming and outgoing messages differently.

## The Complete `ChatMessage` Code Listing

Here's our completed `ChatMessage` component:

[77] http://momentjs.com/

**code/rxjs/chat/app/ts/components/ChatWindow.ts**

```
1  import {
2    Component,
3    OnInit,
4    ElementRef,
5    ChangeDetectionStrategy
6  } from '@angular/core';
7  import {
8    MessagesService,
9    ThreadsService,
10   UserService
11 } from '../services/services';
12 import {Observable} from 'rxjs';
13 import {User, Thread, Message} from '../models';
14
15 @Component({
16   inputs: ['message'],
17   selector: 'chat-message',
18   template: `
19   <div class="msg-container"
20        [ngClass]="{'base-sent': !incoming, 'base-receive': incoming}">
21
22     <div class="avatar"
23          *ngIf="!incoming">
24       <img src="{{message.author.avatarSrc}}">
25     </div>
26
27     <div class="messages"
28       [ngClass]="{'msg-sent': !incoming, 'msg-receive': incoming}">
29       <p>{{message.text}}</p>
30       <p class="time">{{message.author.name}} • {{message.sentAt | fromNow}}</p>
31     </div>
32
33     <div class="avatar"
34          *ngIf="incoming">
35       <img src="{{message.author.avatarSrc}}">
36     </div>
37   </div>
38   `
39 })
40 export class ChatMessage implements OnInit {
41   message: Message;
```

```
42    currentUser: User;
43    incoming: boolean;
44
45    constructor(private userService: UserService) {
46    }
47
48    ngOnInit(): void {
49      this.userService.currentUser
50        .subscribe(
51          (user: User) => {
52            this.currentUser = user;
53            if (this.message.author && user) {
54              this.incoming = this.message.author.id !== user.id;
55            }
56          });
57    }
58
59  }
60
61  @Component({
62    selector: 'chat-window',
63    changeDetection: ChangeDetectionStrategy.OnPush,
64    template: `
65      <div class="chat-window-container">
66        <div class="chat-window">
67          <div class="panel-container">
68            <div class="panel panel-default">
69
70              <div class="panel-heading top-bar">
71                <div class="panel-title-container">
72                  <h3 class="panel-title">
73                    <span class="glyphicon glyphicon-comment"></span>
74                    Chat - {{currentThread.name}}
75                  </h3>
76                </div>
77                <div class="panel-buttons-container">
78                  <!-- you could put minimize or close buttons here -->
79                </div>
80              </div>
81
82              <div class="panel-body msg-container-base">
83                <chat-message
```

```
84                 *ngFor="let message of messages | async"
85                 [message]="message">
86             </chat-message>
87           </div>
88 
89           <div class="panel-footer">
90             <div class="input-group">
91               <input type="text"
92                      class="chat-input"
93                      placeholder="Write your message here..."
94                      (keydown.enter)="onEnter($event)"
95                      [(ngModel)]="draftMessage.text" />
96               <span class="input-group-btn">
97                 <button class="btn-chat"
98                   (click)="onEnter($event)"
99                   >Send</button>
100              </span>
101            </div>
102          </div>
103
104        </div>
105      </div>
106    </div>
107  </div>
108  `
109 })
110 export class ChatWindow implements OnInit {
111   messages: Observable<any>;
112   currentThread: Thread;
113   draftMessage: Message;
114   currentUser: User;
115 
116   constructor(private messagesService: MessagesService,
117               private threadsService: ThreadsService,
118               private userService: UserService,
119               private el: ElementRef) {
120   }
121 
122   ngOnInit(): void {
123     this.messages = this.threadsService.currentThreadMessages;
124 
125     this.draftMessage = new Message();
```

```
126
127     this.threadsService.currentThread.subscribe(
128       (thread: Thread) => {
129         this.currentThread = thread;
130       });
131
132     this.userService.currentUser
133       .subscribe(
134         (user: User) => {
135           this.currentUser = user;
136         });
137
138     this.messages
139       .subscribe(
140         (messages: Array<Message>) => {
141           setTimeout(() => {
142             this.scrollToBottom();
143           });
144         });
145   }
146
147   onEnter(event: any): void {
148     this.sendMessage();
149     event.preventDefault();
150   }
151
152   sendMessage(): void {
153     let m: Message = this.draftMessage;
154     m.author = this.currentUser;
155     m.thread = this.currentThread;
156     m.isRead = true;
157     this.messagesService.addMessage(m);
158     this.draftMessage = new Message();
159   }
160
161   scrollToBottom(): void {
162     let scrollPane: any = this.el
163       .nativeElement.querySelector('.msg-container-base');
164     scrollPane.scrollTop = scrollPane.scrollHeight;
165   }
166
167 }
```

## The `ChatNavBar` Component

The last component we have to talk about is the `ChatNavBar`. In the nav-bar we'll show an unread messages count to the user.

**The Unread Count in the `ChatNavBar` Component**

The best way to try out the unread messages count is to use the "Waiting Bot". If you haven't already, try sending the message '3' to the Waiting Bot and then switch to another window. The Waiting Bot will then wait 3 seconds before sending you a message and you will see the unread messages counter increment.

### The `ChatNavBar @Component`

First we define a pretty plain `@Component` configuration:

**code/rxjs/chat/app/ts/components/ChatNavBar.ts**

```
6  @Component({
7    selector: 'nav-bar',
```

### The `ChatNavBar` Controller

The only thing the `ChatNavBar` controller needs to keep track of is the `unreadMessagesCount`. This is slightly more complicated than it seems on the surface.

The most straightforward way would be to simply listen to `messagesService.messages` and sum the number of `Messages` where `isRead` is false. This works fine for all messages outside of the current thread. However new messages in the current thread aren't guaranteed to be marked as read by the time `messages` emits new values.

The safest way to handle this is to combine the `messages` and `currentThread` streams and make sure we don't count any messages that are part of the current thread.

We do this using the `combineLatest` operator, which we've already used earlier in the chapter:

**code/rxjs/chat/app/ts/components/ChatNavBar.ts**

```
26  export class ChatNavBar implements OnInit {
27    unreadMessagesCount: number;
28  
29    constructor(private messagesService: MessagesService,
30                private threadsService: ThreadsService) {
31    }
32  
33    ngOnInit(): void {
34      this.messagesService.messages
35        .combineLatest(
36          this.threadsService.currentThread,
37          (messages: Message[], currentThread: Thread) =>
38            [currentThread, messages] )
39  
40        .subscribe(([currentThread, messages]: [Thread, Message[]]) => {
41          this.unreadMessagesCount =
42            _.reduce(
43              messages,
44              (sum: number, m: Message) => {
45                let messageIsInCurrentThread: boolean = m.thread &&
46                  currentThread &&
47                  (currentThread.id === m.thread.id);
48                if (m && !m.isRead && !messageIsInCurrentThread) {
49                  sum = sum + 1;
50                }
51                return sum;
52              },
53              0);
54        });
55    }
56  }
```

If you're not an expert in TypeScript you might find the above syntax a little bit hard to parse. In the `combineLatest` callback function we're returning an array with `currentThread` and `messages` as its two elements.

Then we `subscribe` to that stream and we're *destructuring* those objects in the function call. Next we `reduce` over the `messages` and count the number of messages that are unread and not in the current thread.

### The `ChatNavBar template`

In our view, the only thing we have left to do is display our `unreadMessagesCount`:

**code/rxjs/chat/app/ts/components/ChatNavBar.ts**

```
 6 @Component({
 7   selector: 'nav-bar',
 8   template: `
 9   <nav class="navbar navbar-default">
10     <div class="container-fluid">
11       <div class="navbar-header">
12         <a class="navbar-brand" href="https://ng-book.com/2">
13           <img src="${require('images/logos/ng-book-2-minibook.png')}"/>
14            ng-book 2
15         </a>
16       </div>
17       <p class="navbar-text navbar-right">
18         <button class="btn btn-primary" type="button">
19           Messages <span class="badge">{{unreadMessagesCount}}</span>
20         </button>
21       </p>
22     </div>
23   </nav>
24   `
```

### The Completed `ChatNavBar`

Here's the full code listing for `ChatNavBar`:

**code/rxjs/chat/app/ts/components/ChatNavBar.ts**

```
 1 import {Component, OnInit} from '@angular/core';
 2 import {MessagesService, ThreadsService} from '../services/services';
 3 import {Message, Thread} from '../models';
 4 import * as _ from 'underscore';
 5 
 6 @Component({
 7   selector: 'nav-bar',
 8   template: `
 9   <nav class="navbar navbar-default">
10     <div class="container-fluid">
11       <div class="navbar-header">
12         <a class="navbar-brand" href="https://ng-book.com/2">
```

```
13            <img src="${require('images/logos/ng-book-2-minibook.png')}"/>
14             ng-book 2
15          </a>
16        </div>
17        <p class="navbar-text navbar-right">
18          <button class="btn btn-primary" type="button">
19            Messages <span class="badge">{{unreadMessagesCount}}</span>
20          </button>
21        </p>
22      </div>
23    </nav>
24    `
25  })
26  export class ChatNavBar implements OnInit {
27    unreadMessagesCount: number;
28
29    constructor(private messagesService: MessagesService,
30                private threadsService: ThreadsService) {
31    }
32
33    ngOnInit(): void {
34      this.messagesService.messages
35        .combineLatest(
36          this.threadsService.currentThread,
37          (messages: Message[], currentThread: Thread) =>
38            [currentThread, messages] )
39
40        .subscribe(([currentThread, messages]: [Thread, Message[]]) => {
41          this.unreadMessagesCount =
42            _.reduce(
43              messages,
44              (sum: number, m: Message) => {
45                let messageIsInCurrentThread: boolean = m.thread &&
46                  currentThread &&
47                  (currentThread.id === m.thread.id);
48                if (m && !m.isRead && !messageIsInCurrentThread) {
49                  sum = sum + 1;
50                }
51                return sum;
52              },
53              0);
54        });
```

```
55    }
56  }
```

## Summary

There we go, if we put them all together we've got a fully functional chat app!

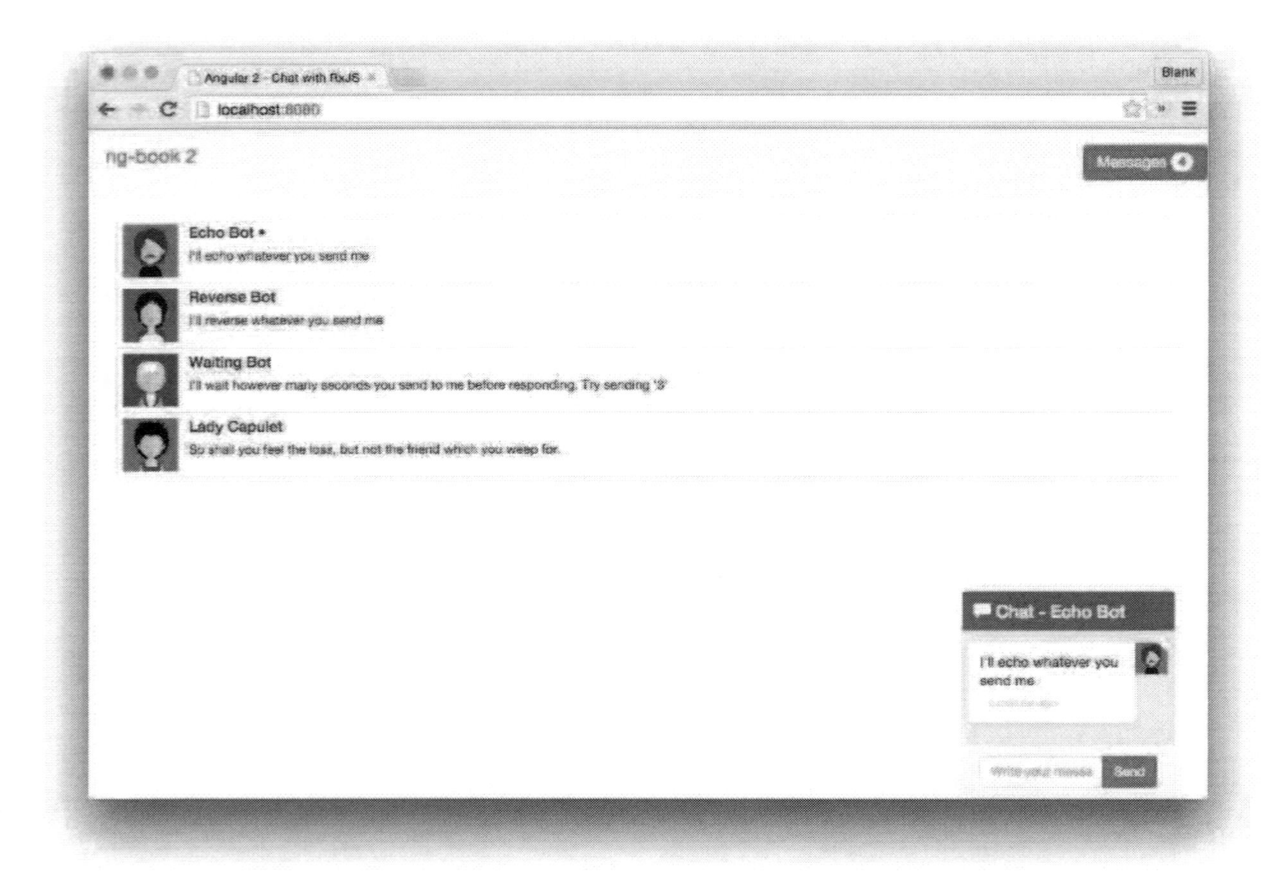


**Completed Chat Application**

If you checkout `code/rxjs/chat/app/ts/ChatExampleData.ts` you'll see we've written a handful of bots for you that you can chat with. Here's a code excerpt from the Reverse Bot:

```
1  let rev: User = new User("Reverse Bot", require("images/avatars/female-avatar-4.\
2  png"));
3  let tRev: Thread = new Thread("tRev", rev.name, rev.avatarSrc);
```

**code/rxjs/chat/app/ts/ChatExampleData.ts**

```
88      messagesService.messagesForThreadUser(tRev, rev)
89        .forEach( (message: Message): void => {
90          messagesService.addMessage(
91            new Message({
92              author: rev,
93              text: message.text.split('').reverse().join(''),
94              thread: tRev
95            })
96          );
97        },
```

Above you can see that we've subscribed to the messages for the "Reverse Bot" by using `messagesForThreadUser`. Try writing a few bots of your own.

# Next Steps

Some ways to improve this chat app would be to become stronger at RxJS and then hook it up to an actual API. We'll talk about how to make API requests in the HTTP Chapter. For now, enjoy your fancy chat application!

# Introduction to Redux with TypeScript

In this chapter and the next we'll be looking at a data-architecture pattern called Redux. In this chapter we're going to discuss the ideas behind Redux, build our own mini version, and then hook it up to Angular. In the next chapter we'll use Redux to build a bigger application.

In most of our projects so far, we've managed state in a fairly direct way: We tend to grab data from services and render them in components, passing values down the component tree along the way.

Managing our apps in this way works fine for smaller apps, but as our apps grow, having multiple components manage different parts of the state becomes cumbersome. For instance, passing all of our values down our component tree suffers from the following downsides:

**Intermediate property passing** - In order to get state to any component we have to pass the values down through `inputs`. This means we have many intermediate components passing state that it isn't directly using or concerned about

**Inflexible refactoring** - Because we're passing `inputs` down through the component tree, we're introducing a coupling between parent and child components that often isn't necessary. This makes it more difficult to put a child component somewhere else in the hierarchy because we have to change all of the new parents to pass the state

**State tree and DOM tree don't match** - The "shape" of our state often doesn't match the "shape" of our view/component hierarchy. By passing all data through the component tree via `props` we run into difficulties when we need to reference data in a far branch of the tree

**State throughout our app** - If we manage state via components, it's difficult to get a snapshot of the total state of our app. This can make it hard to know which component "owns" a particular bit of data, and which components are concerned about changes

Pulling data out of our components and into services helps a lot. At least if services are the "owners" of our data, we have a better idea of where to put things. But this opens a new question: what are the best practices for "service-owned" data? Are there any patterns we can follow? In fact, there are.

In this chapter, we're going to discuss a data-architecture pattern called *Redux* which was designed to help with these issues. We'll implement our own version of Redux which will store **all of our state in a single place**. This idea of holding **all** of our application's state in one place might sound a little crazy, but the results are surprisingly delightful.

# Redux

If you haven't heard of Redux yet you can read a bit about it on the official website[78]. Web application data architecture is evolving and the traditional ways of structuring data aren't quite adequate for large web apps. Redux has been extremely popular because it's both powerful and easy to understand.

Data architecture can be a complex topic and so Redux's best feature is probably its simplicity. If you strip Redux down to the essential core, Redux is fewer than 100 lines of code.

We can build rich, easy to understand, web apps by using Redux as the backbone of our application. But first, let's walk through how to write a minimal Redux and later we'll work out patterns that emerge as we work out these ideas in a larger app.

There are several attempts to use Redux or create a Redux-inspired system that works with Angular. Two notable examples are:

- ngrx/store[79] and
- angular2-redux[80]

`ngrx` is a Redux-inspired architecture that is heavily observables-based. `angular2-redux` uses Redux itself as a dependency, and adds some Angular helpers (dependency-injection, observable wrappers).

Here we're not going to use either. Instead, we're going to use Redux directly in order to show the concepts without introducing a new dependency. That said, both of these libraries may be helpful to you when writing your apps.

## Redux: Key Ideas

The key ideas of Redux are this:

- All of your application's data is in a single data structure called the *state* which is held in the *store*
- Your app reads the **state** from this **store**
- This **store** is never mutated directly
- User interaction (and other code) fires *actions* which describe what happened
- A *new state* is created by combining he **old state** and the **action** by a function called the *reducer*.

[78] http://redux.js.org/
[79] https://github.com/ngrx/store
[80] https://github.com/InfomediaLtd/angular2-redux

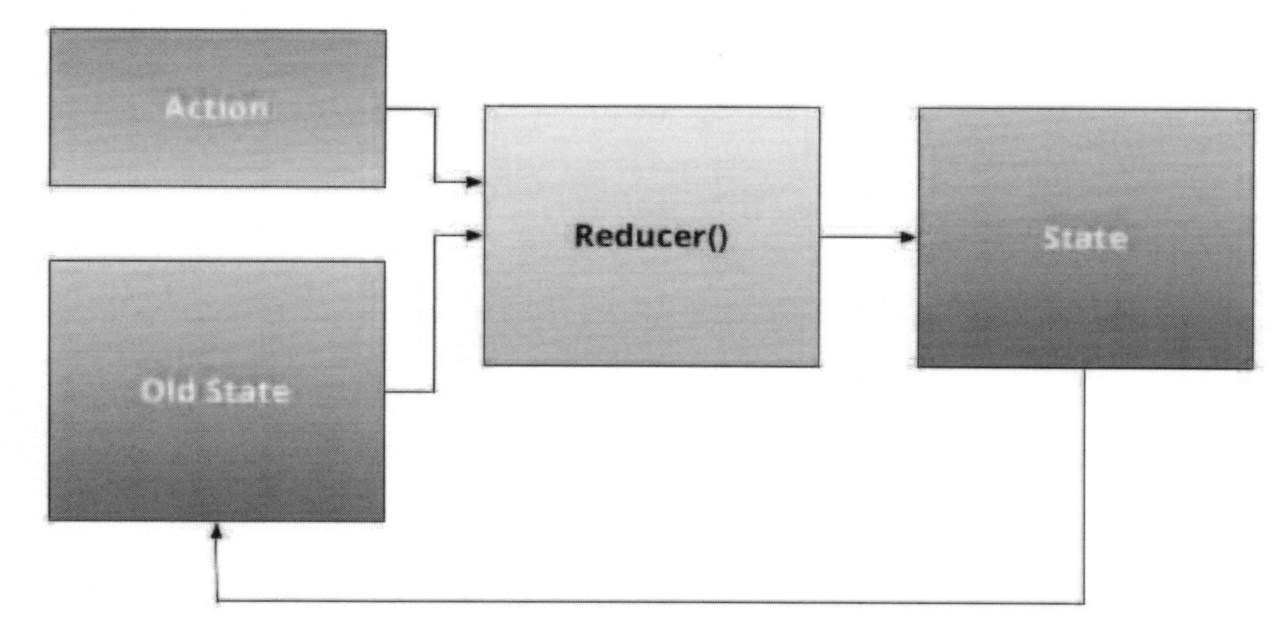


Redux Core

If the above bullet list isn't clear yet, don't worry about it - putting these ideas into practice is the goal of the rest of this chapter.

# Core Redux Ideas

## What's a *reducer*?

Let's talk about the *reducer* first. Here's the idea of a *reducer*: it takes the *old state* and an *action* and returns a *new state*.

A reducer must be a **pure function**[81]. That is:

1. It must not mutate the current state directly
2. It must not use any data outside of its arguments

Put another way, a pure function will always **return the same value, given the same set of arguments**. And a pure function won't call any functions which have an effect on the outside world, e.g. no database calls, no HTTP calls, and no mutating outside data structures.

Reducers should always treat the current state as **read-only**. A reducer **does not change the state** instead, it **returns a new state**. (Often this new state will start with a copy of old state, but let's not get ahead of ourselves.)

Let's define our very first reducer. Remember, there are three things involved:

[81]https://en.wikipedia.org/wiki/Pure_function

1. An `Action`, which defines what to do (with optional arguments)
2. The `state`, which stores *all* of the data in our application
3. The `Reducer` which takes the `state` and the `Action` and returns a new state.

## Defining `Action` and `Reducer` Interfaces

Since we're using TypeScript we want to make sure this whole process is typed, so let's setup an interface for our `Action` and our `Reducer`:

### The `Action` Interface

Our `Action` interface looks like this:

**code/redux/angular2-redux-chat/minimal/tutorial/01-identity-reducer.ts**

```
1  interface Action {
2    type: string;
3    payload?: any;
4  }
```

Notice that our `Action` has two fields:

1. `type` and
2. `payload`

The `type` will be an identifying string that describes the action like `INCREMENT` or `ADD_USER`. The `payload` can be an object of any kind. The `?` on `payload?` means that this field is optional.

### The `Reducer` Interface

Our `Reducer` interface looks like this:

**code/redux/angular2-redux-chat/minimal/tutorial/01-identity-reducer.ts**

```
6  interface Reducer<T> {
7    (state: T, action: Action): T;
8  }
```

Our `Reducer` is using a feature of TypeScript called *generics*. In this case type `T` is the type of the `state`. Notice that we're saying that a valid `Reducer` has a function which takes a `state` (of type `T`) and an `action` and returns a new `state` (also of type `T`).

## Creating Our First Reducer

The simplest possible reducer returns the state itself. (You might call this the *identity* reducer because it applies the identity function[82] on the state. This is the default case for all reducers, as we will soon see).

**code/redux/angular2-redux-chat/minimal/tutorial/01-identity-reducer.ts**

```
10 let reducer: Reducer<number> = (state: number, action: Action) => {
11   return state;
12 };
```

Notice that this `Reducer` makes the generic type concrete to `number` by the syntax `Reducer<number>`. We'll define more sophisticated states beyond a single number soon.

We're not using the `Action` yet, but let's try this `Reducer` just the same.

**Running the examples in this section**

You can find the code for this chapter in the folder `code/redux`. If the example is runnable you will see the filename the code is from above each code box.

In this first section, these examples are run **outside of the browser and run by node.js**. Because we're using TypeScript in these examples, you should run them using the commandline tool `ts-node`, (instead of `node` directly).

You can install `ts-node` by running:

```
1 npm install -g ts-node
```

Or by doing an `npm install` in the `code/redux/angular2-redux-chat` directory and then calling `./node_modules/.bin/ts-node --noProject`

For instance, to run the example above you might type (not including the `$`):

```
1 $ cd code/redux/angular2-redux-chat/minimal/tutorial
2 $ ../../node_modules/.bin/ts-node --noProject 01-identity-reducer.ts
```

Use this same procedure for the rest of the code in this chapter until we instruct you to switch to your browser.

## Running Our First Reducer

Let's put it all together and run this reducer:

[82] https://en.wikipedia.org/wiki/Identity_function

**code/redux/angular2-redux-chat/minimal/tutorial/01-identity-reducer.ts**

```
1  interface Action {
2    type: string;
3    payload?: any;
4  }
5
6  interface Reducer<T> {
7    (state: T, action: Action): T;
8  }
9
10 let reducer: Reducer<number> = (state: number, action: Action) => {
11   return state;
12 };
13
14 console.log( reducer(0, null) ); // -> 0
```

And run it:

```
1  $ cd code/redux/angular2-redux-chat/minimal/tutorial
2  $ ../../node_modules/.bin/ts-node --noProject 01-identity-reducer.ts
3  0
```

It seems almost silly to have that as a code example, but it teaches us our first principle of reducers:

**By default, reducers return the original state.**

In this case, we passed a state of the number `0` and a `null` action. The result from this reducer is the state `0`.

But let's do something more interesting and make our state change.

## Adjusting the Counter With *actions*

Eventually our state is going to be **much more** sophisticated than a single number. We're going to be holding the **all** of the data for our app in the `state`, so we'll need better data structure for the state eventually.

That said, using a single number for the state lets us focus on other issues for now. So let's continue with the idea that our `state` is simply a single number that is storing a counter.

Let's say we want to be able to change the `state` number. Remember that in Redux we do not modify the state. Instead, we create *actions* which instruct the *reducer* on how to generate a *new state*.

Let's create an `Action` to change our counter. Remember that the only required property is a `type`. We might define our first action like this:

```
1  let incrementAction: Action = { type: 'INCREMENT' }
```

We should also create a second action that instructs our reducer to make the counter smaller with:

```
1  let decrementAction: Action = { type: 'DECREMENT' }
```

Now that we have these actions, let's try using them in our reducer:

**code/redux/angular2-redux-chat/minimal/tutorial/02-adjusting-reducer.ts**

```
10  let reducer: Reducer<number> = (state: number, action: Action) => {
11    if (action.type === 'INCREMENT') {
12      return state + 1;
13    }
14    if (action.type === 'DECREMENT') {
15      return state - 1;
16    }
17    return state;
18  };
```

And now we can try out the whole reducer:

**code/redux/angular2-redux-chat/minimal/tutorial/02-adjusting-reducer.ts**

```
20  let incrementAction: Action = { type: 'INCREMENT' };
21
22  console.log( reducer(0, incrementAction )); // -> 1
23  console.log( reducer(1, incrementAction )); // -> 2
24
25  let decrementAction: Action = { type: 'DECREMENT' };
26
27  console.log( reducer(100, decrementAction )); // -> 99
```

Neat! Now the new value of the state is returned according to which action we pass into the reducer.

## Reducer `switch`

Instead of having so many `if` statements, the common practice is to convert the reducer body to a `switch` statement:

**code/redux/angular2-redux-chat/minimal/tutorial/03-adjusting-reducer-switch.ts**

```
10 let reducer: Reducer<number> = (state: number, action: Action) => {
11   switch (action.type) {
12   case 'INCREMENT':
13     return state + 1;
14   case 'DECREMENT':
15     return state - 1;
16   default:
17     return state; // <-- dont forget!
18   }
19 };
20 
21 let incrementAction: Action = { type: 'INCREMENT' };
22 console.log(reducer(0, incrementAction)); // -> 1
23 console.log(reducer(1, incrementAction)); // -> 2
24 
25 let decrementAction: Action = { type: 'DECREMENT' };
26 console.log(reducer(100, decrementAction)); // -> 99
27 
28 // any other action just returns the input state
29 let unknownAction: Action = { type: 'UNKNOWN' };
30 console.log(reducer(100, unknownAction)); // -> 100
```

Notice that the `default` case of the `switch` returns the original `state`. This ensures that if an unknown action is passed in, there's no error and we get the original `state` unchanged.

Q: **Wait, all of my application state is in one giant `switch` statement?**

A: Yes and no.

If this is your first exposure to Redux reducers it might feel a little weird to have all of your application state changes be the result of a giant `switch`. There are two things you should know:

1. Having your state changes centralized in one place can help a **ton** in maintaining your program, particularly because it's easy to track down where the changes are happening when they're all together. (Furthermore, you can easily locate what state changes as the result of any action because you can search your code for the token specified for that action's `type`)
2. You can (and often do) break your reducers down into several sub-reducers which each manage a different branch of the state tree. We'll talk about this later.

## Action "Arguments"

In the last example our actions contained only a `type` which told our reducer either to increment or decrement the state.

But often changes in our app can't be described by a single value - instead we need parameters to describe the change. This is why we have the `payload` field in our `Action`.

In this counter example, say we wanted to add `9` to the counter. One way to do this would be to send 9 `INCREMENT` actions, but that wouldn't be very efficient, especially if we wanted to add, say, 9000.

Instead, let's add a `PLUS` action that will use the `payload` parameter to send a number which specifies how much we want to add to the counter. Defining this action is easy enough:

```
1   let plusSevenAction = { type: 'PLUS', payload: 7 };
```

Next, to support this action, we add a new `case` to our reducer that will handle a `'PLUS'` action:

**code/redux/angular2-redux-chat/minimal/tutorial/04-plus-action.ts**

```
10  let reducer: Reducer<number> = (state: number, action: Action) => {
11    switch (action.type) {
12    case 'INCREMENT':
13      return state + 1;
14    case 'DECREMENT':
15      return state - 1;
16    case 'PLUS':
17      return state + action.payload;
18    default:
19      return state;
20    }
21  };
```

`PLUS` will add whatever number is in the `action.payload` to the `state`. We can try it out:

**code/redux/angular2-redux-chat/minimal/tutorial/04-plus-action.ts**

```
23  console.log( reducer(3, { type: 'PLUS', payload: 7}) );    // -> 10
24  console.log( reducer(3, { type: 'PLUS', payload: 9000}) ); // -> 9003
25  console.log( reducer(3, { type: 'PLUS', payload: -2}) );   // -> 1
```

In the first line we take the state `3` and `PLUS` a payload of `7`, which results in `10`. Neat! However, notice that while we're passing in a `state`, it doesn't really ever *change*. That is, we're not storing the result of our reducer's changes and reusing it for future actions.

# Storing Our State

Our reducers are pure functions, and do not change the world around them. The problem is, in our app, things *do* change. Specifically, our state changes and we need to keep the new state somewhere.

In Redux, we keep our state in the *store*. The store has the responsibility of **running the reducer and then keeping the new state.** Let's take a look at a minimal store:

**code/redux/angular2-redux-chat/minimal/tutorial/05-minimal-store.ts**

```
10 class Store<T> {
11   private _state: T;
12 
13   constructor(
14     private reducer: Reducer<T>,
15     initialState: T
16   ) {
17     this._state = initialState;
18   }
19 
20   getState(): T {
21     return this._state;
22   }
23 
24   dispatch(action: Action): void {
25     this._state = this.reducer(this._state, action);
26   }
27 }
```

Notice that our `Store` is generically typed - we specify the type of the *state* with generic type `T`. We store the state in the private variable `_state`.

We also give our `Store` a `Reducer`, which is also typed to operate on `T`, the state type this is because **each store is tied to a specific reducer**. We store the `Reducer` in the private variable `reducer`.

In Redux, we generally have 1 store and 1 top-level reducer per application.

Let's take a closer look at each method of our `State`:

- In our `constructor` we set the `_state` to the initial state.
- `getState()` simply returns the current `_state`

- `dispatch` takes an action, sends it to the reducer and then **updates the value of `_state`** with the return value

Notice that **`dispatch` doesn't return anything**. It's only *updating* the store's state (once the result returns). This is an important principle of Redux: dispatching actions is a "fire-and-forget" maneuver. **Dispatching actions is not a direct manipulation of the state, and it doesn't return the new state.**

When we dispatch actions, we're sending off a notification of what happened. If we want to know what the current state of the system is, we have to check the state of the store.

## Using the Store

Let's try using our store:

**code/redux/angular2-redux-chat/minimal/tutorial/05-minimal-store.ts**

```
43  // create a new store
44  let store = new Store<number>(reducer, 0);
45  console.log(store.getState()); // -> 0
46
47  store.dispatch({ type: 'INCREMENT' });
48  console.log(store.getState()); // -> 1
49
50  store.dispatch({ type: 'INCREMENT' });
51  console.log(store.getState()); // -> 2
52
53  store.dispatch({ type: 'DECREMENT' });
54  console.log(store.getState()); // -> 1
```

We start by creating a new `Store` and we save this in `store`, which we can use to get the current state and dispatch actions.

The state is set to `0` initially, and then we `INCREMENT` twice and `DECREMENT` once and our final state is `1`.

## Being Notified with `subscribe`

It's great that our `Store` keeps track of what changed, but in the above example we have to *ask* for the state changes with `store.getState()`. It would be nice for us to know immediately when a new action was dispatched so that we could respond. To do this we can implement the Observer pattern - that is, we'll register a callback function that will *subscribe* to all changes.

Here's how we want it to work:

1. We will register a *listener* function using `subscribe`
2. When `dispatch` is called, we will iterate over all listeners and call them, which is the notification that the state has changed.

## Registering Listeners

Our listener callbacks are a going to be a function that takes *no arguments*. Let's define an interface that makes it easy to describe this:

**code/redux/angular2-redux-chat/minimal/tutorial/06-store-w-subscribe.ts**

```
10 interface ListenerCallback {
11   (): void;
12 }
```

After we subscribe a listener, we might want to unsubscribe as well, so lets define the interface for an *unsubscribe* function as well:

**code/redux/angular2-redux-chat/minimal/tutorial/06-store-w-subscribe.ts**

```
14 interface UnsubscribeCallback {
15   (): void;
16 }
```

Not much going on here - it's another function that takes no arguments and has no return value. But by defining these types it makes our code clearer to read.

Our store is going to keep a list of `ListenerCallbacks` let's add that to our `Store`:

**code/redux/angular2-redux-chat/minimal/tutorial/06-store-w-subscribe.ts**

```
18 class Store<T> {
19   private _state: T;
20   private _listeners: ListenerCallback[] = [];
```

Now we want to be able to add to that list of `_listeners` with a `subscribe` function:

**code/redux/angular2-redux-chat/minimal/tutorial/06-store-w-subscribe.ts**

```
38   subscribe(listener: ListenerCallback): UnsubscribeCallback {
39     this._listeners.push(listener);
40     return () => { // returns an "unsubscribe" function
41       this._listeners = this._listeners.filter(l => l !== listener);
42     };
43   }
```

`subscribe` accepts a `ListenerCallback` (i.e. a function with no arguments and no return value) and returns an `UnsubscribeCallback` (the same signature). Adding the new listener is easy: we `push` it on to the `_listeners` array.

The return value is a function which will update the list of `_listeners` to be the list of `_listeners` without the `listener` we just added. That is, it returns the `UnsubscribeCallback` that we can use to remove this listener from the list.

### Notifying Our Listeners

Whenever our state changes, we want to call these listener functions. What this means is, whenever we `dispatch` a new action, whenever the state changes, we want to call all of the listeners:

**code/redux/angular2-redux-chat/minimal/tutorial/06-store-w-subscribe.ts**

```
33   dispatch(action: Action): void {
34     this._state = this.reducer(this._state, action);
35     this._listeners.forEach((listener: ListenerCallback) => listener());
36   }
```

### The Complete Store

We'll try this out below, but before we do that, here's the complete code listing for our new `Store`:

**code/redux/angular2-redux-chat/minimal/tutorial/06-store-w-subscribe.ts**

```
18 class Store<T> {
19   private _state: T;
20   private _listeners: ListenerCallback[] = [];
21 
22   constructor(
23     private reducer: Reducer<T>,
24     initialState: T
25   ) {
26     this._state = initialState;
27   }
28 
29   getState(): T {
30     return this._state;
31   }
32 
33   dispatch(action: Action): void {
34     this._state = this.reducer(this._state, action);
35     this._listeners.forEach((listener: ListenerCallback) => listener());
36   }
37 
38   subscribe(listener: ListenerCallback): UnsubscribeCallback {
39     this._listeners.push(listener);
40     return () => { // returns an "unsubscribe" function
41       this._listeners = this._listeners.filter(l => l !== listener);
42     };
43   }
44 }
```

### Trying Out `subscribe`

Now that we can `subscribe` to changes in our store, let's try it out:

**code/redux/angular2-redux-chat/minimal/tutorial/06-store-w-subscribe.ts**

```
61 let store = new Store<number>(reducer, 0);
62 console.log(store.getState()); // -> 0
63 
64 // subscribe
65 let unsubscribe = store.subscribe(() => {
66   console.log('subscribed: ', store.getState());
67 });
68 
69 store.dispatch({ type: 'INCREMENT' }); // -> subscribed: 1
70 store.dispatch({ type: 'INCREMENT' }); // -> subscribed: 2
71 
72 unsubscribe();
73 store.dispatch({ type: 'DECREMENT' }); // (nothing logged)
74 
75 // decrement happened, even though we weren't listening for it
76 console.log(store.getState()); // -> 1
```

Above we subscribe to our store and in the callback function we'll log `subscribed:` and then the current store state.

Notice that the listener function is **not** given the current state as an argument. This might seem like an odd choice, but because there are some nuances to deal with, it's easier to think of *the notification of state changed* as separate from *the current state.* Without digging too much into the weeds, you can read more about this choice here[83], here[84], and here[85].

We store the `unsubscribe` callback and then notice that after we call `unsubscribe()` our log message isn't called. We can still dispatch actions, we just won't see the results until we ask the store for them.

[83]https://github.com/reactjs/redux/issues/1707
[84]https://github.com/reactjs/redux/issues/1513
[85]https://github.com/reactjs/redux/issues/303

If you're the type of person who likes RxJS and Observables, you might notice that implementing our own subscription listeners could also be implemented using RxJS. You could rewrite our `Store` to use Observables instead of our own subscriptions.

In fact, we've already done this for you and you can find the sample code in the file `code/redux/angular2-redux-chat/minimal/tutorial/06b-rx-store.ts`.

Using RxJS for the `Store` is an interesting and powerful pattern if you're willing to us RxJS for the backbone of our application data.

Here we're not going to use Observables very heavily, particularly because we want to discuss Redux itself and how to think about data architecture with a single state tree. Redux itself is powerful enough to use in our applications without Observables.

Once you get the concepts of using "straight" Redux, adding in Observables isn't difficult (if you already understand RxJS, that is). For now, we're going to use "straight" Redux and we'll give you some guidance on some Observable-based Redux-wrappers at the end.

## The Core of Redux

The above store is the essential core of Redux. Our reducer takes the current state and action and returns a new state, which is held by the store.

There are obviously many more things that we need to add to build a large, production web app. However, all of the new ideas that we'll cover are patterns that flow from building on this simple idea of an immutable, central store of state. If you understand the ideas presented above, you would be likely to invent many of the patterns (and libraries) you find in more advanced Redux apps.

There's still a lot for us to cover about day-to-day use of redux though. For instance, we need to know:

- How to carefully handle more complex data structures in our state
- How to be notified when our state changes without having to poll the state (with subscriptions)
- How to intercept our dispatch for debugging (a.k.a. middleware)
- How to compute derived values (with *selectors*)
- How to split up large reducers into more manageable, smaller ones (and recombine them)
- How to deal with asynchronous data

We'll explain on each of these issues and describe common patterns over the rest of this chapter and the next.

Let's first deal with handling more complex data structures in our state. To do that, we're going to need an example that's more interesting than a counter. Let's start building a chat app where users can send each other messages.

# A Messaging App

In our messaging app, as in all Redux apps, there are three main parts to the data model:

1. The state
2. The actions
3. The reducer

## Messaging App `state`

The `state` in our counter app was a single number. However in our messaging app, the `state` is going to be **an object**.

This state object will have a single property, `messages`. `messages` will be an array of strings, with each string representing an individual message in the application. For example:

```
1 // an example `state` value
2 {
3   messages: [
4     'here is message one',
5     'here is message two'
6   ]
7 }
```

We can define the type for the app's state like this:

**code/redux/angular2-redux-chat/minimal/tutorial/07-messages-reducer.ts**

```
7 interface AppState {
8   messages: string[];
9 }
```

## Messaging App `actions`

Our app will process two actions: `ADD_MESSAGE` and `DELETE_MESSAGE`.

The `ADD_MESSAGE` action object will always have the property `message`, the message to be added to the state. The `ADD_MESSAGE` action object has this shape:

```
{
  type: 'ADD_MESSAGE',
  message: 'Whatever message we want here'
}
```

The DELETE_MESSAGE action object will delete a specified message from the state. A challenge here is that we have to be able to specify *which message* we want to delete.

If our messages were objects, we could assign each message an id property when it is created. However, to simplify this example, our messages are just simple strings, so we'll have to get a handle to the message another way. The easiest way for now is to just use the index of the message in the array (as a proxy for the ID).

With that in mind, the DELETE_MESSAGE action object has this shape:

```
{
  type: 'DELETE_MESSAGE',
  index: 2                // <- or whatever index is appropriate
}
```

We can define the types for these actions by using the `interface ... extends` syntax in TypeScript:

**code/redux/angular2-redux-chat/minimal/tutorial/07-messages-reducer.ts**

```
interface AddMessageAction extends Action {
  message: string;
}

interface DeleteMessageAction extends Action {
  index: number;
}
```

In this way our AddMessageAction is able to specify a message and the DeleteMessageAction will specify an index.

## Messaging App reducer

Remember that our reducer needs to handle two actions: ADD_MESSAGE and DELETE_MESSAGE. Let's talk about these individually.

### Reducing ADD_MESSAGE

**code/redux/angular2-redux-chat/minimal/tutorial/07-messages-reducer.ts**

```
19 let reducer: Reducer<AppState> =
20   (state: AppState, action: Action): AppState => {
21   switch (action.type) {
22   case 'ADD_MESSAGE':
23     return {
24       messages: state.messages.concat(
25         (<AddMessageAction>action).message
26       ),
27     };
```

We start by switching on the `action.type` and handling the `ADD_MESSAGE` case.

**TypeScript objects already have a type, so why are we adding a `type` field?**

There are many different ways we might choose to handle this sort of “polymorphic dispatch”. Keeping a string in a `type` field (where `type` means “action-type”) is a straightforward, portable way we can use to distinguish different types of actions and handle them in one reducer. In part, it means that you don’t *have* to create a new `interface` for every action.

That said, it would be more satisfying to be able to use reflection to switch on the concrete type. While this might become possible with more advanced type guards[86], this isn’t currently possible in today’s TypeScript.

Broadly speaking, types are a compile-time construct and this code is compiled down to JavaScript and we can lose some of the typing metadata.

That said, if switching on a `type` field bothers you and you’d like to use language features directly, you could use the decoration reflection metadata[87]. For now, a simple `type` field will suffice.

## Adding an Item Without Mutation

When we handle an `ADD_MESSAGE` action, we need to add the given message to the state. As will all reducer handlers, we need to **return a new state**. Remember that our reducers must be *pure* and not mutate the old state.

What would be the problem with the following code?

[86]https://basarat.gitbooks.io/typescript/content/docs/types/typeGuard.html
[87]http://blog.wolksoftware.com/decorators-metadata-reflection-in-typescript-from-novice-to-expert-part-4

```
1  case 'ADD_MESSAGE':
2    state.messages.push( action.message );
3    return { messages: messages };
4    // ...
```

The problem is that this code **mutates** the `state.messages` array, which changes our old state! Instead what we want to do is create a *copy* of the `state.messages` array and add our new message to the copy.

**code/redux/angular2-redux-chat/minimal/tutorial/07-messages-reducer.ts**

```
22    case 'ADD_MESSAGE':
23      return {
24        messages: state.messages.concat(
25          (<AddMessageAction>action).message
26        ),
27      };
```

The syntax `<AddMessageAction>action` will cast our `action` to the more specific type. That is, notice that our reducer takes the more general type `Action`, which does not have the `message` field. If we leave off the cast, then the compiler will complain that `Action` does not have a field `message`.

Instead, we know that we have an `ADD_MESSAGE` action so we cast it to an `AddMessageAction`. We use parenthesis to make sure the compiler knows that we want to cast `action` and not `action.message`.

Remember that the reducer **must return a new `AppState`**. When we return an object from our reducer it must match the format of the `AppState` that was input. In this case we only have to keep the key `messages`, but in more complicated states we have more fields to worry about.

## Deleting an Item Without Mutation

Remember that when we handle the `DELETE_MESSAGE` action we are passing the index of the item in the array as the faux ID. (Another common way of handling the same idea would be to pass a real item ID.) Again, because we do not want to mutate the old `messages` array, we need to handle this case with care:

**code/redux/angular2-redux-chat/minimal/tutorial/07-messages-reducer.ts**

```
28    case 'DELETE_MESSAGE':
29      let idx = (<DeleteMessageAction>action).index;
30      return {
31        messages: [
32          ...state.messages.slice(0, idx),
33          ...state.messages.slice(idx + 1, state.messages.length)
34        ]
```

Here we use the `slice` operator twice. First we take all of the items up until the item we are removing. And we concatenate the items that come after.

There are four common non-mutating operations:

- Adding an item to an array
- Removing an item from an array
- Adding / changing a key in an object
- Removing a key from an object

The first two (array) operations we just covered. We'll talk more about the object operations further down, but for now know that a common way to do this is to use `Object.assign`. As in:

```
1  Object.assign({}, oldObject, newObject)
2                // <-------<-------------
```

You can think of `Object.assign` as merging objects in from the right into the object on the left. `newObject` is merged into `oldObject` which is merged into `{}`. This way all of the fields in `oldObject` will be kept, except for where the field exists in `newObject`. Neither `oldObject` nor `newObject` will be mutated.

Of course, handling all of this on your own takes great care and it is easy to make a mistake. This is one of the reasons many people use Immutable.js[88], which is a set of data structures that help enforce immutability.

## Trying Out Our Actions

Now let's try running our actions:

[88] https://facebook.github.io/immutable-js/

**code/redux/angular2-redux-chat/minimal/tutorial/07-messages-reducer.ts**

```
42  let store = new Store<AppState>(reducer, { messages: [] });
43  console.log(store.getState()); // -> { messages: [] }
44
45  store.dispatch({
46    type: 'ADD_MESSAGE',
47    message: 'Would you say the fringe was made of silk?'
48  } as AddMessageAction);
49
50  store.dispatch({
51    type: 'ADD_MESSAGE',
52    message: 'Wouldnt have no other kind but silk'
53  } as AddMessageAction);
54
55  store.dispatch({
56    type: 'ADD_MESSAGE',
57    message: 'Has it really got a team of snow white horses?'
58  } as AddMessageAction);
59
60  console.log(store.getState());
61  // ->
62  // { messages:
63  //    [ 'Would you say the fringe was made of silk?',
64  //      'Wouldnt have no other kind but silk',
65  //      'Has it really got a team of snow white horses?' ] }
```

Here we start with a new store and we call `store.getState()` and see that we have an empty `messages` array.

Next we add three messages[89] to our store. For each message we specify the `type` as `ADD_MESSAGE` and we cast each object to an `AddMessageAction`.

Finally we log the new state and we can see that `messages` contains all three messages.

Our three `dispatch` statements are a bit ugly for two reasons:

1. we manually have to specify the `type` string each time. We could use a constant, but it would be nice if we didn't have to do this and
2. we're manually casting to an `AddMessageAction`

Instead of creating these objects as an object directly we should create a *function* that will create these objects. This idea of writing a function to create actions is so common in Redux that the pattern has a name: *Action Creators*.

[89] https://en.wikipedia.org/wiki/The_Surrey_with_the_Fringe_on_Top

## Action Creators

Instead of creating the ADD_MESSAGE actions directly as objects, let's create a function to do this for us:

**code/redux/angular2-redux-chat/minimal/tutorial/08-action-creators.ts**

```
19 class MessageActions {
20   static addMessage(message: string): AddMessageAction {
21     return {
22       type: 'ADD_MESSAGE',
23       message: message
24     };
25   }
26   static deleteMessage(index: number): DeleteMessageAction {
27     return {
28       type: 'DELETE_MESSAGE',
29       index: index
30     };
31   }
32 }
```

Here we've created a class with two static methods addMessage and deleteMessage. They return an AddMessageAction and a DeleteMessageAction respectively.

You definitely don't *have* to use static methods for your action creators. You could use plain functions, functions in a namespace, even instance methods on an object, etc. The key idea is to keep them organized in a way that makes them easy to use.

Now let's use our new action creators:

**code/redux/angular2-redux-chat/minimal/tutorial/08-action-creators.ts**

```
55 let store = new Store<AppState>(reducer, { messages: [] });
56 console.log(store.getState()); // -> { messages: [] }
57
58 store.dispatch(
59   MessageActions.addMessage('Would you say the fringe was made of silk?'));
60
61 store.dispatch(
62   MessageActions.addMessage('Wouldnt have no other kind but silk'));
63
64 store.dispatch(
```

```
65   MessageActions.addMessage('Has it really got a team of snow white horses?'));
66
67 console.log(store.getState());
68 // ->
69 // { messages:
70 //    [ 'Would you say the fringe was made of silk?',
71 //      'Wouldnt have no other kind but silk',
72 //      'Has it really got a team of snow white horses?' ] }
```

This feels much nicer!

An added benefit is that if we eventually decided to change the format of our messages, we could do it without having to update all of our `dispatch` statements. For instance, say we wanted to add the time each message was created. We could add a `created_at` field to `addMessage` and now all `AddMessageActions` will be given a `created_at` field:

```
1 class MessageActions {
2   static addMessage(message: string): AddMessageAction {
3     return {
4       type: 'ADD_MESSAGE',
5       message: message,
6       // something like this
7       created_at: new Date()
8     };
9   }
10  // ....
```

## Using Real Redux

Now that we've built our own mini-redux you might be asking, "What do I need to do to use the *real* Redux?" Thankfully, not very much. Let's update our code to use the real Redux now!

If you haven't already, you'll want to run `npm install` in the `code/redux/angular2-redux-chat/minimal/tutorial` directory.

The first thing we need to do is import `Action`, `Reducer`, and `Store` from the `redux` package. We're also going to import a helper method `createStore` while we're at it:

**code/redux/angular2-redux-chat/minimal/tutorial/09-real-redux.ts**

```
1  import {
2    Action,
3    Reducer,
4    Store,
5    createStore
6  } from 'redux';
```

Next, instead of specifying our initial state when we create the *store* instead we're going to let the *reducer* create the initial state. Here we'll do this as the default argument to the reducer. This way if there is no state passed in (e.g. the first time it is called at initialization) we will use the initial state:

**code/redux/angular2-redux-chat/minimal/tutorial/09-real-redux.ts**

```
35 let initialState: AppState = { messages: [] };
36 
37 let reducer: Reducer<AppState> =
38   (state: AppState = initialState, action: Action) => {
```

What's neat about this is that the rest of our reducer stays the same!

The last thing we need to do is create the store using the `createStore` helper method from Redux:

**code/redux/angular2-redux-chat/minimal/tutorial/09-real-redux.ts**

```
58 let store: Store<AppState> = createStore<AppState>(reducer);
```

After that, everything else just works!

**code/redux/angular2-redux-chat/minimal/tutorial/09-real-redux.ts**

```
58 let store: Store<AppState> = createStore<AppState>(reducer);
59 console.log(store.getState()); // -> { messages: [] }
60 
61 store.dispatch(
62   MessageActions.addMessage('Would you say the fringe was made of silk?'));
63 
64 store.dispatch(
65   MessageActions.addMessage('Wouldnt have no other kind but silk'));
66 
67 store.dispatch(
68   MessageActions.addMessage('Has it really got a team of snow white horses?'));
69 
```

```
70 console.log(store.getState());
71 // ->
72 // { messages:
73 //    [ 'Would you say the fringe was made of silk?',
74 //      'Wouldnt have no other kind but silk',
75 //      'Has it really got a team of snow white horses?' ] }
```

Now that we have a handle on using Redux in isolation, the next step is to hook it up to our web app. Let's do that now.

## Using Redux in Angular

In the last section we walked through the core of Redux and showed how to create reducers and use stores to manage our data in isolation. Now it's time to level-up and integrate Redux with our Angular components.

In this section we're going to create a minimal Angular app that contains just a counter which we can increment and decrement with a button.

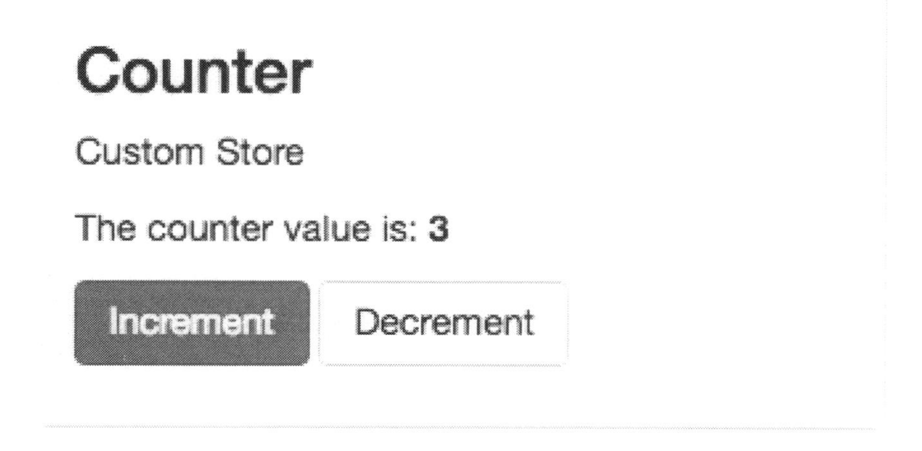


**Counter App**

By using such a small app we can focus on the integration points between Redux and Angular and then we can move on to a larger app in the next section. But first, let's see how to build this counter app!

Here we are going to be integrating Redux directly with Angular without any helper libraries in-between. There are several open-source libraries with the goal of making this process easier, and you can find them in the references section below.

That said, it can be much easier to use those libraries once you understand what is going on underneath the hood, which is what we work through here.

## Planning Our App

If you recall, the three steps to planning our Redux apps are to:

1. Define the structure of our central app state
2. Define actions that will change that state and
3. Define a reducer that takes the old state and an action and returns a new state.

For this app, we're just going to increment and decrement a counter. We did this in the last section, and so our actions, store, and reducer will all be very familiar.

The other thing we need to do when writing Angular apps is decide where we will create components. In this app, we'll have a top-level `CounterApp` which will have one component, the `CounterComponent` which contains the view we see in the screenshot.

At a high level we're going to do the following:

1. Create our `Store` and make it accessible to our whole app via dependency injection
2. Subscribe to changes to the `Store` and display them in our components
3. When something changes (a button is pressed) we will dispatch an action to the `Store`.

Enough planning, let's look at how this works in practice!

## Setting Up Redux

We start by importing a few things we'll need along the way:

**code/redux/angular2-redux-chat/minimal/app.ts**

```
 9  import {
10    Component
11  } from '@angular/core';
12  import { NgModule } from "@angular/core";
13  import { BrowserModule } from "@angular/platform-browser";
14  import { platformBrowserDynamic } from "@angular/platform-browser-dynamic";
15  import {
16    createStore,
17    Store,
18    StoreEnhancer
19  } from 'redux';
20  import { counterReducer } from './counter-reducer';
```

We're importing `Store` (the class) and `createStore` (the helper creation function), which we've used before. We're also importing a new class called `StoreEnhancer` - more on that in a minute.

We import our reducer from `counter-reducer.ts` and our state interface `AppState` from `app-state.ts`.

## Defining the Application State

Let's take a look at our `AppState`:

**code/redux/angular2-redux-chat/minimal/app-state.ts**

```
 9  export interface AppState {
10    counter: number;
11  };
```

Here we are defining our core state structure as `AppState` - it is an object with one key, `counter` which is a `number`. In the next example (the chat app) we'll talk about how to have more sophisticated states, but for now this will be fine.

## Defining the Reducers

Next lets define the reducer which will handle incrementing and decrementing the counter in the application state:

**code/redux/angular2-redux-chat/minimal/counter-reducer.ts**

```
 6  import {
 7    INCREMENT,
 8    DECREMENT
 9  } from './counter-action-creators';
10
11  let initialState: AppState = { counter: 0 };
12
13  // Create our reducer that will handle changes to the state
14  export const counterReducer: Reducer<AppState> =
15    (state: AppState = initialState, action: Action): AppState => {
16      switch (action.type) {
17      case INCREMENT:
18        return Object.assign({}, state, { counter: state.counter + 1 });
19      case DECREMENT:
20        return Object.assign({}, state, { counter: state.counter - 1 });
21      default:
```

```
22        return state;
23    }
24  };
```

We start by importing the constants `INCREMENT` and `DECREMENT`, which are exported by our action creators. They're just defined as the strings `'INCREMENT'` and `'DECREMENT'`, but it's nice to get the extra help from the compiler in case we make a typo. We'll look at those action creators in a minute.

The `initialState` is an `AppState` which sets the counter to `0`.

The `counterReducer` handles two actions: `INCREMENT`, which adds `1` to the current counter and `DECREMENT`, which subtracts `1`. Both actions use `Object.assign` to ensure that we don't *mutate* the old state, but instead create a new object that gets returned as the new state.

Since we're here, let's look at the action creators

## Defining Action Creators

Our action creators are functions which return objects that define the action to be taken. `increment` and `decrement` below return an object that defines the appropriate `type`.

**code/redux/angular2-redux-chat/minimal/counter-action-creators.ts**

```
 1  import {
 2    Action,
 3    ActionCreator
 4  } from 'redux';
 5
 6  export const INCREMENT: string = 'INCREMENT';
 7  export const increment: ActionCreator<Action> = () => ({
 8    type: INCREMENT
 9  });
10
11  export const DECREMENT: string = 'DECREMENT';
12  export const decrement: ActionCreator<Action> = () => ({
13    type: DECREMENT
14  });
```

Notice that our action creator functions return the type `ActionCreator<Action>`. `ActionCreator` is a generic class defined by Redux that we use to define functions that create actions. In this case we're using the concrete class `Action`, but we could use a more specific `Action` class, such as `AddMessageAction` that we defined in the last section.

## Creating the Store

Now that we have our reducer and state, we could create our store like so:

```
let store: Store<AppState> = createStore<AppState>(counterReducer);
```

However, one of the awesome things about Redux is that it has a robust set of developer tools. Specifically, there is a Chrome extension[90] that will let us monitor the state of our application and dispatch actions.

**Counter App With Redux Devtools**

What's really neat about the Redux Devtools is that it gives us clear insight to every action that flows through the system and it's affect on the state.

Go ahead and install the Redux Devtools Chrome extension[91] now!

In order to use the Devtools we have to do one thing: add it to our store.

**code/redux/angular2-redux-chat/minimal/app.ts**

```
25  let devtools: StoreEnhancer<AppState> =
26    window['devToolsExtension'] ?
27    window['devToolsExtension']() : f => f;
```

Not everyone who uses our app will necessarily have the Redux Devtools installed. The code above will check for `window.devToolsExtension`, which is defined by Redux Devtools, and if it exists, we will use it. If it doesn't exist, we're just returning an *identity function* (`f => f`) that will return whatever is passed to it.

[90]https://chrome.google.com/webstore/detail/redux-devtools/lmhkpmbekcpmknklioeibfkpmmfibljd?hl=en

[91]https://chrome.google.com/webstore/detail/redux-devtools/lmhkpmbekcpmknklioeibfkpmmfibljd?hl=en

*Middleware* is a term for a function that enhances the functionality of another library. The Redux Devtools is one of many possible middleware libraries for Redux. Redux supports lots of interesting middleware and it's easy to write our own.

You can read more about Redux middleware here[92]

In order to use this `devtools` we pass it as *middleware* to our Redux store:

**code/redux/angular2-redux-chat/minimal/app.ts**

```
29 let store: Store<AppState> = createStore<AppState>(
30   counterReducer,
31   devtools
32 );
```

Now whenever we dispatch an action and change our state, we can inspect it in our browser!

## `CounterApp` Component

Now that we have the Redux core setup, let's turn our attention to our Angular components. Let's create our top-level app component, `CounterApp`. This will be the component we use to `bootstrap` Angular:

**code/redux/angular2-redux-chat/minimal/app.ts**

```
34 @Component({
35   selector: 'minimal-redux-app',
36   template: `
37   <div>
38     <counter-component>
39     </counter-component>
40   </div>
41   `
42 })
43 class CounterApp {
44 }
```

All this component does is create an instance of the `CounterComponent`, which we'll define in a minute. But before we define `CounterComponent`, let's bootstrap our app.

[92]http://redux.js.org/docs/advanced/Middleware.html

# Providing the Store

We're going to use the `CounterApp` as the root component. Remember that since this is a Redux app, we need to make our store instance accessible everywhere in our app. How should we do this? We'll use dependency injection (DI).

If you recall from the dependency injection chapter, when we want to make something available via DI, then we use the `providers` configuration to add it to the list of `providers` in our `NgModule`.

When we provide something to the DI system, we specify two things:

1. the *token* to use to refer this injectable dependency
2. the *way* to inject the dependency

Oftentimes if we want to provide a singleton service we might use the `useClass` option as in:

```
1 { provide: SpotifyService, useClass: SpotifyService }
```

In the case above, we're using the class `SpotifyService` as the *token* in the DI system. The `useClass` option tells Angular to *create an instance* of `SpotifyService` and reuse that instance whenever the `SpotifyService` injection is requested (e.g. maintain a Singleton).

One problem with us using this method is that we don't want Angular to create our store - we did it ourselves above with `createStore`. We just want to use the `store` we've already created.

To do this we'll use the `useValue` option of `provide`. We've done this before with configurable values like `API_URL`:

```
1 { provide: API_URL, useValue: 'http://localhost/api' }
```

The one thing we have left to figure out is what token we want to use to inject. Our `store` is of type `Store<AppState>`:

**code/redux/angular2-redux-chat/minimal/app.ts**

```
29 let store: Store<AppState> = createStore<AppState>(
30   counterReducer,
31   devtools
32 );
```

`Store` is an *interface*, not a class and, unfortunately, we can't use interfaces as a dependency injection key.

If you're interested in *why* we can't use an interface as a DI key, it's because TypeScript interfaces are removed after compilation and not available at runtime.

If you'd like to read more, see here[93], here[94], and here[95].

This means we need to create our own token that we'll use for injecting the store. Thankfully, Angular makes this easy to do. Let's create this token in it's own file so that way we can `import` it from anywhere in our application;

**code/redux/angular2-redux-chat/minimal/app-store.ts**

```
1  import { OpaqueToken } from '@angular/core';
2
3  export const AppStore = new OpaqueToken('App.store');
```

Here we have created a `const AppStore` which uses the `OpaqueToken` class from Angular. `OpaqueToken` is a better choice than injecting a string directly because it helps us avoid collisions.

Now we can use this token `AppStore` with `provide`. Let's do that now.

# Bootstrapping the App

Back in `app.ts`, let's create the `NgModule` we'll use to bootstrap our app:

**code/redux/angular2-redux-chat/minimal/app.ts**

```
46  @NgModule({
47    declarations: [
48      CounterApp,
49      CounterComponent
50    ],
51    imports: [ BrowserModule ],
52    bootstrap: [ CounterApp ],
53    providers: [
54      {provide: AppStore, useValue: store }
55    ]
56
57  })
58  class CounterAppAppModule {}
59
60  platformBrowserDynamic().bootstrapModule(CounterAppAppModule)
```

[93] http://stackoverflow.com/questions/32254952/binding-a-class-to-an-interface
[94] https://github.com/angular/angular/issues/135
[95] http://victorsavkin.com/post/126514197956/dependency-injection-in-angular-1-and-angular-2

Now we are able to get a reference to our Redux store anywhere in our app by injecting `AppStore`. The place we need it most now is our `CounterComponent`.

## The `CounterComponent`

With our setup out of the way, we can start creating our component that actually displays the counter to the user and provides buttons for the user to change the state.

### `imports`

Let's start by looking at the imports:

**code/redux/angular2-redux-chat/minimal/CounterComponent.ts**

```
 8  import {
 9    Component,
10    Inject
11  } from '@angular/core';
12  import { Store } from 'redux';
13  import { AppStore } from './app-store';
14  import { AppState } from './app-state';
15  import * as CounterActions from './counter-action-creators';
```

We import `Store` from Redux as well as our injector token `AppStore`, which will get us a reference to the singleton *instance* of our store. We also import the `AppState` type, which helps us know the structure of the central state.

Lastly, we import our action creators with `* as CounterActions`. This syntax will let us call `CounterActions.increment()` to create an `INCREMENT` action.

## The template

Let's look at the template of our `CounterComponent`:

**Counter App Template**

**code/redux/angular2-redux-chat/minimal/CounterComponent.ts**

```
@Component({
  selector: 'counter-component',
  template: `
   <div class="row">
     <div class="col-sm-6 col-md-4">
       <div class="thumbnail">
         <div class="caption">
           <h3>Counter</h3>
           <p>Custom Store</p>

           <p>
             The counter value is:
             <b>{{ counter }}</b>
           </p>

           <p>
             <button (click)="increment()"
                     class="btn btn-primary">
               Increment
             </button>
             <button (click)="decrement()"
                     class="btn btn-default">
               Decrement
             </button>
           </p>
```

```
41            </div>
42          </div>
43        </div>
44      </div>
45    `
```

The three things to note here are that we're:

1. displaying the value of the counter in `{{ counter }}`
2. calling the `increment()` function in a button and
3. calling the `decrement()` function in a button.

## The `constructor`

Remember that we need this component depends on the `Store`, so we need to inject it in the constructor. This is how we use our custom `AppStore` token to inject a dependency:

**code/redux/angular2-redux-chat/minimal/CounterComponent.ts**

```
48  export default class CounterComponent {
49    counter: number;
50
51    constructor(@Inject(AppStore) private store: Store<AppState>) {
52      store.subscribe(() => this.readState());
53      this.readState();
54    }
55
56    readState() {
57      let state: AppState = this.store.getState() as AppState;
58      this.counter = state.counter;
59    }
60
61    increment() {
62      this.store.dispatch(CounterActions.increment());
63    }
64
65    decrement() {
66      this.store.dispatch(CounterActions.decrement());
67    }
68  }
```

We use the `@Inject` annotation to inject `AppStore` - notice that we define the type of the variable `store` to `Store<AppState>`. Having a different injection token than the type of the dependency injected is a little different than when we use the class as the injection token (and Angular infers what to inject).

We set the `store` to an instance variable (with `private store`). Now that we have the store we can listen for changes. Here we call `store.subscribe` and call `this.readState()`, which we define below.

The store will call `subscribe` only when a new action is dispatched, so in this case we need to make sure we manually call `readState` at least once to ensure that our component gets the initial data.

The method `readState` reads from our store and updates `this.counter` to the current value. Because `this.counter` is a property on this class and bound in the view, Angular will detect when it changes and re-render this component.

We define two helper methods: `increment` and `decrement`, each of which dispatch their respective actions to the store.

## Putting It All Together

Here's the full listing of our `CounterComponent`

**code/redux/angular2-redux-chat/minimal/CounterComponent.ts**

```
8  import {
9    Component,
10   Inject
11 } from '@angular/core';
12 import { Store } from 'redux';
13 import { AppStore } from './app-store';
14 import { AppState } from './app-state';
15 import * as CounterActions from './counter-action-creators';
16 
17 @Component({
18   selector: 'counter-component',
19   template: `
20     <div class="row">
21       <div class="col-sm-6 col-md-4">
22         <div class="thumbnail">
23           <div class="caption">
24             <h3>Counter</h3>
25             <p>Custom Store</p>
26 
27             <p>
```

```
28                The counter value is:
29                <b>{{ counter }}</b>
30              </p>
31 
32              <p>
33                <button (click)="increment()"
34                        class="btn btn-primary">
35                  Increment
36                </button>
37                <button (click)="decrement()"
38                        class="btn btn-default">
39                  Decrement
40                </button>
41              </p>
42            </div>
43          </div>
44        </div>
45      </div>
46    `
47  })
48  export default class CounterComponent {
49    counter: number;
50 
51    constructor(@Inject(AppStore) private store: Store<AppState>) {
52      store.subscribe(() => this.readState());
53      this.readState();
54    }
55 
56    readState() {
57      let state: AppState = this.store.getState() as AppState;
58      this.counter = state.counter;
59    }
60 
61    increment() {
62      this.store.dispatch(CounterActions.increment());
63    }
64 
65    decrement() {
66      this.store.dispatch(CounterActions.decrement());
67    }
68  }
```

Try it out!

```
1 cd code/redux/angular2-redux-chat
2 npm install
3 npm run go
4 open http://localhost:8080/minimal.html
```

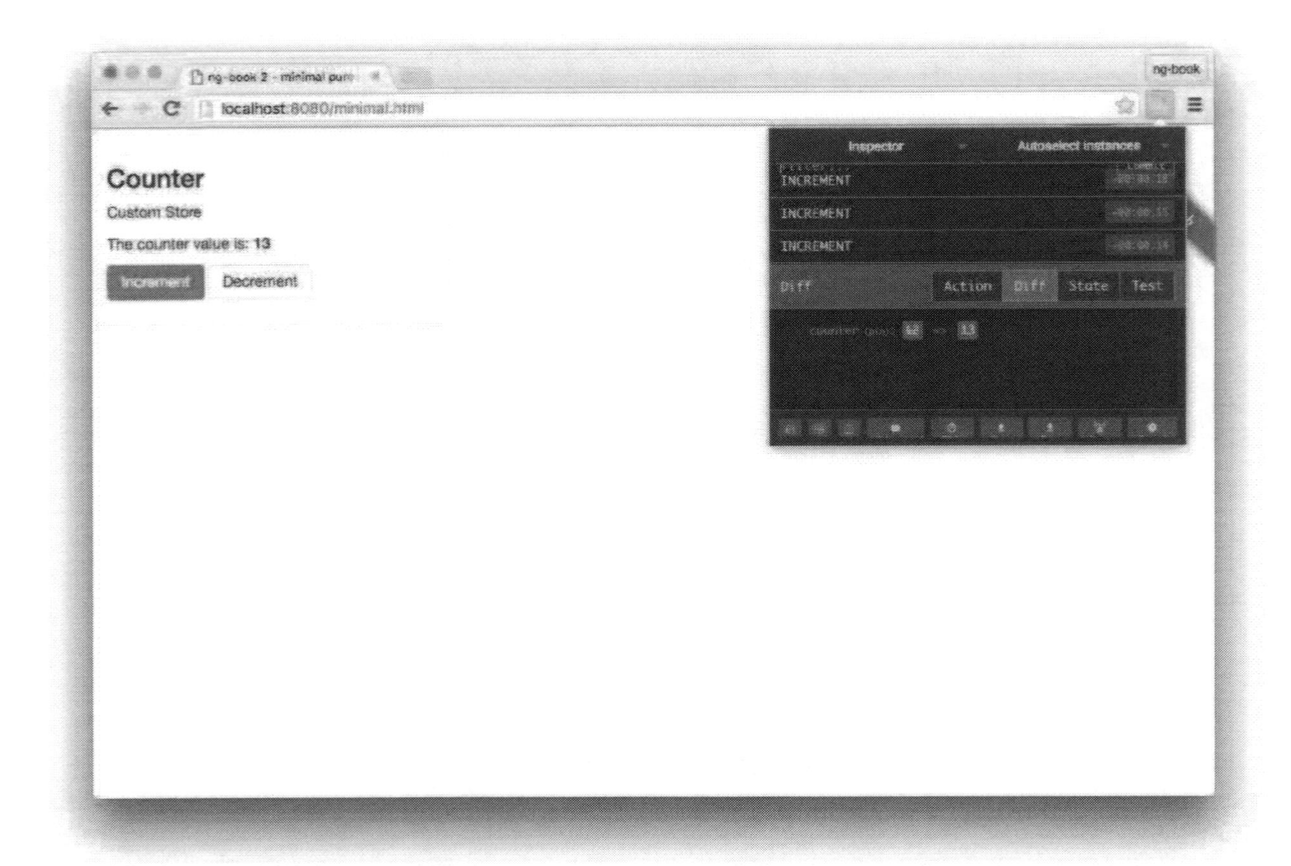


**Working Counter App**

Congratulations! You've created your first Angular and Redux app!

# What's Next

Now that we've built a basic app using Redux and Angular, we should try building a more complicated app. When we build bigger apps we encounter new challenges like:

- How do we combine reducers?

- How do we extract data from different branches of the state?
- How should we organize our Redux code?

In the next chapter, we'll build a chat app which will tackle all of these questions!

# References

If you want to learn more about Redux, here are some good resources:

- Official Redux Website[96]
- This Video Tutorial by Redux's Creator[97]
- Real World Redux[98] (presentation slides)
- The power of higher-order reducers[99]

To learn more about Redux and Angular checkout:

- angular2-redux[100]
- ng2-redux[101]
- ngrx/store[102]

Onward!

[96]http://redux.js.org/
[97]https://egghead.io/courses/getting-started-with-redux
[98]https://speakerdeck.com/chrisui/real-world-redux
[99]http://slides.com/omnidan/hor
[100]https://github.com/InfomediaLtd/angular2-redux
[101]https://github.com/angular-redux/ng2-redux
[102]https://github.com/ngrx/store

# Intermediate Redux in Angular

In the last chapter we learned about Redux, the popular and elegant data architecture. In that chapter, we built an extremely basic app that tied our Angular components and the Redux store together.

In this chapter we're going to take on those ideas and build on them to create a more sophisticated chat app.

Here's a screenshot of the app we're going to build:

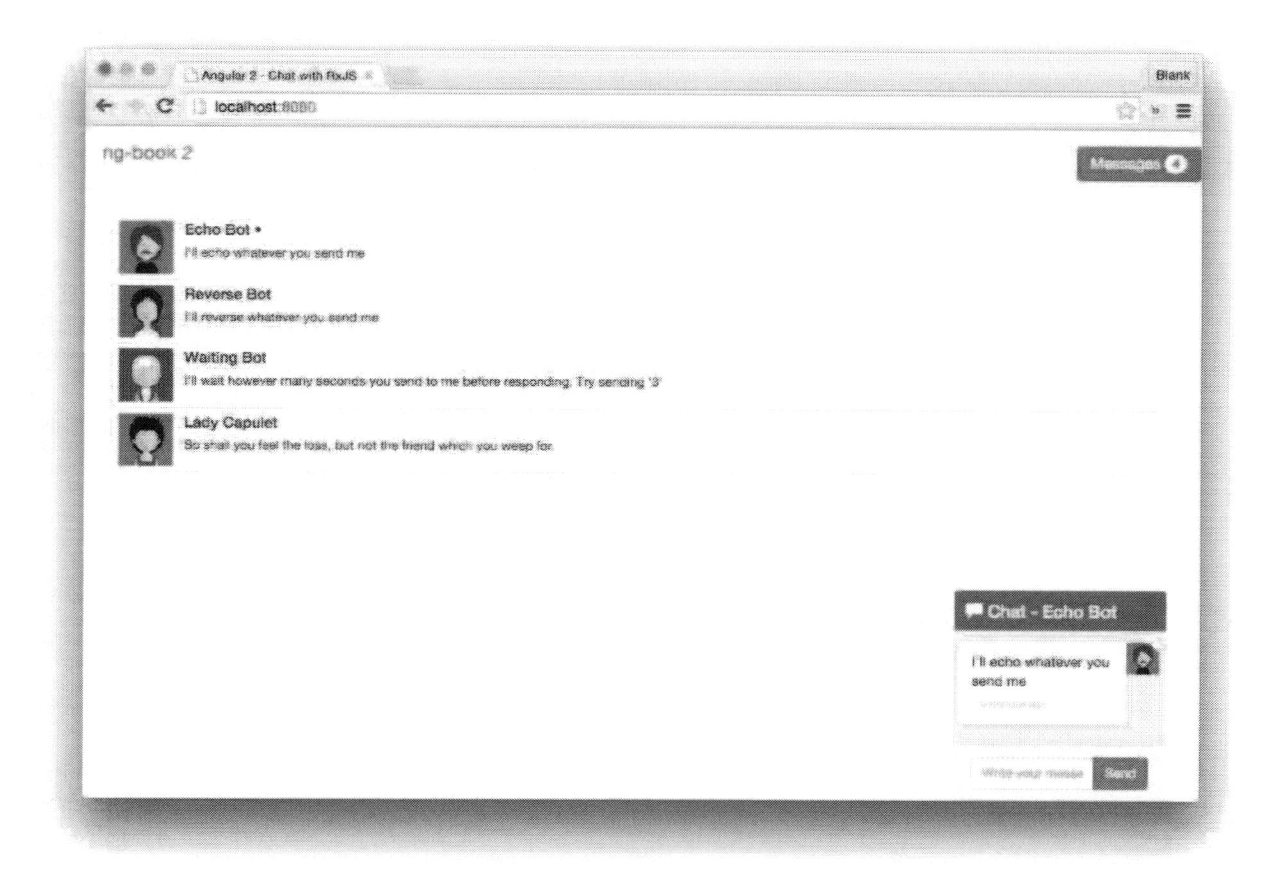


**Completed Chat Application**

## Context For This Chapter

Earlier in this book we built a chat app using RxJS. We're going to be building that same app again only this time with Redux. The point is for you to be able to compare and contrast how the same app works with different data architecture strategies.

You are not required to have read the RxJS chapter in order to work through this one. This chapter stands on its own with regard to the RxJS chapters. If you have read that chapter, you'll be able to skim through some of the sections here where the code is largely the same (for instance, the data models themselves don't change much).

We *do* expect that you've read through the previous Redux chapter or at least have some familiarity with Redux.

# Chat App Overview

In this application we've provided a few bots you can chat with. Open up the code and try it out:

```
1   cd code/redux/angular2-redux-chat
2   npm install
3   npm run go
```

Now open your browser to `http://localhost:8080`.

If the above URL doesn't work, try this URL: `http://localhost:8080/webpack-dev-server/index.html`

Some Windows users may have trouble doing an `npm install` on this repo. If this causes problems for you, make sure you're running these commands inside Cygwin[103].

Notice a few things about this application:

- You can click on the threads to chat with another person
- The bots will send you messages back, depending on their personality
- The unread message count in the top corner stays in sync with the number of unread messages

Let's look at an overview of how this app is constructed. We have

- 3 top-level Angular Components
- 3 models
- and 2 reducers, with their respective action creators

Let's look at them one at a time.

[103] https://www.cygwin.com/

## Components

The page is broken down into three top-level components:

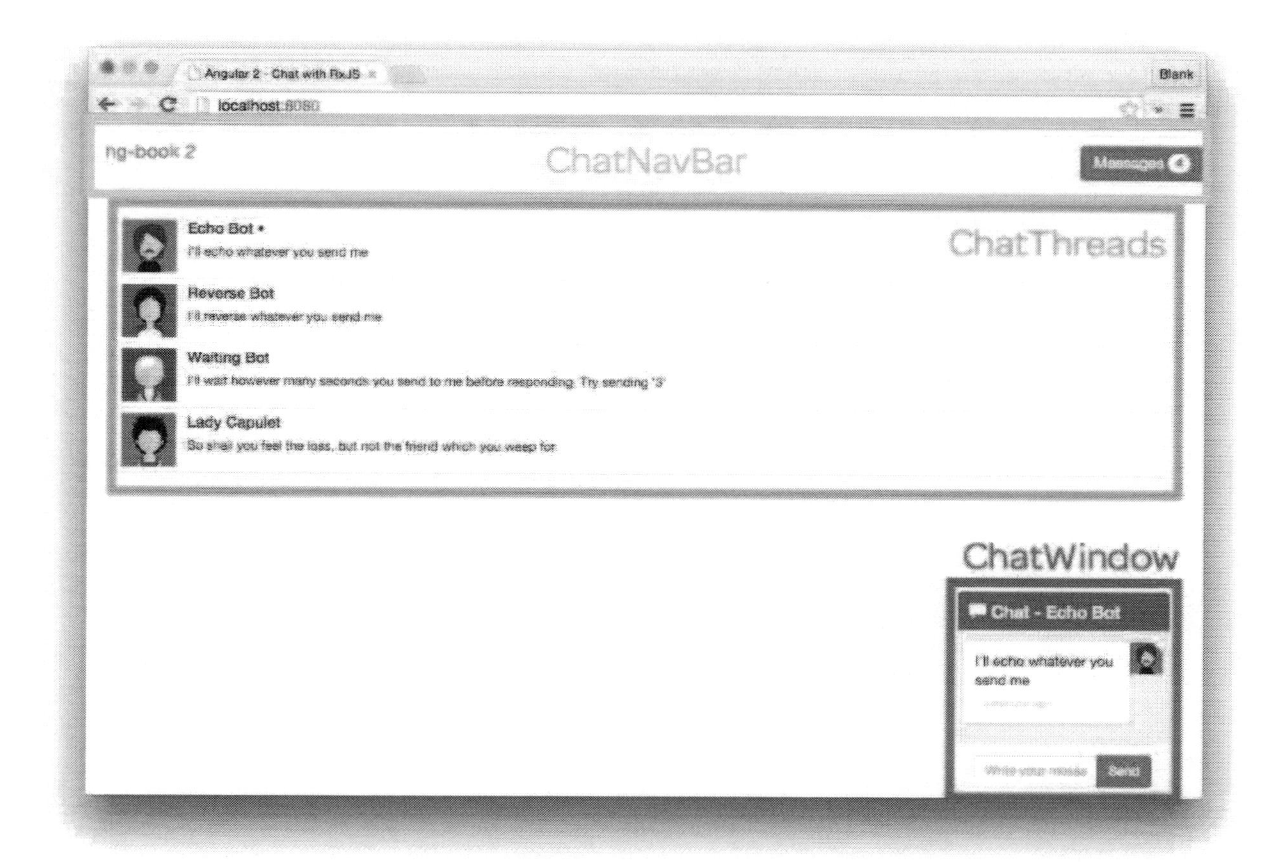


**Redux Chat Top-Level Components**

- `ChatNavBar` - contains the unread messages count
- `ChatThreads` - shows a clickable list of threads, along with the most recent message and the conversation avatar
- `ChatWindow` - shows the messages in the current thread with an input box to send new messages

## Models

This application also has three models:

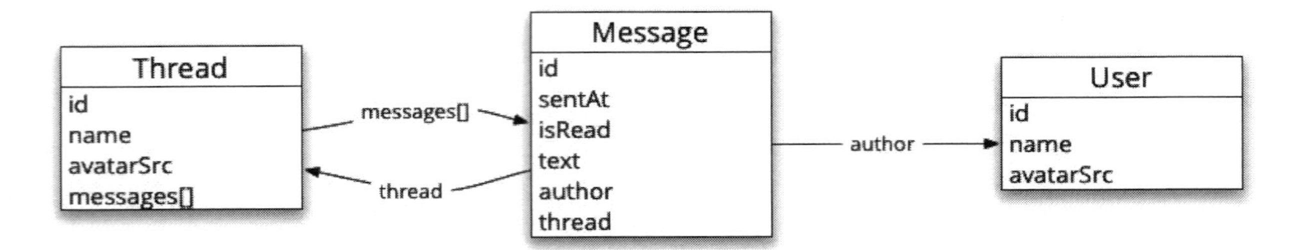


**Redux Chat Models**

- `User` - stores information about a chat participant
- `Message` - stores an individual message
- `Thread` - stores a collection of `Messages` as well as some data about the conversation

## Reducers

In this app, we have two reducers:

- `UsersReducer` - handles information about the current user
- `ThreadsReducer` - handles threads and their messages

## Summary

At a high level our data architecture looks like this:

- All information about the users and threads (which hold messages) are contained in our central store
- Components subscribe to changes in that store and display the appropriate data (unread count, list of threads, the messages themselves
- When the user sends a message, our components dispatch an action to the store

In the rest of this chapter, we're going to go in-depth on how we implement this using Angular and Redux. We'll start by implementing our models, then look at how we create our app state and reducers, and then finally we'll implement the Components.

# Implementing the Models

Let's start with the easy stuff and take a look at the models.

We're going to be specifying each of our model definitions as `interfaces`. This isn't a requirement and you're free to use more elaborate objects if you wish. That said, objects with methods that mutate their internal state can break the functional model that we're striving for.

That is, all mutations to our app state should only be made by the reducers - the objects in the state should be immutable themselves.

So by defining an `interface` for our models,

1. we're able to ensure that the objects we're working with conform to an expected format at compile time and
2. we don't run the risk of someone accidentally adding a method to the model object that would work in an unexpected way.

### User

Our `User` interface has an `id`, `name`, and `avatarSrc`.

**code/redux/angular2-redux-chat/app/ts/models/User.ts**

```
12 export interface User {
13   id: string;
14   name: string;
15   avatarSrc: string;
16   isClient?: boolean;
17 }
```

We also have a boolean `isClient` (the question mark indicates that this field is optional). We will set this value to `true` for the `User` that represents the client, the person using the app.

### Thread

Similarly, `Thread` is also a TypeScript interface:

**code/redux/angular2-redux-chat/app/ts/models/Thread.ts**

```
14 export interface Thread {
15   id: string;
16   name: string;
17   avatarSrc: string;
18   messages: Message[];
19 }
```

We store the `id` of the `Thread`, the `name`, and the current `avatarSrc`. We also expect an array of `Messages` in the `messages` field.

### Message

`Message` is our third and final model `interface`:

**code/redux/angular2-redux-chat/app/ts/models/Message.ts**

```
15 export interface Message {
16   id?: string;
17   sentAt?: Date;
18   isRead?: boolean;
19   thread?: Thread;
20   author: User;
21   text: string;
22 }
```

Each message has:

- `id` - the id of the message
- `sentAt` - when the message was sent
- `isRead` - a boolean indicating that the message was read
- `author` - the `User` who wrote this message
- `text` - the text of the message
- `thread` - a reference to the containing `Thread`

# App State

Now that we have our models, let's talk about the shape of our central state. In the previous chapter, our central state was a single object with the key `counter` which had the value of a `number`. This app, however, is more complicated.

Here's the first part of our app state:

**code/redux/angular2-redux-chat/app/ts/reducers/index.ts**

```
26  export interface AppState {
27    users: UsersState;
28    threads: ThreadsState;
29  }
```

Our `AppState` is also an `interface` and it has two top level keys: `users` and `threads` - these are defined by two more interfaces `UsersState` and `ThreadsState`, which are defined in their respective reducers.

## A Word on Code Layout

This is a common pattern we use in Redux apps: the top level state has a top-level key for each reducer. In our app we're going to keep this top-level reducer in `reducers/index.ts`.

Each reducer will have it's own file. In that file we'll store:

- The `interface` that describes that branch of the state tree
- The value of the initial state, for that branch of the state tree
- The reducer itself
- Any *selectors* that query that branch of the state tree - we haven't talked about *selectors* yet, but we will soon.

The reason we keep all of these different things together is because they all deal with the structure of this branch of the state tree. By putting these things in the same file it's very easy to refactor everything at the same time.

You're free to have multiple layers of nesting, if you so desire. It's a nice way to break up large modules in your app.

## The Root Reducer

Since we're talking about how to split up reducers, let's look at our root reducer now:

**code/redux/angular2-redux-chat/app/ts/reducers/index.ts**

```
26 export interface AppState {
27   users: UsersState;
28   threads: ThreadsState;
29 }
30
31 const rootReducer: Reducer<AppState> = combineReducers<AppState>({
32   users: UsersReducer,
33   threads: ThreadsReducer
34 });
```

Notice the symmetry here - our `UsersReducer` will operate on the `users` key, which is of type `UsersState` and our `ThreadsReducer` will operate on the `threads` key, which is of type `ThreadsState`.

This is made possible by the `combineReducers` function which takes a map of keys and reducers and returns a new reducer that operates appropriately on those keys.

Of course we haven't finished looking at the structure of our `AppState` yet, so let's do that now.

## The `UsersState`

Our `UsersState` holds a reference to the `currentUser`.

**code/redux/angular2-redux-chat/app/ts/reducers/UsersReducer.ts**

```
18 export interface UsersState {
19   currentUser: User;
20 };
21
22 const initialState: UsersState = {
23   currentUser: null
24 };
```

You could imagine that this branch of the state tree could hold information about all of the users, when they were last seen, their idle time, etc. But for now this will suffice.

We'll use `initialState` in our reducer when we define it below, but for now we're just going to set the current user to `null`.

## The `ThreadsState`

Let's look at the `ThreadsState`:

**code/redux/angular2-redux-chat/app/ts/reducers/ThreadsReducer.ts**

```
33 export interface ThreadsEntities {
34   [id: string]: Thread;
35 }
36
37 export interface ThreadsState {
38   ids: string[];
39   entities: ThreadsEntities;
40   currentThreadId?: string;
41 };
42
43 const initialState: ThreadsState = {
44   ids: [],
45   currentThreadId: null,
46   entities: {}
47 };
```

We start by defining an interface called `ThreadsEntities` which is a map of thread `ids` to `Threads`. The idea is that we'll be able to look up any thread by id in this map.

In the `ThreadsState` we're also storing an array of the `ids`. This will store the list of possible ids that we might find in `entities`.

This strategy is used by the commonly-used library normalizr[104]. The idea is that when we standardize how we store entities in our Redux state, we're able to build helper libraries and it's clearer to work with. Instead of wondering what the format is for each tree of the state, when we use `normalizr` a lot of the choices have been made for us and we're able to work more quickly.

I've opted not to teach `normalizr` in this chapter because we're learning so many other things. That said, I would be very likely to use `normalizr` in my production applications.

That said, `normalizr` is totally optional - nothing major changes in our app by not using it.

If you'd like to learn how to use `normalizr`, checkout the official docs[105], this blog post[106], and the thread referenced by Redux creator Dan Abramov here[107]

We store the currently viewed thread in `currentThreadId` - the idea here is that we want to know which thread the user is currently looking at.

We set our `initialState` to "empty" values.

[104]https://github.com/paularmstrong/normalizr

[105]https://github.com/paularmstrong/normalizr

[106]https://medium.com/@mcowpercoles/using-normalizr-js-in-a-redux-store-96ab33991369#.l8ur7ipu6

[107]https://twitter.com/dan_abramov/status/663032263702106112

## Visualizing Our `AppState`

Redux Devtools provides us with a "Chart" view that lets us inspect the state of our app. Here's what mine looks like after being booted with all of the demo data:

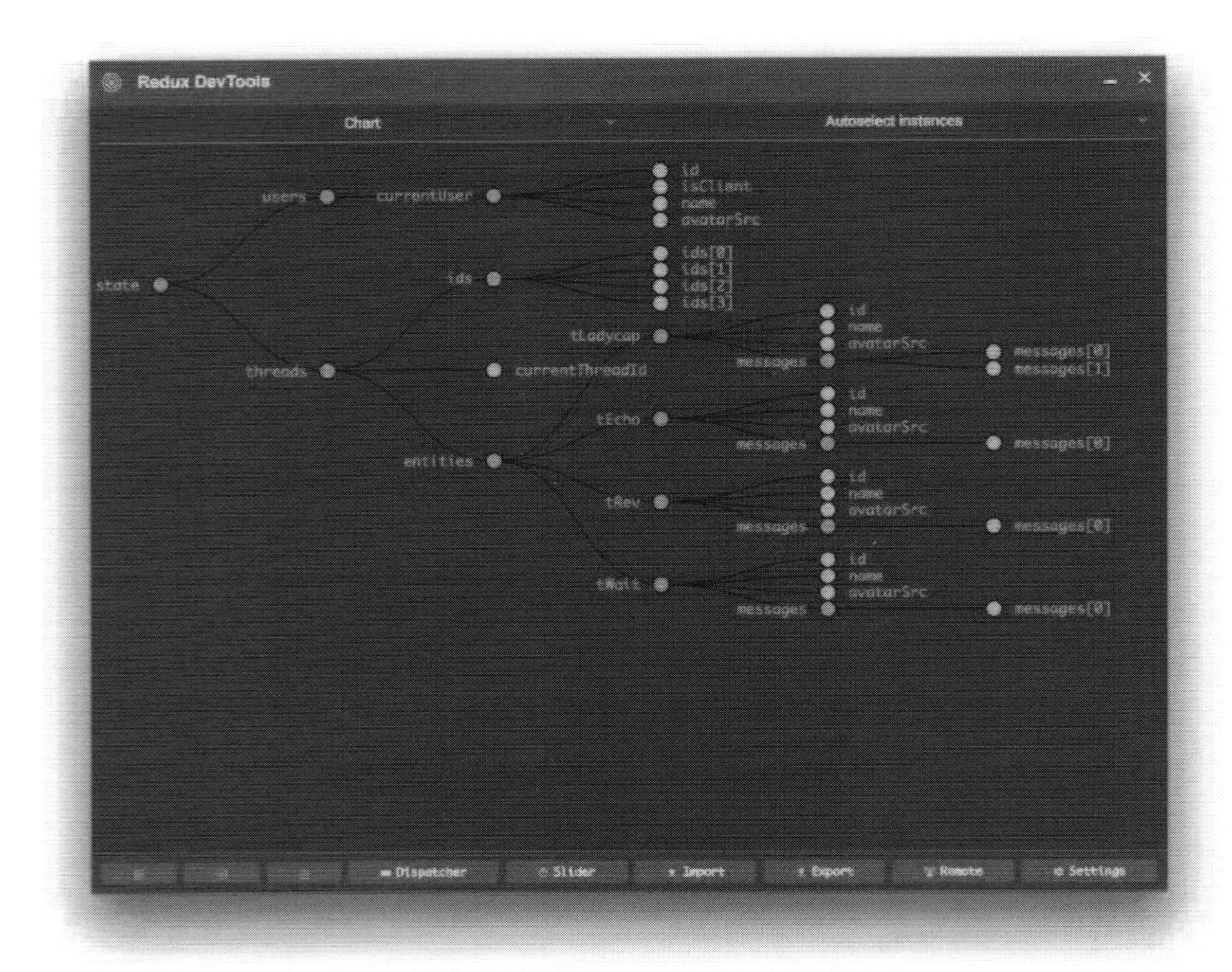


**Redux Chat State Chart**

What's neat is that we can hover over an individual node and see the attributes of that piece of data:

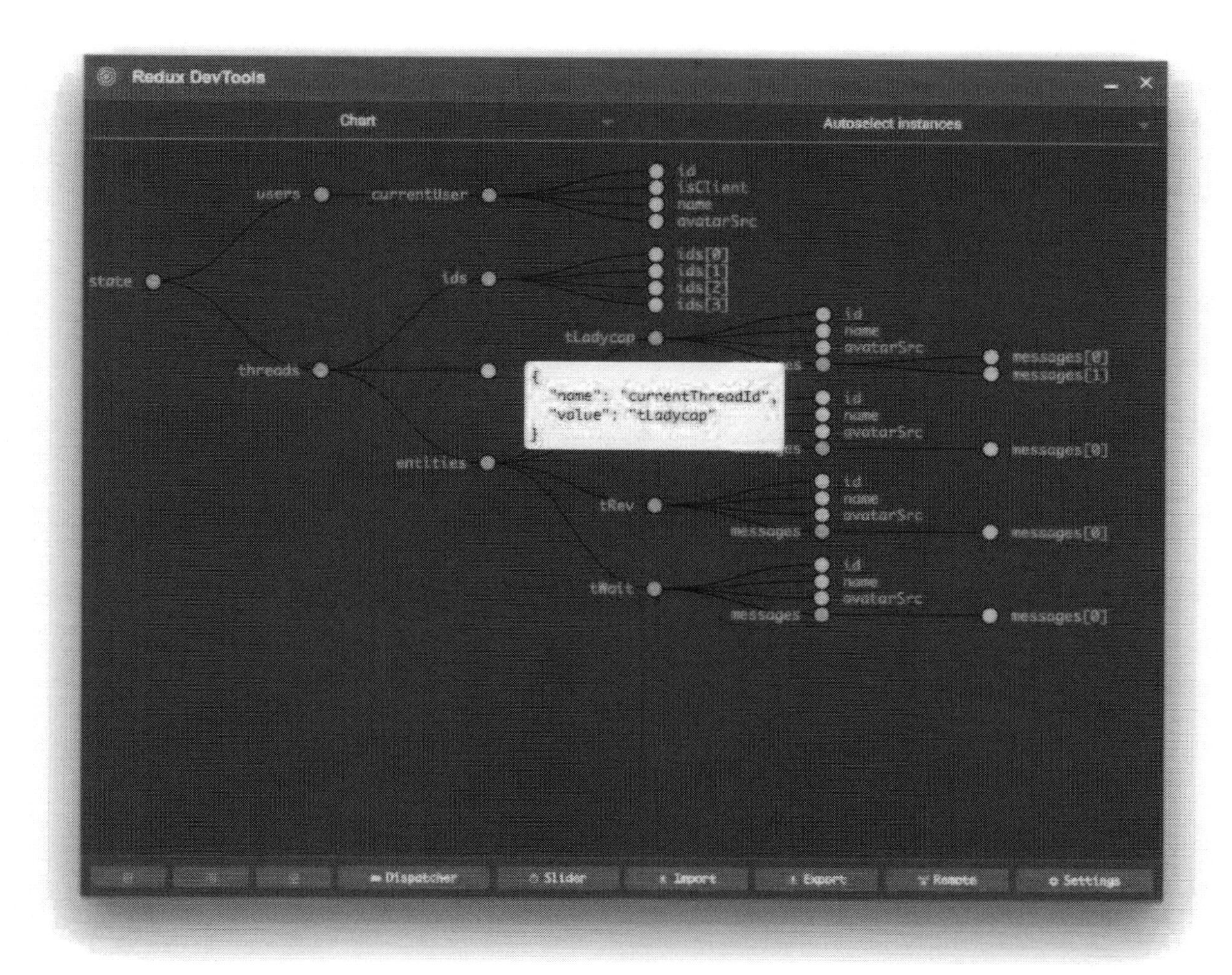


Inspecting the current thread

## Building the Reducers (and Action Creators)

Now that we have our central state, we can start changing it using our reducers!

Since reducers handle actions, we need to know the format of our actions in our reducer. So let's build our action creators at the same time we build our reducers

### Set Current User Action Creators

The `UsersState` stores the current user. This means we need an action to set the current user. We're going to keep our actions in the `actions` folder and name the actions to match their corresponding reducer, in this case `UserActions`.

**code/redux/angular2-redux-chat/app/ts/actions/UserActions.ts**

```
20 export const SET_CURRENT_USER = '[User] Set Current';
21 export interface SetCurrentUserAction extends Action {
22   user: User;
23 }
24 export const setCurrentUser: ActionCreator<SetCurrentUserAction> =
25   (user) => ({
26     type: SET_CURRENT_USER,
27     user: user
28   });
```

Here we define the `const SET_CURRENT_USER`, which we'll use to `switch` on in our reducer.

We also define a new subinterface `SetCurrentUserAction` which extends `Action` to add a `user` property. We'll use the `user` property to indicate *which user* we want to make the current user.

The function `setCurrentUser` is our proper action creator function. It takes `user` as an argument, and returns a `SetCurrentUserAction` which we can give to our reducer.

### `UsersReducer` - Set Current User

Now we turn our attention to our `UsersReducer`:

**code/redux/angular2-redux-chat/app/ts/reducers/UsersReducer.ts**

```
26 export const UsersReducer =
27   function(state: UsersState = initialState, action: Action): UsersState {
28   switch (action.type) {
29     case UserActions.SET_CURRENT_USER:
30     const user: User = (<UserActions.SetCurrentUserAction>action).user;
31       return {
32         currentUser: user
33       };
34     default:
35       return state;
36   }
37 };
```

Our `UsersReducer` takes a `UsersState` as the first argument. Notice that this isn't the `AppState`! Our "child reducer" only works with it's branch of the state tree.

Our `UsersReducer`, like all reducers, returns a new state, in this case it is of type `UsersState`.

Next we `switch` on the `action.type` and we handle the `UserActions.SET_CURRENT_USER`.

In order to set the current user, we need to get the `user` from the incoming action. To do this, we first cast the `action` to `UserActions.SetCurrentUserAction` and then we read the `.user` field.

It might seem a little weird that we originally created a `SetCurrentUserAction` but then now we switch on a `type` string instead of using the type directly.

Indeed, we are fighting TypeScript a little here. We lose interface metadata when the TypeScript is compiled to JavaScript. We could instead try some sort of reflection (through decorator metadata, or looking at a constructor etc.).

While down-casting our `SetCurrentUserAction` to an `Action` on `dispatch` and then re-casting is a bit ugly, it's a straightforward and portable way to handle this "polymorphic dispatch" for this app.

We need to return a new `UsersState`. Since `UsersState` only has one key, we return an object with the `currentUser` set to the incoming action's `user`.

## Thread and Messages Overview

The core of our application is messages in threads. There are three actions we need to support:

1. Adding a new thread to the state
2. Adding messages to a thread
3. Selecting a thread

Let's start by creating a new thread

## Adding a New Thread Action Creators

Here's the action creator for adding a new `Thread` to our state:

**code/redux/angular2-redux-chat/app/ts/actions/ThreadActions.ts**

```
24  export const ADD_THREAD = '[Thread] Add';
25  export interface AddThreadAction extends Action {
26    thread: Thread;
27  }
28  export const addThread: ActionCreator<AddThreadAction> =
29    (thread) => ({
30      type: ADD_THREAD,
31      thread: thread
32    });
```

Notice that this is structurally very similar to our previous action creator. We define a const ADD_-THREAD that we can switch on, a custom Action, and an action creator addThread which generates the Action.

Notice that we don't initialize the Thread itself here - the Thread is accepted as an argument.

## Adding a New Thread Reducer

Now let's start our ThreadsReducer by handling ADD_THREAD:

**code/redux/angular2-redux-chat/app/ts/reducers/ThreadsReducer.ts**

```
53 export const ThreadsReducer =
54   function(state: ThreadsState = initialState, action: Action): ThreadsState {
55   switch (action.type) {
56 
57     // Adds a new Thread to the list of entities
58     case ThreadActions.ADD_THREAD: {
59       const thread = (<ThreadActions.AddThreadAction>action).thread;
60 
61       if (state.ids.includes(thread.id)) {
62         return state;
63       }
64 
65       return {
66         ids: [ ...state.ids, thread.id ],
67         currentThreadId: state.currentThreadId,
68         entities: Object.assign({}, state.entities, {
69           [thread.id]: thread
70         })
71       };
72     }
73 
74     // Adds a new Message to a particular Thread
```

Our ThreadsReducer handles the ThreadsState. When we handle the ADD_THREAD action, we cast the action object back into a ThreadActions.AddThreadAction and then pull the Thread out.

Next we check to see if this new thread.id already appears in the list of state.ids. If it does, then we don't make any changes, but instead return the current state.

However if this thread is new, then we need to add it to our current state.

Remember when we create a new ThreadsState we need to take care to now mutate our old state. This looks more complicated than any state we've done so far, but it's not very different in principle.

We start by adding our `thread.id` to the `ids` array. Here we're using the ES6 spread operator (. . .) to indicate that we want to put all of the existing `state.ids` into this new array and then append `thread.id` to the end.

`currentThreadId` does not change when we add a new thread, so we return the *old* `state.currentThreadId` for this field.

For `entities`, remember that it is an object where the key is the `string` id of each thread and the value is the thread itself. We're using `Object.assign` here to create a new object that merges the old `state.entities` with our newly added `thread` into a new object.

You might be kind of tired of meticulously copying these objects when we need to make changes. That's a common response! In fact, it's easy to make mutations here by accident.

This is why Immutable.js[108] was written. Immutable.js is often used with Redux for this purpose. When we use Immutable, these careful updates are handled for us.

I'd encourage you to take a look at Immutable.js and see if it is a good fit for your reducers.

Now we can add new threads to our central state!

## Adding New Messages Action Creators

Now that we have threads we can start adding messages to them.

Let's define a new action for adding messages:

**code/redux/angular2-redux-chat/app/ts/actions/ThreadActions.ts**

```
34  export const ADD_MESSAGE = '[Thread] Add Message';
35  export interface AddMessageAction extends Action {
36    thread: Thread;
37    message: Message;
38  }
```

The `AddMessageAction` adds a `Message` to a `Thread`.

Here's the action creator for adding a message:

[108]https://facebook.github.io/immutable-js/

**code/redux/angular2-redux-chat/app/ts/actions/ThreadActions.ts**

```
39 export const addMessage: ActionCreator<AddMessageAction> =
40   (thread: Thread, messageArgs: Message): AddMessageAction => {
41     const defaults = {
42       id: uuid(),
43       sentAt: new Date(),
44       isRead: false,
45       thread: thread
46     };
47     const message: Message = Object.assign({}, defaults, messageArgs);
48 
49     return {
50       type: ADD_MESSAGE,
51       thread: thread,
52       message: message
53     };
54   };
```

The `addMessage` action creator accepts a `thread` and an object we use for crafting the message. Notice here that we keep a list of `defaults`. The idea here is that we want to encapsulate creating an id, setting the timestamp, and setting the `isRead` status. Someone who wants to send a message shouldn't have to worry about how the UUIDs are formed, for instance.

That said, maybe the client using this library crafted the message beforehand and if they send a message with an existing id, we want to keep it. To enable this default behavior we merge the `messageArgs` into the `defaults` and copy those values to a new object.

Lastly we return the `ADD_MESSAGE` action with the this `thread` and new `message`.

## Adding A New Message Reducer

Now we will add our `ADD_MESSAGE` handler to our `ThreadsReducer`. When a new message is added, we need to take the thread and add the message to it.

There is one tricky thing we need to handle that may not be obvious at this point: if the `thread` is the "current thread" we need to *mark this message as read.*

The user will always have one thread that is the "current thread" that they're looking at. We're going to say that if a new message is added to the current thread, then it's automatically marked as read.

**code/redux/angular2-redux-chat/app/ts/reducers/ThreadsReducer.ts**

```
75      case ThreadActions.ADD_MESSAGE: {
76        const thread = (<ThreadActions.AddMessageAction>action).thread;
77        const message = (<ThreadActions.AddMessageAction>action).message;
78
79        // special case: if the message being added is in the current thread, then
80        // mark it as read
81        const isRead = message.thread.id === state.currentThreadId ?
82                        true : message.isRead;
83        const newMessage = Object.assign({}, message, { isRead: isRead });
84
85        // grab the old thraed from entities
86        const oldThread = state.entities[thread.id];
87
88        // create a new thread which has our newMessage
89        const newThread = Object.assign({}, oldThread, {
90          messages: [...oldThread.messages, newMessage]
91        });
92
93        return {
94          ids: state.ids, // unchanged
95          currentThreadId: state.currentThreadId, // unchanged
96          entities: Object.assign({}, state.entities, {
97            [thread.id]: newThread
98          })
99        };
100     }
101
102     // Select a particular thread in the UI
```

The code is a bit long because we're being careful not to mutate the original thread, but it is not much different than what we've done so far in principle.

We start by extracting the `thread` and `message`.

Next we mark the message as read, if its part of the "current thread" (we'll look at how to set the current thread next).

Then we grab the `oldThread` and create a `newThread` which has the `newMessage` appended on to the old `messages`.

Finally we return the new `ThreadsState`. The current list of thread `ids` and the `currentThreadId` are unchanged by adding a message, so we pass the old values here. The only thing we change is that we update `entities` with our `newThread`.

Now let's implement the last part of our data backbone: selecting a thread.

## Selecting A Thread Action Creators

Our user can have multiple chat sessions in progress at the same time. However, we only have one chat window (where the user can read and send messages). When the user clicks on a thread, we want to show that thread's messages in the chat window.

**Selecting A Thread**

We need to keep track of which thread is the currently selected thread. To do that, we'll use the `currentThreadId` property in the `ThreadsState`.

Let's create the actions for this:

**code/redux/angular2-redux-chat/app/ts/actions/ThreadActions.ts**

```
56 export const SELECT_THREAD = '[Thread] Select';
57 export interface SelectThreadAction extends Action {
58   thread: Thread;
59 }
60 export const selectThread: ActionCreator<SelectThreadAction> =
61   (thread) => ({
62     type: SELECT_THREAD,
63     thread: thread
64   });
```

There's nothing conceptually new in this action: we've got a new type of `SELECT_THREAD` and we pass the `Thread` that we're selecting as an argument.

## Selecting A Thread Reducer

To select a `thread` we need to do two things:

1. set `currentThreadId` to the selected `thread`'s `id`
2. mark all messages in that `thread` as read

Here's the code for that reducer:

**code/redux/angular2-redux-chat/app/ts/reducers/ThreadsReducer.ts**

```
103      case ThreadActions.SELECT_THREAD: {
104        const thread = (<ThreadActions.SelectThreadAction>action).thread;
105        const oldThread = state.entities[thread.id];
106
107        // mark the messages as read
108        const newMessages = oldThread.messages.map(
109          (message) => Object.assign({}, message, { isRead: true }));
110
111        // give them to this new thread
112        const newThread = Object.assign({}, oldThread, {
113          messages: newMessages
114        });
115
116        return {
117          ids: state.ids,
118          currentThreadId: thread.id,
119          entities: Object.assign({}, state.entities, {
120            [thread.id]: newThread
121          })
122        };
123      }
124
125      default:
126        return state;
127    }
128  };
```

We start by getting the `thread`-to-select and then using that `thread.id` to get the current `Thread` that exists in `state` to get the values.

This maneuver is a bit defensive. Why not just use the `thread` that is passed in? That might be the right design decision for some apps. In this case we protect against some external mutation of `thread` by reading the last known values of that thread in `state.entities`.

Next we create a copy of all of the old messages and set them as `isRead: true`. Then we assign those new read messages to `newThread`.

Finally we return our new `ThreadsState`.

### Reducers Summary

We did it! Above is everything we need for the backbone of our data architecture.

To recap, we have a `UsersReducer` which maintains the current user. We have a `ThreadsReducer` which manages:

- The list of threads
- The messages in those threads
- The currently selected thread

We can derive everything else that we need (e.g. the unread count) from these pieces of data.

Now we need to hook them up to our components!

## Building the Angular Chat App

As we mentioned earlier in the chapter, the page is broken down into three top-level components:

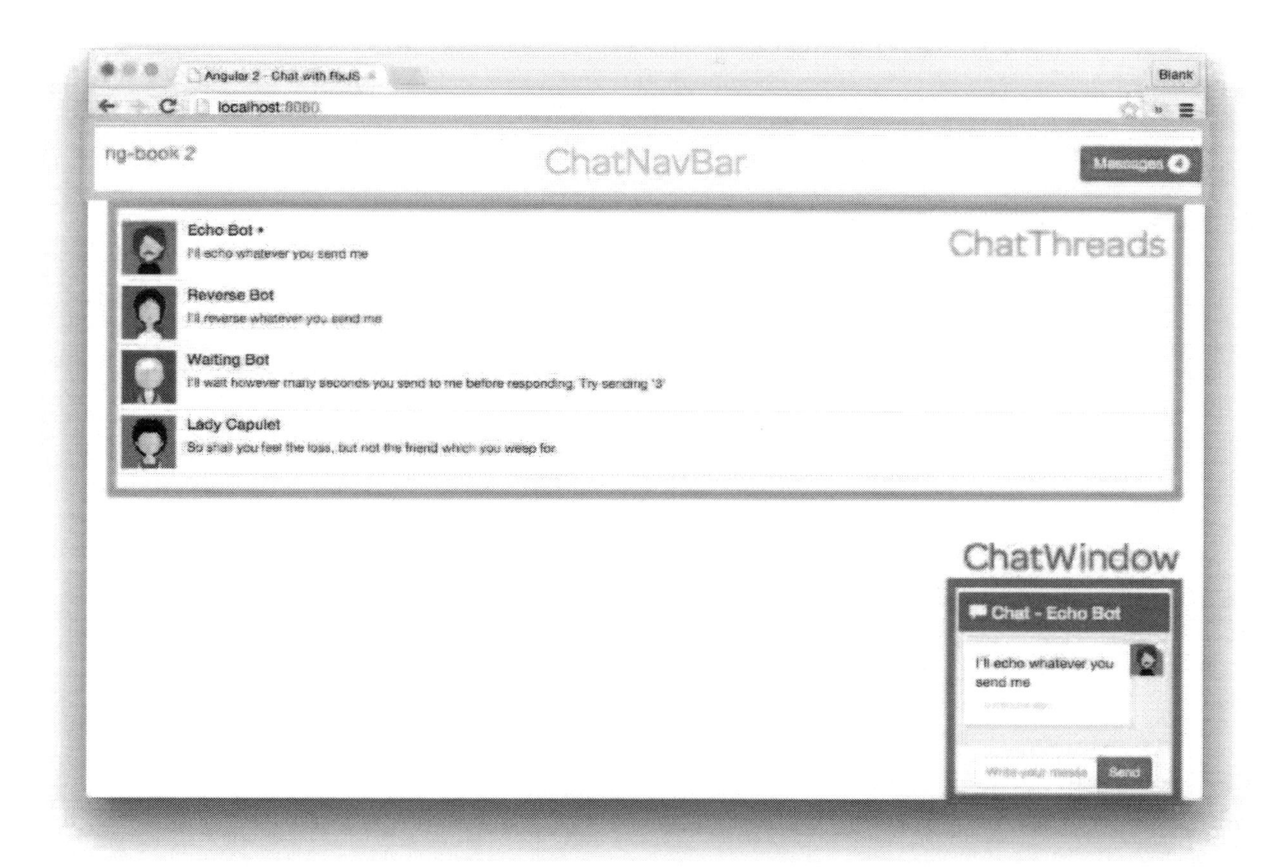


**Redux Chat Top-Level Components**

- `ChatNavBar` - contains the unread messages count
- `ChatThreads` - shows a clickable list of threads, along with the most recent message and the conversation avatar
- `ChatWindow` - shows the messages in the current thread with an input box to send new messages

We're going to bootstrap our app much like we did in the last chapter. We're going to initialize our Redux store at the top of the app and provide it via Angular's dependency injection system (take a look at the previous chapter if this looks unfamiliar):

**code/redux/angular2-redux-chat/app/ts/app.ts**

```
58 let store: Store<AppState> = createStore<AppState>(
59   reducer,
60   compose(devtools)
61 );
62
63 @NgModule({
64   declarations: [
65     ChatApp,
66     ChatPage,
67     ChatThreads,
68     ChatNavBar,
69     ChatWindow,
70     ChatThread,
71     ChatMessage,
72     FromNowPipe
73   ],
74   imports: [
75     BrowserModule,
76     FormsModule
77   ],
78   bootstrap: [ ChatApp ],
79   providers: [
80     { provide: AppStore, useFactory: () => store }
81   ]
82 })
83 class ChatAppModule {}
84
85 platformBrowserDynamic().bootstrapModule(ChatAppModule)
```

## The top-level `ChatApp`

Our `ChatApp` component is the top-level component. It doesn't do much other than render the `ChatPage`.

**code/redux/angular2-redux-chat/app/ts/app.ts**

```
40  @Component({
41    selector: 'chat-app',
42    template: `
43    <div>
44      <chat-page></chat-page>
45    </div>
46    `
47  })
48  class ChatApp {
49    constructor(@Inject(AppStore) private store: Store<AppState>) {
50      ChatExampleData(store);
51    }
52  }
```

For this app the bots operate on data on the client and are not connected to a server. The function `ChatExampleData()` sets up the initial data for the app. We won't be covering this code in detail in the book, so feel free to look at the code on disk if you want to learn more about how it works.

We're not using a router in this app, but if we were, we would put it here at the top level of the app. For now, we're going to create a `ChatPage` which will render the bulk of our app.

We don't have any other pages in this app, but it's a good idea to give each page it's own component in case we add some in the future.

## The `ChatPage`

Our chat page renders our three main components:

- `ChatNavBar`
- `ChatThreads` and
- `ChatWindow`

Here it is in code:

**code/redux/angular2-redux-chat/app/ts/pages/ChatPage.ts**

```
18 @Component({
19   selector: 'chat-page',
20   template: `
21   <div>
22     <chat-nav-bar></chat-nav-bar>
23     <div class="container">
24       <chat-threads></chat-threads>
25       <chat-window></chat-window>
26     </div>
27   </div>
28   `
29 })
30 export default class ChatPage {
31 }
```

For this app we are using a design pattern called *container components* and these three components are all container components. Let's talk about what that means.

## Container vs. Presentational Components

It is hard to reason about our apps if there is data spread throughout all of our components. However, our apps are dynamic - they need to be populated with runtime data and they need to be responsive to user interaction.

One of the patterns that has emerged in managing this tension is the idea of presentational vs. container components. The idea is this:

1. You want to minimize the number of components which interact with outside data sources. (e.g. APIs, the Redux Store, Cookies etc.)
2. Therefore deliberately put data access into "container" components and
3. Require purely 'functional' presentation components to have all of their properties (inputs and outputs) managed by container components.

The great thing about this design is that presentational components are predictable. They're reusable because they don't make assumptions about your overall data-architecture, they only give requirements for their own use.

But even beyond reuse, they're predictable Given the same inputs, they always return the same outputs (e.g. render the same way).

If you squint, you can see that the philosophy that requires reducers to be pure functions is the same that requires presentational components be 'pure components'

It would be great if our entire app could be all presentational components, but of course, the real world has messy, changing data. So we try to put this complexity of adapting our real-world data into our container components.

If you're an advanced programmer you may see that there is a loose analogy between MVC and container/presentation components. That is, the presentational component is sort of a "view" of data that is passed in. A container component is sort of a "controller" in that it takes the "model" (the data from the rest of the app) and adapts it for the presentational components.

That said, if you haven't been programming very long, take this analogy with a grain of salt as Angular components are already a view and a controller themselves.

In our app the container components are going to be the components which interact with the store. This means our container components will be anything that:

1. Reads data from the store
2. Subscribes to the store for changes
3. Dispatches actions to the store

Our three main components are container components and anything below them will be presentational (i.e. functional / pure / not interact with the store).

Let's build our first container component, the nav bar.

## Building the `ChatNavBar`

In the nav bar we'll show an unread messages count to the user.

**The Unread Count in the `ChatNavBar` Component**

The best way to try out the unread messages count is to use the "Waiting Bot". If you haven't already, try sending the message '3' to the Waiting Bot and then switch to another window. The Waiting Bot will then wait 3 seconds before sending you a message and you will see the unread messages counter increment.

Let's look at the component code first:

**code/redux/angular2-redux-chat/app/ts/containers/ChatNavBar.ts**

```
24 @Component({
25   selector: 'chat-nav-bar',
26   template: `
27   <nav class="navbar navbar-default">
28     <div class="container-fluid">
29       <div class="navbar-header">
30         <a class="navbar-brand" href="https://ng-book.com/2">
31           <img src="${require('images/logos/ng-book-2-minibook.png')}"/>
32            ng-book 2
33         </a>
34       </div>
35       <p class="navbar-text navbar-right">
36         <button class="btn btn-primary" type="button">
37           Messages <span class="badge">{{ unreadMessagesCount }}</span>
38         </button>
39       </p>
40     </div>
41   </nav>
42   `
43 })
44 export default class ChatNavBar  {
45   unreadMessagesCount: number;
46 
47   constructor(@Inject(AppStore) private store: Store<AppState>) {
48     store.subscribe(() => this.updateState());
49     this.updateState();
50   }
51 
52   updateState() {
53     this.unreadMessagesCount = getUnreadMessagesCount(this.store.getState());
54   }
55 }
```

Our template gives us the DOM structure and CSS necessary for rending a nav bar (these CSS-classes come from the CSS framework Bootstrap).

The only variable we're showing in this template us `unreadMessagesCount`.

Our `ChatNavBar` has `unreadMessagesCount` as an instance variable. This number will be set to the sum of unread messages in all threads.

Notice in our `constructor` we do three things:

1. `Inject` our store
2. Subscribe to any changes in the store
3. Call `this.updateState()`

We call `this.updateState()` after `subscribe` because we want to make sure this component is initialized with the most recent data. `subscribe` will only be called if something changes *after* this component is initialized.

`updateState()` is the most interesting function - we set `unreadMessagesCount` to the value of the function `getUnreadMessagesCount`. What is `getUnreadMessagesCount` and where did it come from?

`getUnreadMessagesCount` is a new concept called *selectors*.

## Redux Selectors

Thinking about our `AppState`, how might we go about getting the unread messages count? How about something like this:

```
1  // get the state
2  let state = this.store.getState();
3
4  // get the threads state
5  let threadsState = state.threads;
6
7  // get the entities from the threads
8  let threadsEntities = threadsState.entities;
9
10 // get all of the threads from state
11 let allThreads = Object.keys(threadsEntities)
12               .map((threadId) => entities[threadId]);
13
14 // iterate over all threads and ...
15 let unreadCount = allThreads.reduce(
16       (unreadCount: number, thread: Thread) => {
17         // foreach message in that thread
18         thread.messages.forEach((message: Message) => {
19           if (!message.isRead) {
20             // if it's unread, increment unread count
21             ++unreadCount;
22           }
23         });
24         return unreadCount;
25       },
26       0);
```

Should we put this logic in the `ChatNavBar` component? There's two problems with that approach:

1. This chunk of code reaches deep into our `AppState`. A better approach would be to co-locate this logic next to where the state itself is written.
2. What if we need the unread count somewhere else in the app? How could we share this logic?

Solving these problems is the idea behind *selectors*.

**Selectors are functions that take a part of the state and return a value.**

Let's take a look at how to make a few selectors.

## Threads Selectors

Let's start with an easy one. Say we have our `AppState` and we want to get the `ThreadsState`:

**code/redux/angular2-redux-chat/app/ts/reducers/ThreadsReducer.ts**

```
130 export const getThreadsState = (state): ThreadsState => state.threads;
```

Pretty easy, right? Here we're saying, given the top-level `AppState`, we can find the `ThreadsState` at `state.threads`.

Let's say that we want to get the current thread. We could do it like this:

```
1 const getCurrentThread = (state: AppState): Thread => {
2   let currentThreadId = state.threads.currentThreadId;
3   return state.threads.entities[currentThreadId];
4 }
```

For this small example, this selector works fine. But it's worth thinking about how we can make our selectors maintainable as the app grows. It would be nice if we could use selectors to query other selectors. It also would be nice to be able to specify a selector that has multiple selectors as a dependency.

This is what the reselect[109] library provides. With `reselect` we can create small, focused selectors and then combine them together into bigger functionality.

Let's look at how we will get the current thread using `createSelector` from `reselect`.

[109] https://github.com/reactjs/reselect#createselectorinputselectors--inputselectors-resultfunc

**code/redux/angular2-redux-chat/app/ts/reducers/ThreadsReducer.ts**

```
132 export const getThreadsEntities = createSelector(
133   getThreadsState,
134   ( state: ThreadsState ) => state.entities );
```

We start by writing `getThreadsEntities`. `getThreadsEntities` uses `createSelector` and passes two arguments:

1. `getThreadsState`, the selector we defined above and
2. A callback function which will receive *the value of the selector in #1* and return the value we want to select.

This might seem like a lot of overhead to call `state.entities`, but it sets us up for a much more maintainable selectors down the line. Let's look at `getCurrentThread` using `createSelector`:

**code/redux/angular2-redux-chat/app/ts/reducers/ThreadsReducer.ts**

```
155 export const getCurrentThread = createSelector(
156   getThreadsEntities,
157   getThreadsState,
158   ( entities: ThreadsEntities, state: ThreadsState ) =>
159     entities[state.currentThreadId] );
```

Notice here that we're citing **two** selectors as dependencies: `getThreadsEntities` and `getThreadsState` - when these selectors resolve they become the arguments to the callback function. We can then combine them together to return the selected thread.

## Unread Messages Count Selector

Now that we understand how selectors work, let's create a selector that will get the number of unread messages. If you look at our first attempt at unread messages above, we can see that each variable could instead become it's own selector (`getThreadsState`, `getThreadsEntities`, etc.)

Here's a selector that will get all `Threads`:

**code/redux/angular2-redux-chat/app/ts/reducers/ThreadsReducer.ts**

```
136 export const getAllThreads = createSelector(
137   getThreadsEntities,
138   ( entities: ThreadsEntities ) => Object.keys(entities)
139                         .map((threadId) => entities[threadId]));
```

And then given all of the threads, we can get the sum of the unread messages over all threads:

**code/redux/angular2-redux-chat/app/ts/reducers/ThreadsReducer.ts**

```
141 export const getUnreadMessagesCount = createSelector(
142   getAllThreads,
143   ( threads: Thread[] ) => threads.reduce(
144       (unreadCount: number, thread: Thread) => {
145         thread.messages.forEach((message: Message) => {
146           if (!message.isRead) {
147             ++unreadCount;
148           }
149         });
150         return unreadCount;
151       },
152       0));
```

Now that we have this selector, we can use it to get the number of unread messages in our `ChatNavBar` (and anywhere else in our app where we might need it).

## Building the `ChatThreads` Component

Next let's build our thread list in the `ChatThreads` component.

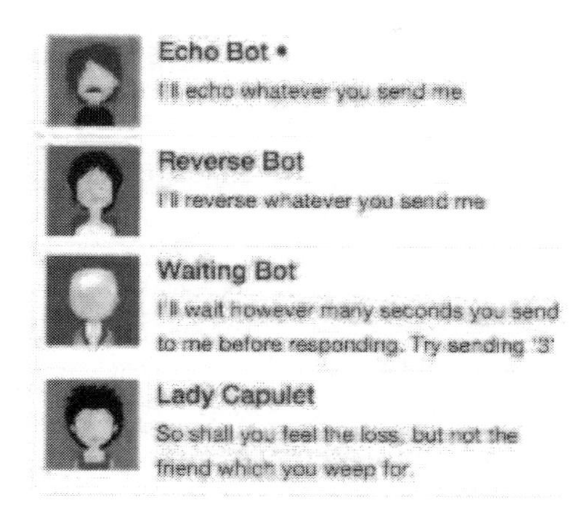


**Time Ordered List of Threads**

## `ChatThreads` Controller

Let's take a look at our component controller `ChatThreads` before we look at the template:

**code/redux/angular2-redux-chat/app/ts/containers/ChatThreads.ts**

```
48 export default class ChatThreads {
49   threads: Thread[];
50   currentThreadId: string;
51 
52   constructor(@Inject(AppStore) private store: Store<AppState>) {
53     store.subscribe(() => this.updateState());
54     this.updateState();
55   }
56 
57   updateState() {
58     let state = this.store.getState();
59 
60     // Store the threads list
61     this.threads = getAllThreads(state);
62 
63     // We want to mark the current thread as selected,
64     // so we store the currentThreadId as a value
65     this.currentThreadId = getCurrentThread(state).id;
66   }
67 
68   handleThreadClicked(thread: Thread) {
```

```
69      this.store.dispatch(ThreadActions.selectThread(thread));
70    }
71  }
```

We're storing two instance variables on this component:

- `threads` - the list of `Threads`
- `currentThreadId` - the current thread (conversation) that the user is participating in

In our `constructor` we keep a reference to the Redux store and subscribe to updates. When the store changes, we call `updateState()`.

`updateState()` keeps our instance variables in sync with the Redux store. Notice that we're using two selectors:

- `getAllThreads` and
- `getCurrentThread`

which keep their respective instance variables up to date.

The one new idea we've added is an event handler: `handleThreadClicked`. `handleThreadClicked` will dispatch the `selectThread` action. The idea here is that when a thread is clicked on, we'll tell our store to set this new thread as the selected thread and the rest of the application should update in turn.

### `ChatThreads template`

Let's look at the `ChatThreads template` and its configuration:

**code/redux/angular2-redux-chat/app/ts/containers/ChatThreads.ts**

```
30   */
31  @Component({
32    selector: 'chat-threads',
33    template: `
34    <!-- conversations -->
35    <div class="row">
36      <div class="conversation-wrap">
37        <chat-thread
38             *ngFor="let thread of threads"
39             [thread]="thread"
40             [selected]="thread.id === currentThreadId"
```

```
41            (onThreadSelected)="handleThreadClicked($event)">
42        </chat-thread>
43      </div>
44    </div>
45    `
```

In our template we're using ngFor to iterate over our threads. We're using a new directive to render the individual threads called ChatThread.

ChatThread is a *presentational* component. We **won't** be able to access the store in ChatThread, neither for fetching data nor dispatching actions. Instead, we're going to pass everything we need to this component through inputs and handle any interaction through outputs.

We'll look at the implementation of ChatThread next, but look at the inputs and outputs we have in this template first.

- We're sending the input [thread] with the individual thread
- On the input [selected] we're passing a *boolean* which indicates if this thread (thread.id) is the "current" thread (currentThreadId)
- If the thread is clicked, we will emit the output event (onThreadSelected) - when this happens we'll call handleThreadClicked() (which dispatches a thread selected event to the store).

Let's dig in to the ChatThread component.

## The Single ChatThread Component

The ChatThread component will be used to display a **single thread** in the list of threads. Remember that ChatThread is a *presentational component* - it doesn't manipulate any data that isn't given to it directly.

Because it is a presentational component, we're going to store it in the app/ts/components directory.

Here's the component controller code:

**code/redux/angular2-redux-chat/app/ts/components/ChatThread.ts**

```
43 export default class ChatThread {
44   thread: Thread;
45   selected: boolean;
46   onThreadSelected: EventEmitter<Thread>;
47 
48   constructor() {
49     this.onThreadSelected = new EventEmitter<Thread>();
50   }
51 
52   clicked(event: any): void {
53     this.onThreadSelected.emit(this.thread);
54     event.preventDefault();
55   }
56 }
```

The main thing to look at here is the `onThreadSelected` `EventEmitter`. If you haven't used `EventEmitters` much, the idea is that it's an implementation of the observer pattern. We use it as the "output channel" for this component - when we want to send data we call `onThreadSelected.emit` and pass whatever data we want along with it.

In this case, we want to emit the current thread as the argument to the `EventEmitter`. When this element is clicked, we will call `onThreadSelected.emit(this.thread)` which will trigger the callback in our parent (`ChatThreads`) component.

### `ChatThread` `@Component` and `template`

Here's the code for our `@Component` annotation and `template`:

**code/redux/angular2-redux-chat/app/ts/components/ChatThread.ts**

```
21 @Component({
22   inputs: ['thread', 'selected'],
23   selector: 'chat-thread',
24   outputs: ['onThreadSelected'],
25   template: `
26   <div class="media conversation">
27     <div class="pull-left">
28       <img class="media-object avatar"
29            src="{{thread.avatarSrc}}">
30     </div>
31     <div class="media-body">
```

```
32        <h5 class="media-heading contact-name">{{thread.name}}
33          <span *ngIf="selected">&bull;</span>
34        </h5>
35        <small class="message-preview">
36          {{thread.messages[thread.messages.length - 1].text}}
37        </small>
38      </div>
39      <a (click)="clicked($event)" class="div-link">Select</a>
40    </div>
41    `
```

Here is where we specify our `inputs` of `thread` and `selected`, as well as the output of `onThreadS-elected`.

Notice that in our view we've got some straight-forward bindings like `{{thread.avatarSrc}}`, `{{thread.name}}`. In the `message-preview` tag we've got the following:

```
1    {{ thread.messages[thread.messages.length - 1].text }}
```

This gets the last message in the thread and displays the text of that message. The idea is we are showing a preview of the most recent message in that thread.

We've got an `*ngIf` which will show the `•` symbol only if this is the selected thread.

Lastly, we're binding to the `(click)` event to call our `clicked()` handler. Notice that when we call `clicked` we're passing the argument `$event`. This is a special variable provided by Angular that describes the event. We use that in our `clicked` handler by calling `event.preventDefault();`. This makes sure that we don't navigate to a different page.

# Building the `ChatWindow` Component

The `ChatWindow` is the most complicated component in our app. Let's take it one section at a time:

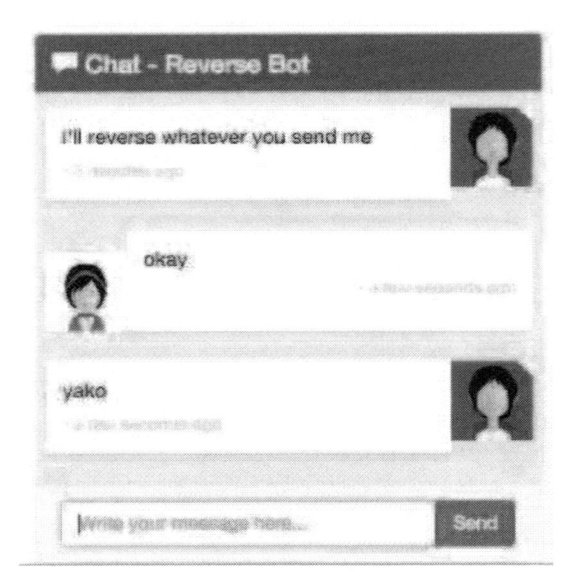


**The Chat Window**

Our `ChatWindow` class has three properties: `currentThread` (which holds `messages`), `draftMessage`, and `currentUser`:

**code/redux/angular2-redux-chat/app/ts/containers/ChatWindow.ts**

```
81  export default class ChatWindow {
82    currentThread: Thread;
83    draftMessage: { text: string };
84    currentUser: User;
```

Here's a diagram of where each one is used:

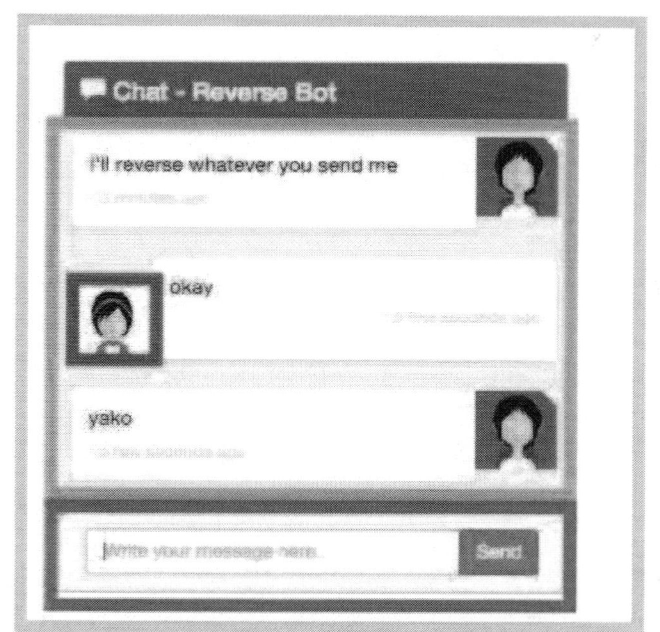


**Chat Window Properties**

In our `constructor` we're going to inject two things:

**code/redux/angular2-redux-chat/app/ts/containers/ChatWindow.ts**

```
86    constructor(@Inject(AppStore) private store: Store<AppState>,
87                private el: ElementRef) {
88      store.subscribe(() => this.updateState() );
89      this.updateState();
90      this.draftMessage = { text: '' };
91    }
```

The first is our Redux Store. The second, `el` is an `ElementRef` which we can use to get access to the host DOM element. We'll use that when we scroll to the bottom of the chat window when we create and receive new messages.

In our constructor we subscribe to our store, as we have in our other container components.

The next thing we do is to set a default `draftMessage` with an empty string for the `text`. We'll use the `draftMessage` to keep track of the input box as the user is typing their message.

#### `ChatWindow updateState()`

When the store changes we will update the instance variables for this component:

**code/redux/angular2-redux-chat/app/ts/containers/ChatWindow.ts**

```
93   updateState() {
94     let state = this.store.getState();
95     this.currentThread = getCurrentThread(state);
96     this.currentUser = getCurrentUser(state);
97     this.scrollToBottom();
98   }
```

Here we store the current thread and the current user. If a new message comes in, we also want to scroll to the bottom of the window. It's a bit coarse to call `scrollToBottom` here, but it's a simple way to make sure that the user doesn't have to scroll manually each time there is a new message (or they switch to a new thread).

#### ChatWindow scrollToBottom()

To scroll to the bottom of the chat window, we're going to use the `ElementRef el` that we saved in the constructor. To make this element scroll, we're going to set the `scrollTop` property of our host element:

**code/redux/angular2-redux-chat/app/ts/containers/ChatWindow.ts**

```
100   scrollToBottom(): void {
101     let scrollPane: any = this.el
102       .nativeElement.querySelector('.msg-container-base');
103     if (scrollPane) {
104       setTimeout(() => scrollPane.scrollTop = scrollPane.scrollHeight);
105     }
106   }
```

**Why do we have the `setTimeout`?**

If we call `scrollToBottom` immediately when we get a new message then what happens is we scroll to the bottom before the new message is rendered. By using a `setTimeout` we're telling Javascript that we want to run this function when it is finished with the current execution queue. This happens **after** the component is rendered, so it does what we want.

#### ChatWindow sendMessage

When we want to send a new message, we'll do it by taking:

- The current thread
- The current user
- The draft message text

And then dispatching a new `addMessage` action on the store. Here's what it looks like in code:

**code/redux/angular2-redux-chat/app/ts/containers/ChatWindow.ts**

```
108   sendMessage(): void {
109     this.store.dispatch(ThreadActions.addMessage(
110       this.currentThread,
111       {
112         author: this.currentUser,
113         isRead: true,
114         text: this.draftMessage.text
115       }
116     ));
117     this.draftMessage = { text: '' };
118   }
```

The `sendMessage` function above takes the `draftMessage`, sets the `author` and `thread` using our component properties. Every message we send has "been read" already (we wrote it) so we mark it as read.

After we dispatch the message, we create a new `Message`** and set that new `Message` to `this.draftMessage`. This will clear the search box, and by creating a new object we ensure we don't mutate the message that was sent to the store.

#### `ChatWindow` `onEnter`

In our view, we want to send the message in two scenarios

1. the user hits the "Send" button or
2. the user hits the Enter (or Return) key.

Let's define a function that will handle both events:

**code/redux/angular2-redux-chat/app/ts/containers/ChatWindow.ts**

```
120   onEnter(event: any): void {
121     this.sendMessage();
122     event.preventDefault();
123   }
```

We create this `onEnter` event handler as a separate function from `sendMessage` because `onEnter` will accept an `event` as an argument and then call `event.preventDefault()`. This way we *could* call `sendMessage` in scenarios other than in response to a browser event. In this case, we're not really calling `sendMessage` in any other situation, but I find that it's nice to separate the event handler from the function that 'does the work'.

That is, a `sendMessage` function that also 1. requires an event to be passed to it and 2. handles that event is feels like a function that may be handling too many concerns.

Now that we've handled the controller code, let's look at the `template`

#### ChatWindow template

We start our template by opening the panel tags: and showing the chat name in the header:

**code/redux/angular2-redux-chat/app/ts/containers/ChatWindow.ts**

```
33 @Component({
34   selector: 'chat-window',
35   template: `
36     <div class="chat-window-container">
37       <div class="chat-window">
38         <div class="panel-container">
39           <div class="panel panel-default">
40
41             <div class="panel-heading top-bar">
42               <div class="panel-title-container">
43                 <h3 class="panel-title">
44                   <span class="glyphicon glyphicon-comment"></span>
45                   Chat - {{currentThread.name}}
46                 </h3>
47               </div>
48               <div class="panel-buttons-container" >
49                 <!-- you could put minimize or close buttons here -->
50               </div>
51             </div>
```

```
52
53                <div class="panel-body msg-container-base">
```

Next we show the list of messages. Here we use ngFor to iterate over our list of messages. We'll describe the individual chat-message component in a minute.

**code/redux/angular2-redux-chat/app/ts/containers/ChatWindow.ts**

```
54                  <chat-message
55                       *ngFor="let message of currentThread.messages"
56                       [message]="message">
57                  </chat-message>
58                </div>
59
60                <div class="panel-footer">
```

Lastly we have the message input box and closing tags:

**code/redux/angular2-redux-chat/app/ts/containers/ChatWindow.ts**

```
61                  <div class="input-group">
62                    <input type="text"
63                           class="chat-input"
64                           placeholder="Write your message here..."
65                           (keydown.enter)="onEnter($event)"
66                           [(ngModel)]="draftMessage.text" />
67                    <span class="input-group-btn">
68                      <button class="btn-chat"
69                         (click)="onEnter($event)"
70                         >Send</button>
71                    </span>
72                  </div>
73                </div>
74
75              </div>
76            </div>
77          </div>
78        </div>
79      `
80  })
81  export default class ChatWindow {
```

The message input box is the most interesting part of this view, so let's talk about two interesting properties: 1. (keydown.enter) and 2. [(ngModel)].

## Handling keystrokes

Angular provides a straightforward way to handle keyboard actions: we bind to the event on an element. In this case, we're binding to `keydown.enter` which says if "Enter" is pressed, call the function in the expression, which in this case is `onEnter($event)`.

**code/redux/angular2-redux-chat/app/ts/containers/ChatWindow.ts**

```
63                    class="chat-input"
64                    placeholder="Write your message here..."
65                    (keydown.enter)="onEnter($event)"
66                    [(ngModel)]="draftMessage.text" />
67            <span class="input-group-btn">
```

## Using ngModel

As we've talked about before, we don't generally use two-way data binding as the crux of our data architecture (like we might have in Angular 1). This is particularly true when we're using Redux which is strictly a one-way data flow.

However it can be very useful to have a two-way binding between a component and its view. As long as the side-effects are kept local to the component, it can be a very convenient way to keep a component property in sync with the view.

In this case, we're establishing a two-way bind **between the value of the input tag and `draftMessage.text`**. That is, if we type into the `input` tag, `draftMessage.text` will automatically be set to the value of that `input`. Likewise, if we were to update `draftMessage.text` in our code, the value in the `input` tag would change in the view.

## Clicking "Send"

On our "Send" button we bind the `(click)` property to the `onEnter` function of our component:

**code/redux/angular2-redux-chat/app/ts/containers/ChatWindow.ts**

```
69                  (click)="onEnter($event)"
70                  >Send</button>
71            </span>
```

We're using the same `onEnter` function to handle the events which should send the draft message for both the button and hitting the enter button.

# The `ChatMessage` Component

Instead of putting the rendering code for each individual message in this component, instead we're going to create another *presentational component* `ChatMessage`.

Tip: If you're using `ngFor` that's a good indication you should create a new component.

Each `Message` is rendered by the `ChatMessage` component.

The `ChatMessage` Component

This component is relatively straightforward. The main logic here is rendering a slightly different view depending on if the message was authored by the current user. If the `Message` was **not** written by the current user, then we consider the message `incoming`.

## Setting `incoming`

Remember that each `ChatMessage` belongs to one `Message`. So in `ngOnInit` we will subscribe to the `currentUser` stream and set `incoming` depending on if this `Message` was written by the current user:

**code/redux/angular2-redux-chat/app/ts/components/ChatMessage.ts**

```
45 export default class ChatMessage implements OnInit {
46   message: Message;
47   incoming: boolean;
48
49   ngOnInit(): void {
50     this.incoming = !this.message.author.isClient;
51   }
52 }
```

### The `ChatMessage template`

In our `template` we have two interesting ideas:

1. the `FromNowPipe`
2. `[ngClass]`

First, here's the code:

**code/redux/angular2-redux-chat/app/ts/components/ChatMessage.ts**

```
19  */
20  @Component({
21   inputs: ['message'],
22   selector: 'chat-message',
23   template: `
24   <div class="msg-container"
25        [ngClass]="{'base-sent': !incoming, 'base-receive': incoming}">
26
27     <div class="avatar"
28          *ngIf="!incoming">
29       <img src="{{message.author.avatarSrc}}">
30     </div>
31
32     <div class="messages"
33       [ngClass]="{'msg-sent': !incoming, 'msg-receive': incoming}">
34       <p>{{message.text}}</p>
35       <p class="time">{{message.sender}} • {{message.sentAt | fromNow}}</p>
36     </div>
37
38     <div class="avatar"
```

```
39          *ngIf="incoming">
40        <img src="{{message.author.avatarSrc}}">
41      </div>
42    </div>
43    `
```

The `FromNowPipe` is a pipe that casts our `Messages` sent-at time to a human-readable "x seconds ago" message. You can see that we use it by: `{{message.sentAt | fromNow}}`

`FromNowPipe` uses the excellent `moment.js`[110] library. You can read the source of the `FromNowPipe` in `code/redux/angular2-redux-chat/app/ts/util/FromNowPipe.ts`

We also make extensive use of `ngClass` in this view. The idea is, when we say:

```
1     [ngClass]="{'msg-sent': !incoming, 'msg-receive': incoming}"
```

We're asking Angular to apply the `msg-receive` class if `incoming` is truthy (and apply `msg-sent` if `incoming` is falsey).

By using the `incoming` property, we're able to display incoming and outgoing messages differently.

## Summary

There we go, if we put them all together we've got a fully functional chat app!

[110] http://momentjs.com/

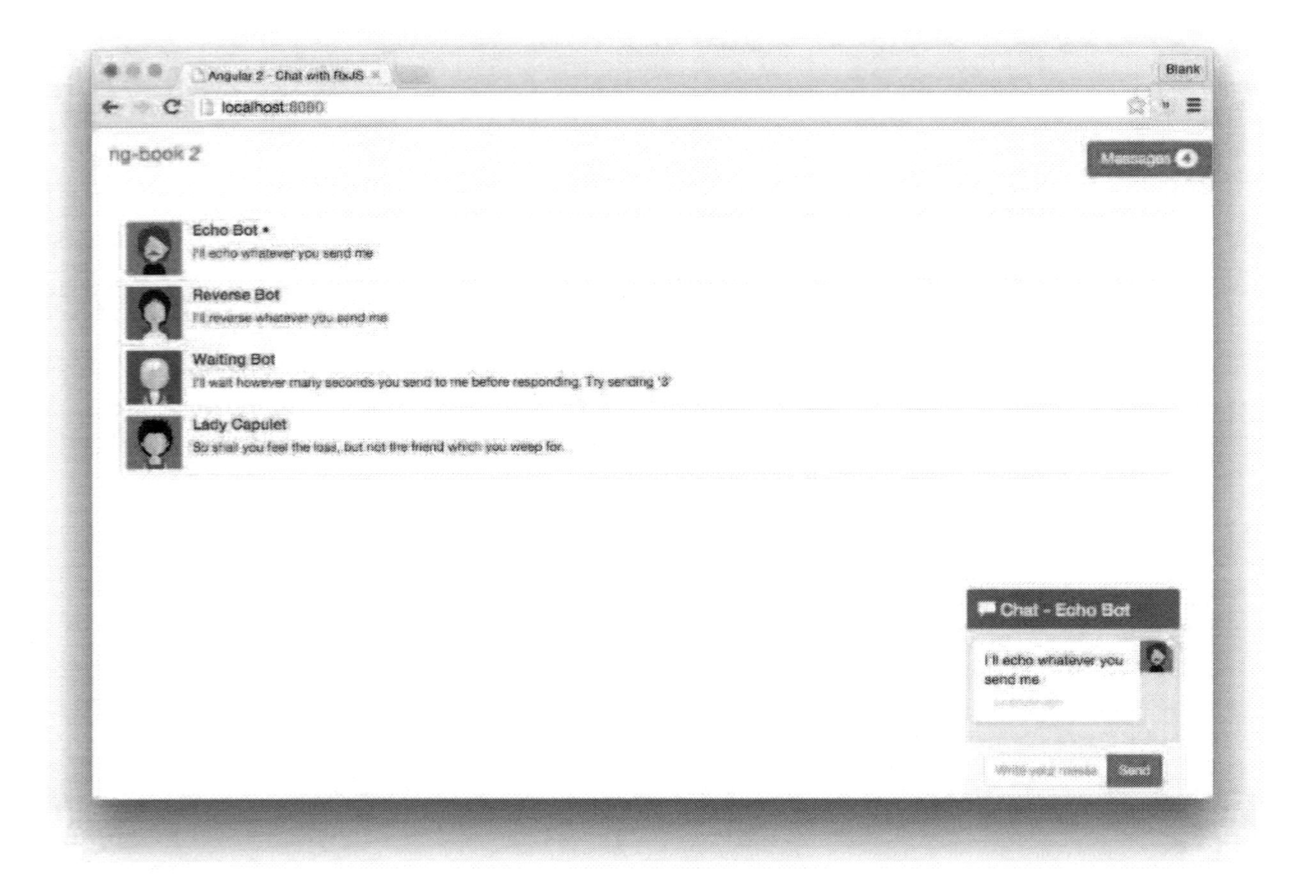


**Completed Chat Application**

If you checkout `code/redux/angular2-redux-chat/app/ts/ChatExampleData.ts` you'll see we've written a handful of bots for you that you can chat with. Checkout the code and try writing a few bots of your own!

# Advanced Components

Throughout this book, we've learned how to use Angular's built-in directives and how to create components of our own. In this chapter we'll take a deep dive into **advanced** features we can use to make components.

In this chapter we'll learn the following concepts:

- Styling components (with encapsulation)
- Modifying host DOM elements
- Modifying templates with *content projection*
- Accessing neighbor directives
- Using lifecycle hooks
- Detecting changes

## Styling

Angular provides a mechanism for specifying component-specific styles. CSS stands for *cascading style sheet*, but sometimes we **don't** want the cascade. Instead we want to provide styles for a component that won't leak out into the rest of our page.

Angular provides two attributes that allow us to define CSS classes for our component.

To define the style for our component, we use the View attribute `styles` to define in-line styles, or `styleUrls`, to use external CSS files. We can also declare those attributes directly on the `Component` annotation.

Let's write a component that uses inline styles:

**code/advanced_components/app/ts/styling/styling.ts**

```
5  @Component({
6    selector: 'inline-style',
7    styles: [`
8    .highlight {
9      border: 2px solid red;
10     background-color: yellow;
11     text-align: center;
12     margin-bottom: 20px;
13   }
```

```
  `],
  template: `
  <h4 class="ui horizontal divider header">
    Inline style example
  </h4>

  <div class="highlight">
    This uses component <code>styles</code>
    property
  </div>
  `
})
class InlineStyle {
}
```

In this example we defined the styles we want to use by declaring the `.highlight` class as an item on the array on the `styles` parameter.

Further on in the template we reference that class on the div using `<div class="highlight">`.

And the result is exactly what we expect - a div with a red border and yellow background:

Inline style example

This uses component `styles` property

**Example of component using styles**

Another way to declare CSS classes is to use the `styleUrls` property. This allows us to declare our CSS on an external file and just reference them from the component.

Let's write another component that uses this, but first let's create a file called `external.css` with the following class:

**code/advanced_components/app/ts/styling/external.css**

```
.highlight {
  border: 2px dotted red;
  text-align: center;
  margin-bottom: 20px;
}
```

Then we can write the code that references it:

**code/advanced_components/app/ts/styling/styling.ts**

```
29  @Component({
30    selector: 'external-style',
31    styleUrls: [externalCSSUrl],
32    template: `
33    <h4 class="ui horizontal divider header">
34      External style example
35    </h4>
36
37    <div class="highlight">
38      This uses component <code>styleUrls</code>
39      property
40    </div>
41    `
42  })
43  class ExternalStyle {
44  }
```

And when we load the page, we see our div with a dotted border:

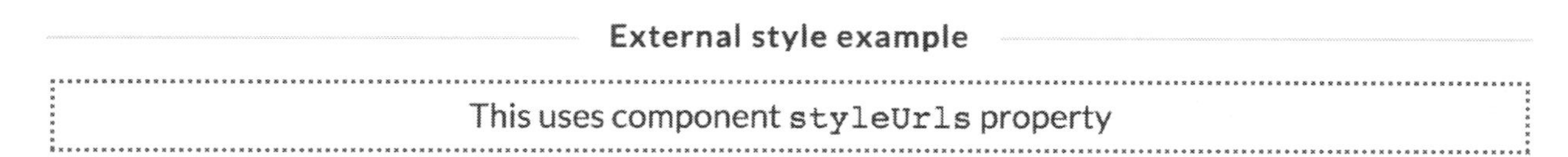


Example of component using styleUrls

## View (Style) Encapsulation

One interesting thing about this example is that both components define a class called `highlight` with different properties, but the attributes of one didn't leak into the other.

This happens because Angular styles are **encapsulated by the component context** by default. If we inspect the page and expand the `<head>`, we'll notice that Angular injected a `<style>` tag with our style:

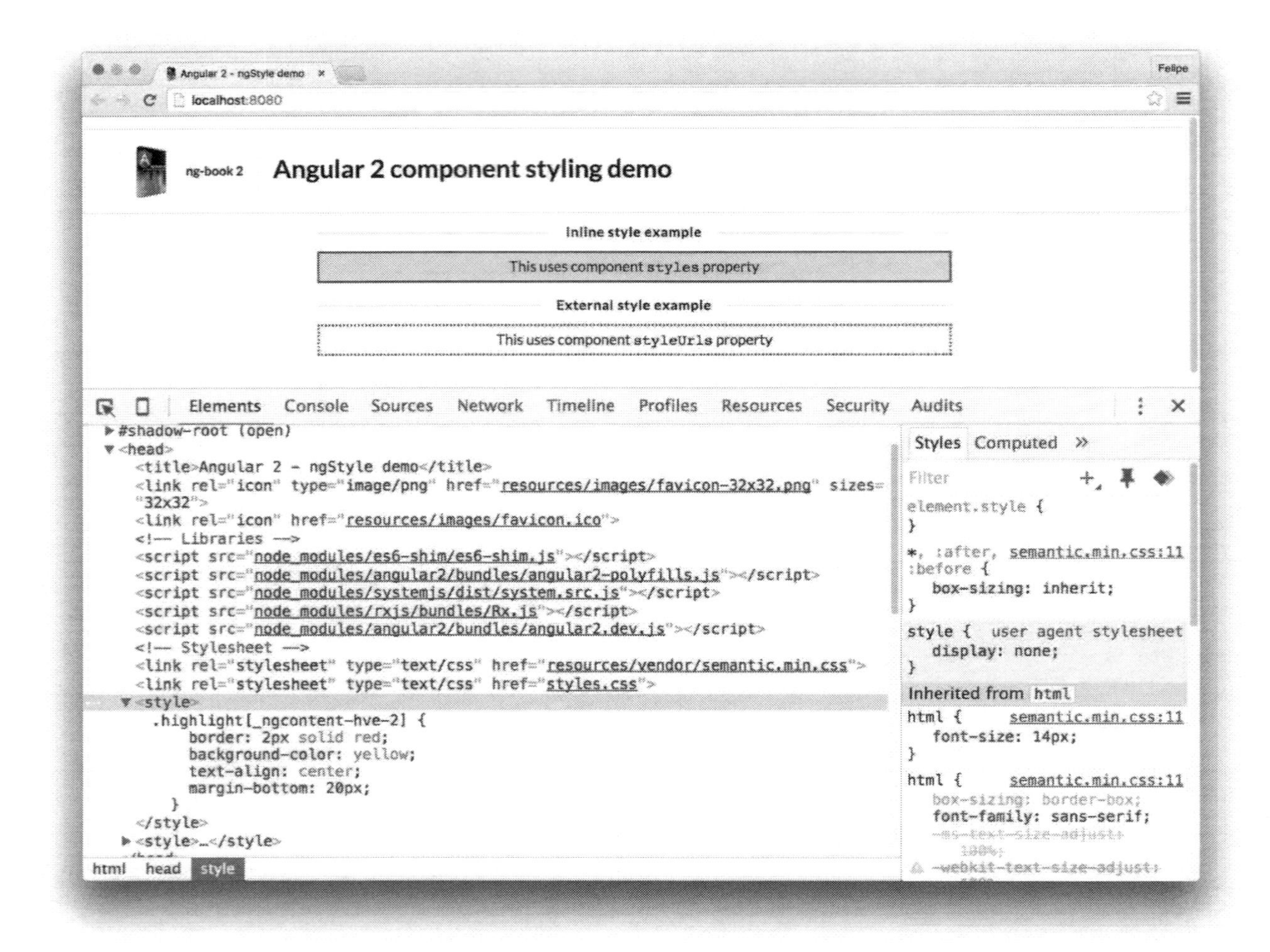


**Injected style**

You'll also notice that the CSS class has been scoped with _ngcontent-hve-2:

```
1 .highlight[\_ngcontent-hve-2] {
2   border: 2px solid red;
3   background-color: yellow;
4   text-align: center;
5   margin-bottom: 20px;
6 }
```

And if we check how our `<div>` is rendered, you'll find that _ng-content-hve-2 was added:

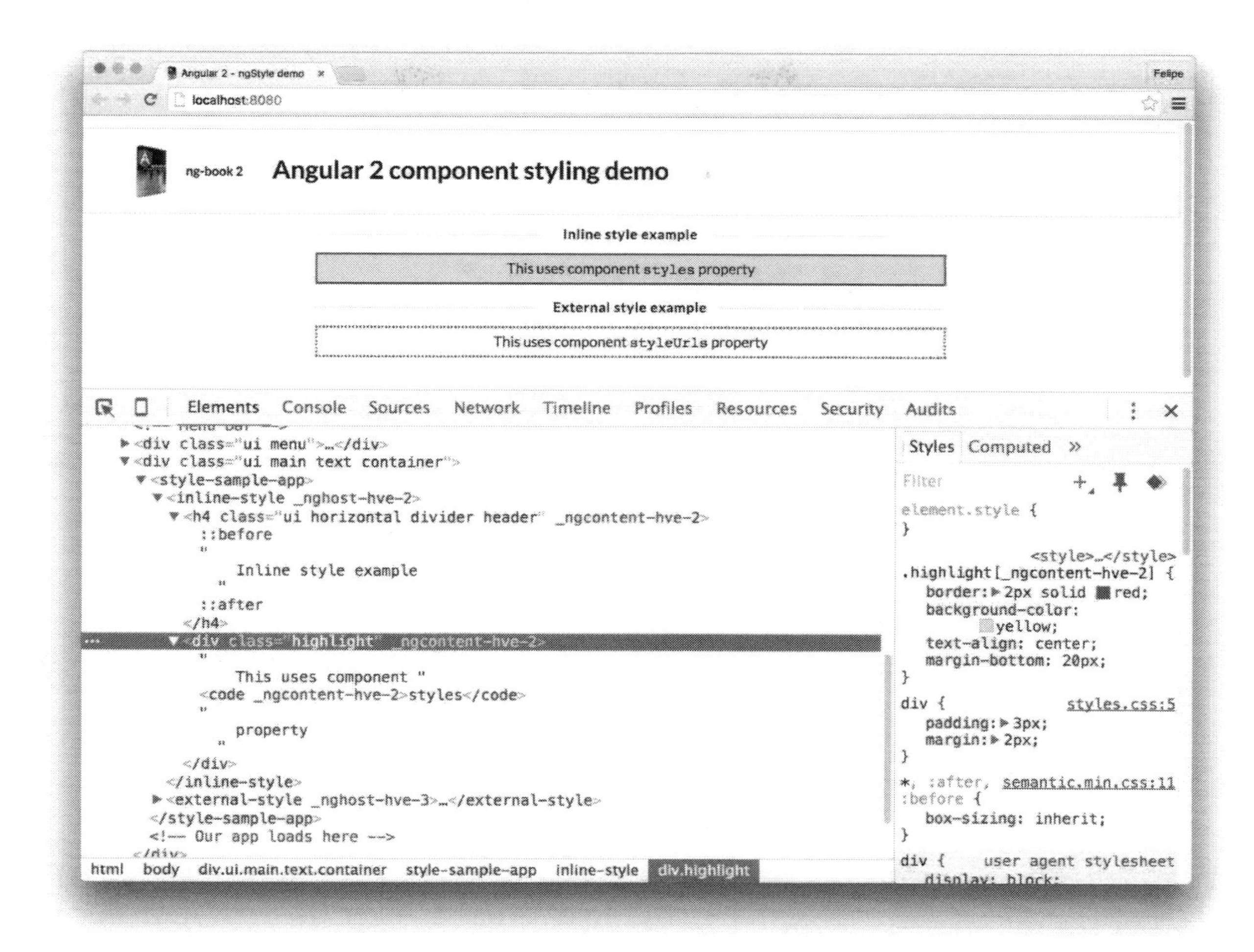


**Injected style**

The same thing happens for our external style:

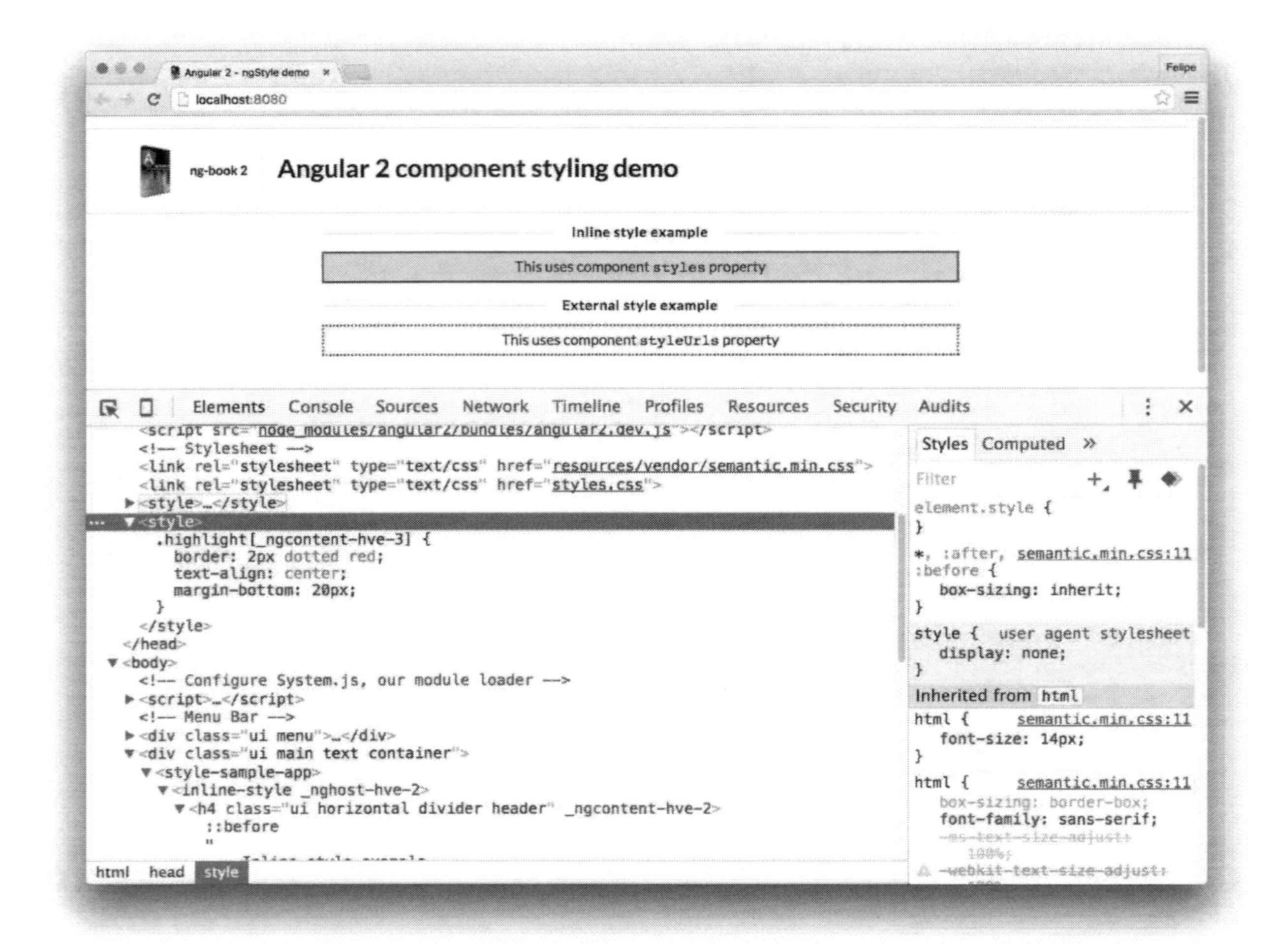


**External style**

and:

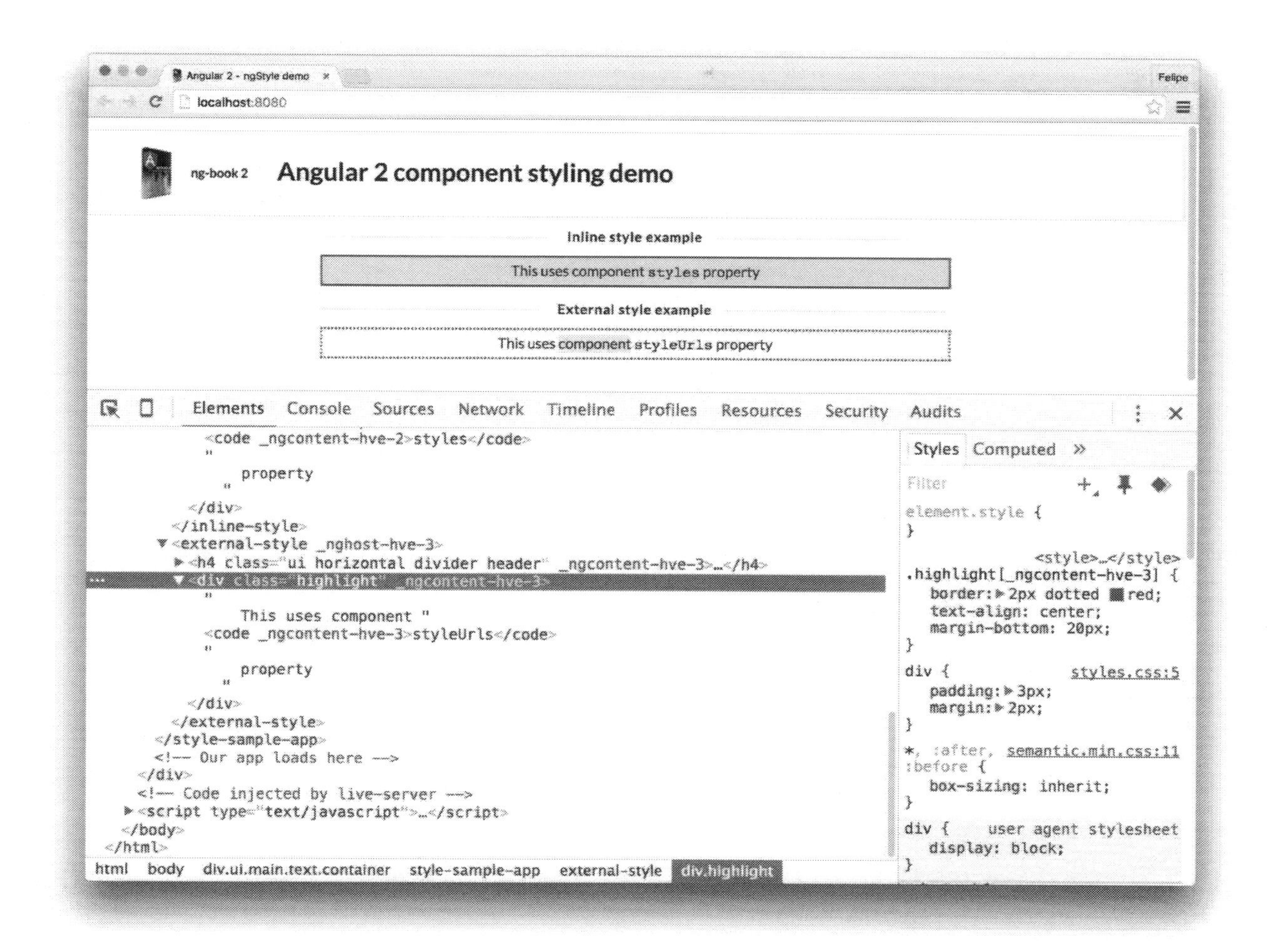


**External style**

Angular allows us to change this behavior, by using the `encapsulation` property.

This property can have the following values, defined by the `ViewEncapsulation` enum:

- **Emulated** - this is the default option and it will encapsulate the styles using the technique we just explained above
- **Native** - with this option, Angular will use the Shadow DOM (more on this below)
- **None** - with this option set, Angular won't encapsulate the styles at all, allowing them to leak to other elements on the page

## Shadow DOM Encapsulation

You might be wondering: what is the point of using the Shadow DOM? By using the Shadow DOM the component we **use a unique DOM tree that is hidden from the other elements on the page**. This allows styles defined within that element to be invisible to the rest of the page.

For a deep dive into Shadow DOM, please check this guide by Eric Bidelman[111].

Let's create another component that uses the **Native** encapsulation (Shadow DOM) to understand how this works:

**code/advanced_components/app/ts/styling/styling.ts**

```
46  @Component({
47    selector: `native-encapsulation`,
48    styles: [`
49    .highlight {
50      text-align: center;
51      border: 2px solid black;
52      border-radius: 3px;
53      margin-botton: 20px;
54    }`],
55    template: `
56    <h4 class="ui horizontal divider header">
57      Native encapsulation example
58    </h4>
59
60    <div class="highlight">
61      This component uses <code>ViewEncapsulation.Native</code>
62    </div>
63    `,
64    encapsulation: ViewEncapsulation.Native
65  })
66  class NativeEncapsulation {
67  }
```

In this case, if we inspect the source code, we'll see:

[111] http://www.html5rocks.com/en/tutorials/webcomponents/shadowdom/

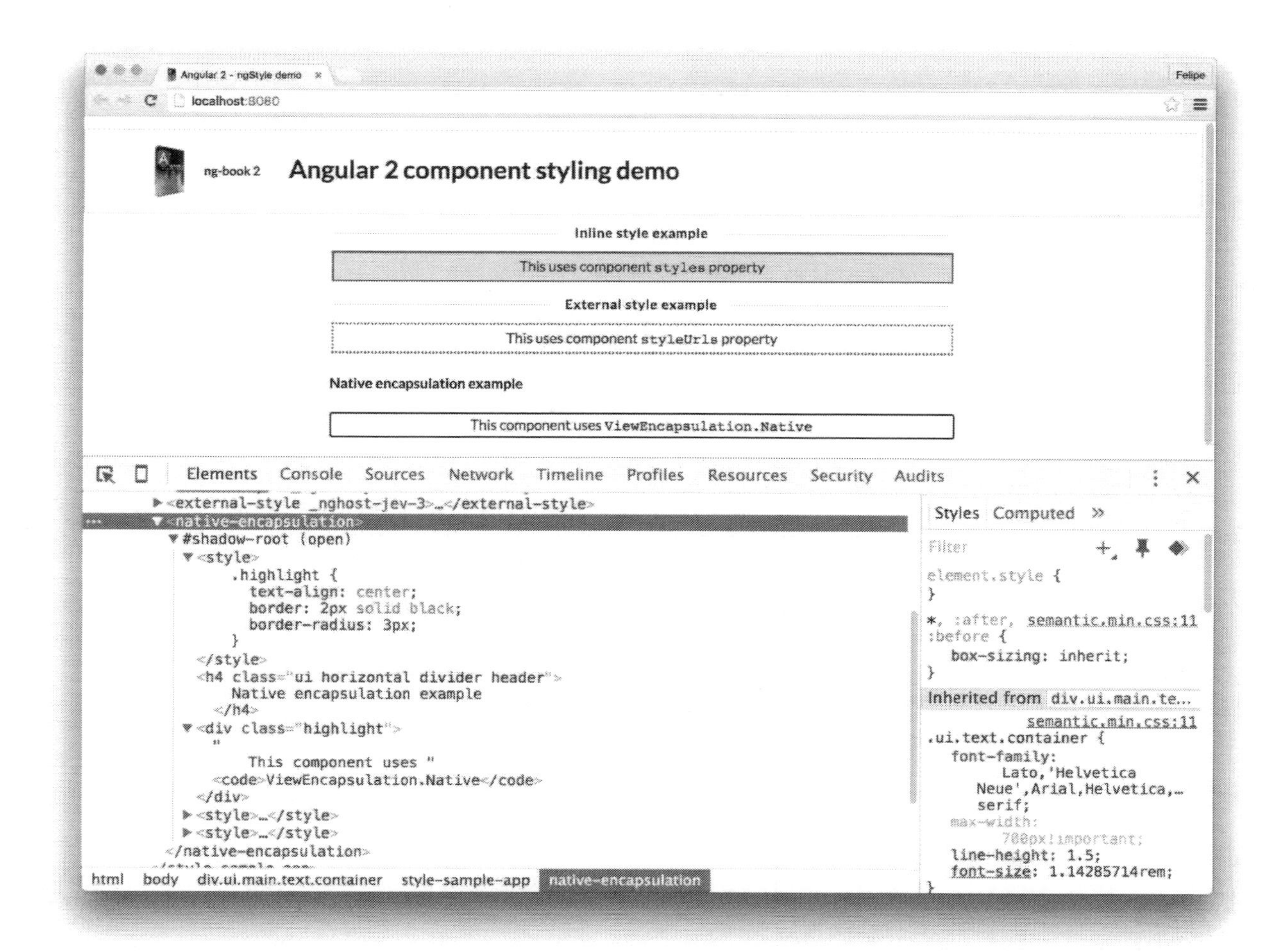


**Native encapsulation**

Everything inside the `#shadow-root` element has been encapsulated and isolated from the rest of the page.

## No Encapsulation

Finally, if we create a component that specifies `ViewEncapsulation.None`, no style encapsulation will be added:

**code/advanced_components/app/ts/styling/styling.ts**

```
69 @Component({
70   selector: `no-encapsulation`,
71   styles: [`
72   .highlight {
73     border: 2px dashed red;
74     text-align: center;
75     margin-bottom: 20px;
76   }
77   `],
78   template: `
79   <h4 class="ui horizontal divider header">
80     No encapsulation example
81   </h4>
82 
83   <div class="highlight">
84     This component uses <code>ViewEncapsulation.None</code>
85   </div>
86   `,
87   encapsulation: ViewEncapsulation.None
88 })
89 class NoEncapsulation {
90 }
```

When we inspect the element:

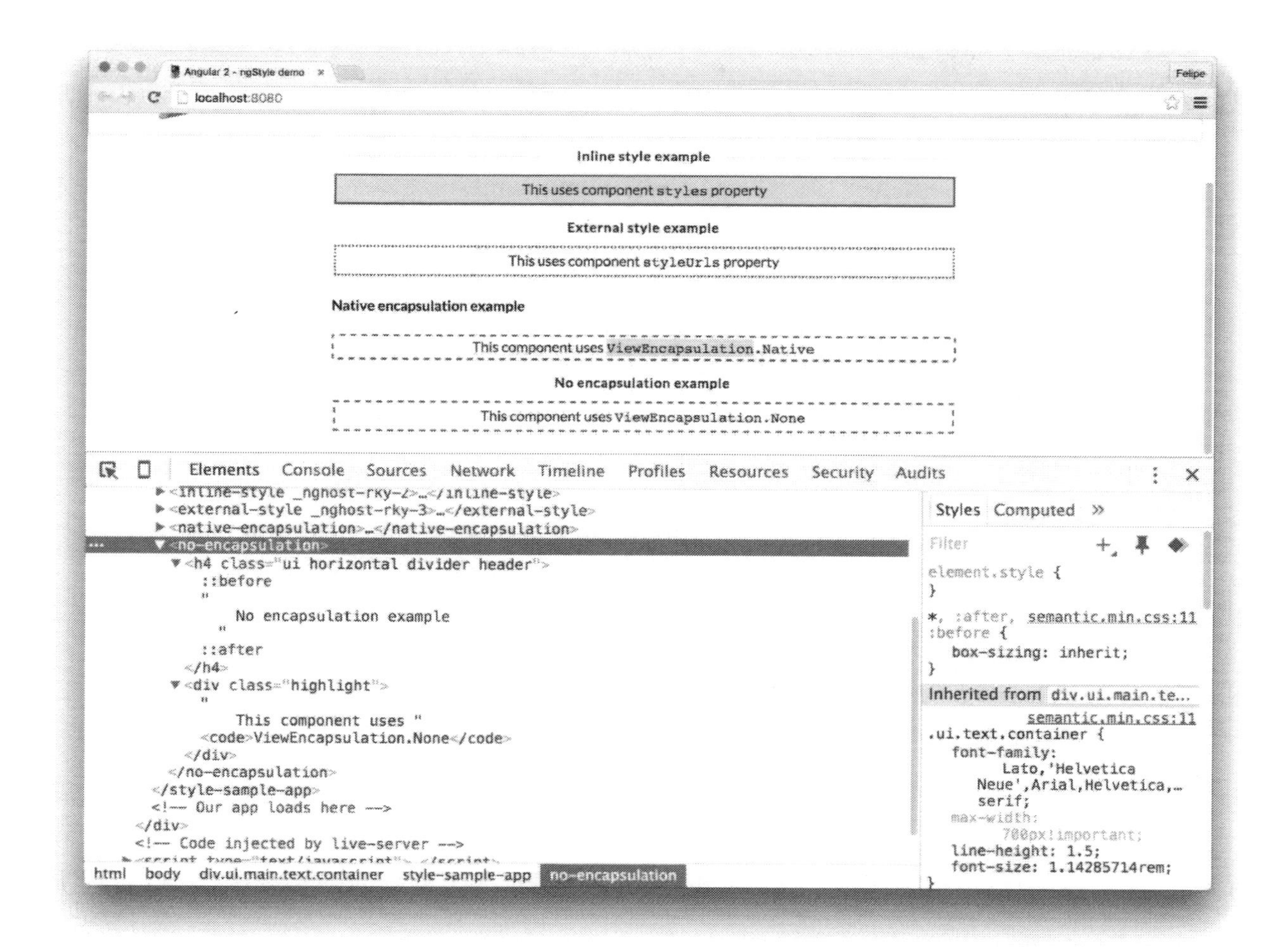


**No encapsulation**

We can see that nothing was injected on the HTML. Also on the header we can find that the `<style>` tag was also injected exactly like we defined on the `styles` parameter:

```
1  .highlight {
2    border: 2px dashed red;
3    text-align: center;
4    margin-bottom: 20px;
5  }
```

One side-effect of using `ViewEncapsulation.None` is that, since we don't have any encapsulation, this style "leaks" into other components. If we check the picture above, the `ViewEncapsulation.Native` component style was affected by this new component's style. But sometimes this can be exactly what you want.

You can comment out the `<no-encapsulation></no-encapsulation>` code on the `StyleSampleApp` template to see the difference.

# Creating a Popup - Referencing and Modifying Host Elements

The *host element* is the element to which the directive or component is bound. Sometimes we have a component that needs to attach markup or behavior to its host element.

In this example, we're going to create a `Popup` directive that will attach behavior to its host element which will display a message when clicked.

### Components vs. Directives - What's the difference?

Components and directives are closely related, but they are slightly different.

You may have heard that "components are directives with a view". This isn't exactly true. Components come with functionality that makes it easy to add views, but directives can have views too. In fact, **components are implemented with directives**.

One great example of a directive that renders a conditional view is `NgIf`.

But we can attach behaviors to an element **without a template** by using a *directive*.

Think of it this way: Components are Directives and Components always have a view. Directives may or may not have a view.

If you choose to render a view (a template) in your Directive, you can have more control over how that template is rendered. We'll talk more about how to use that control later in this chapter.

## Popup Structure

Now let's write our first directive. We want this directive to **show an alert when we click a DOM element** that includes the attribute `popup`. The message displayed will be identified by the element's `message` attribute.

Here's what we want it to look like:

```
1   <element popup message="Some message"></element>
```

In order to make this directive work, there are a couple of things we need to do:

- receive the `message` attribute *from* the host
- be notified when the host element is clicked

Let's start coding our directive:

**code/advanced_components/app/ts/host/steps/host_01.ts**

```
4   @Directive({
5     selector: '[popup]'
6   })
7   class Popup {
8     constructor() {
9       console.log('Directive bound');
10    }
11  }
```

We use the `Directive` annotation and set the `selector1 parameter to` [popup]‘. This will make this directive bind to any elements that define the popup attribute.

Now let’s create an app that has an element that has the `popup` attribute:

**code/advanced_components/app/ts/host/steps/host_01.ts**

```
13  @Component({
14    selector: 'host-sample-app',
15    template: `
16    <div class="ui message" popup>
17      <div class="header">
18        Learning Directives
19      </div>
20
21      <p>
22        This should use our Popup diretive
23      </p>
24    </div>
25    `
26  })
27  export class HostSampleApp1 {
28  }
```

When we run this application, we expect the message `Directive bound` to be logged on the console, indicating we have successfully bound to the first `<div>` in our template:

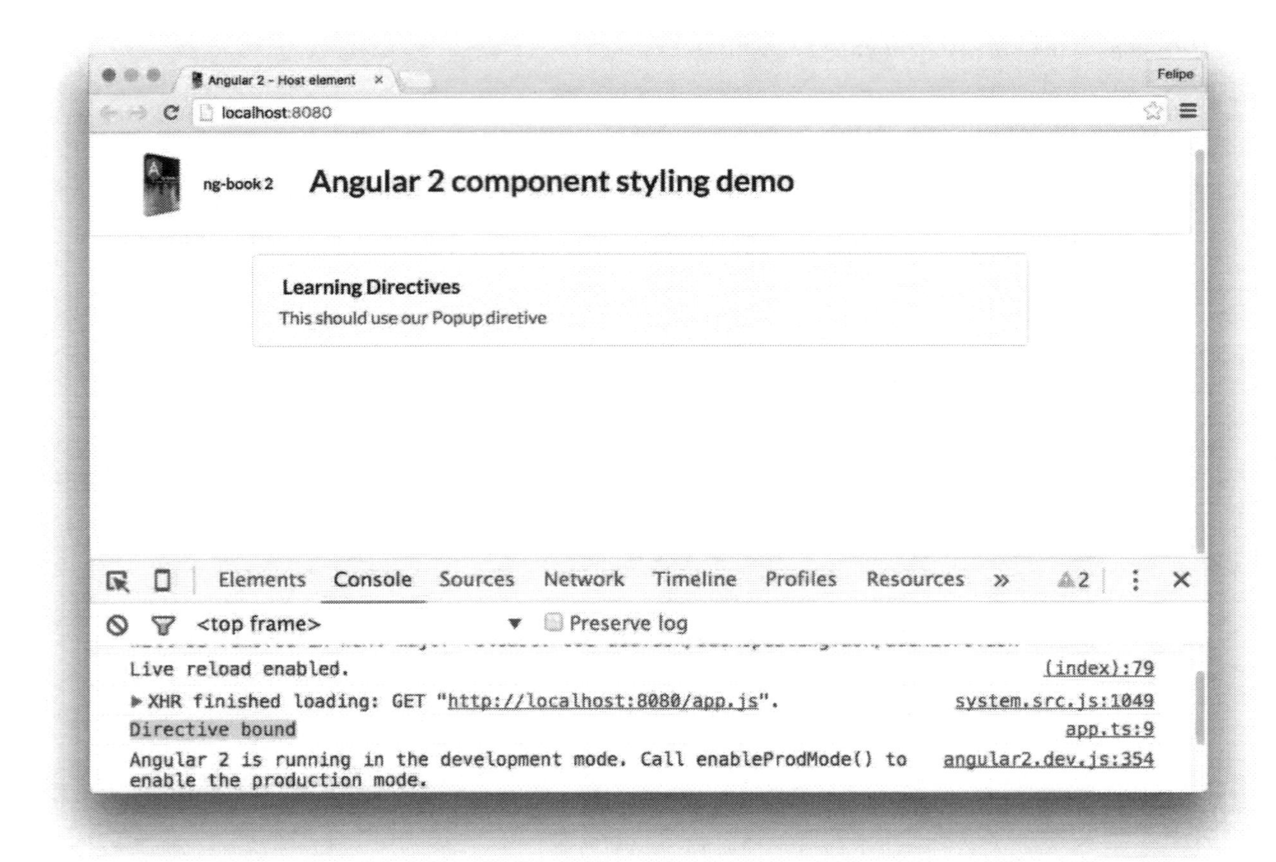


**Binding to host element**

## Using `ElementRef`

If we want to learn more about the host element a directive is bound to, we can use the built-in `ElementRef` class.

This class holds the information about a given Angular element, including the native DOM element using the `nativeElement` property.

In order to see the elements our directive is binding to, we can change our directive constructor to receive the `ElementRef` and log it to the console:

**code/advanced_components/app/ts/host/steps/host_02.ts**

```
 4  @Directive({
 5    selector: '[popup]'
 6  })
 7  class Popup {
 8    constructor(_elementRef: ElementRef) {
 9      console.log(_elementRef);
10    }
11  }
```

We can also add a second element to the page that uses our directive, so we can see two different `ElementRefs` logged to the console:

**code/advanced_components/app/ts/host/steps/host_02.ts**

```
13  @Component({
14    selector: 'host-sample-app',
15    template: `
16    <div class="ui message" popup>
17      <div class="header">
18        Learning Directives
19      </div>
20
21      <p>
22        This should use our Popup diretive
23      </p>
24    </div>
25
26    <i class="alarm icon" popup></i>
27    `
28  })
29  export class HostSampleApp2 {
30  }
```

When we run our app now, we can see two different ElementRefs: one with `div.ui.message` and the other with `i.alarm.icon`. This means that the directive was successfully bound to two different host elements:

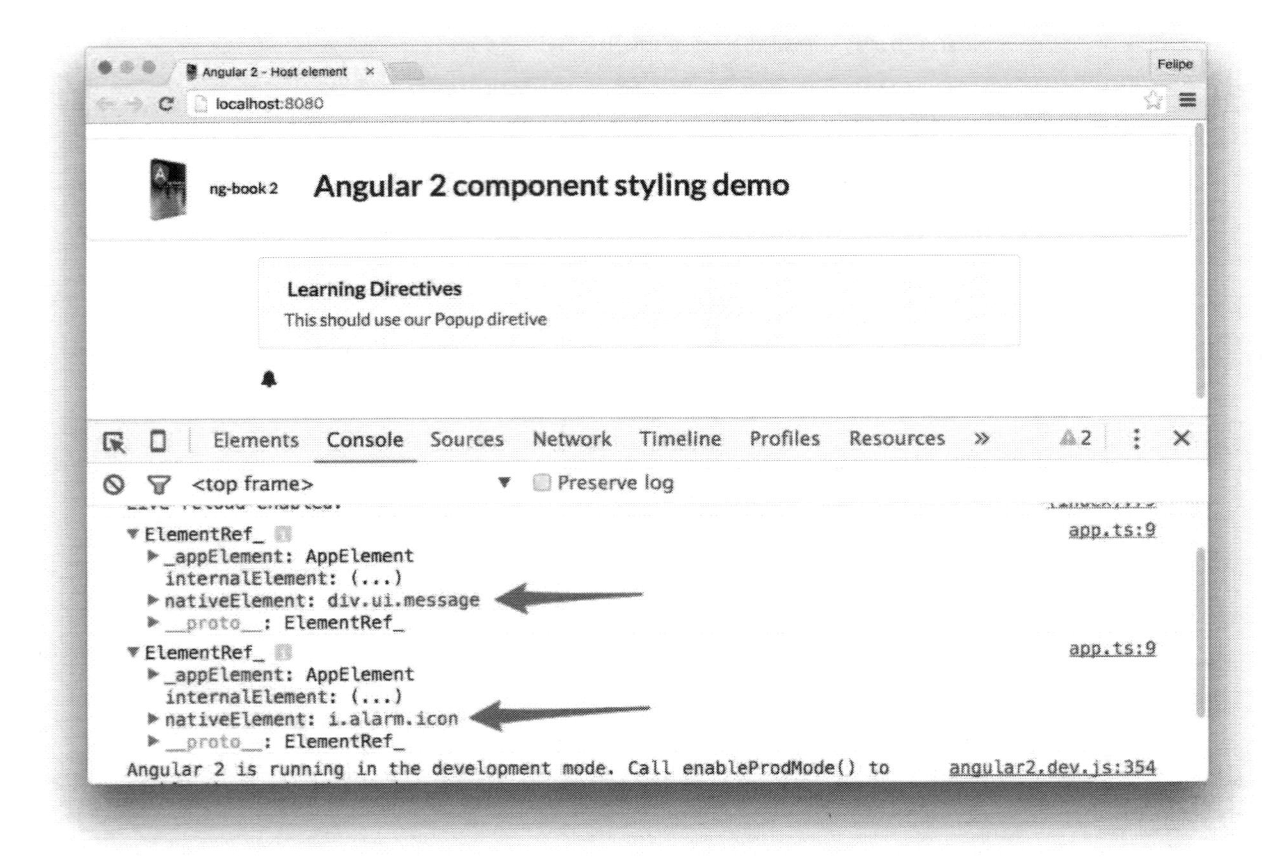


**ElementRefs**

## Binding to the `host`

Moving on, our next goal is to do something when the host element is clicked.

We learned before that the way we bind events in elements in Angular is using the `(event)` syntax.

In order to bind events of the host element, we must do something very similar, but using the `host` attribute of the directive. **The `host` attribute allows a directive to change the attributes and behaviors of its host element**.

We also want the host element to define what message will pop up when the element is clicked, using the `message` attribute.

In order to do that we use something we've used many times before: we add an `inputs` attribute to the directive.

Here's what our directive annotation looks like with those additions:

**code/advanced_components/app/ts/host/steps/host_03.ts**

```
 4  @Directive({
 5    selector: '[popup]',
 6    inputs: ['message'],
 7    host: {
 8      '(click)': 'displayMessage()'
 9    }
10  })
```

We're saying that we expect to receive an input called `message` and that when the host element is clicked we'll call the directive's `displayMessage` method.

We need to change our `Popup` class code by:

1. Adding a new `message` field to receive the input and
2. Creating the `displayMessage` function which will display the message the host element defines

Here's how we do it:

**code/advanced_components/app/ts/host/steps/host_03.ts**

```
11  class Popup {
12    message: String;
13  
14    constructor(_elementRef: ElementRef) {
15      console.log(_elementRef);
16    }
17  
18    displayMessage(): void {
19      alert(this.message);
20    }
21  }
```

And finally, we need to change our app template a bit to add the message we want displayed for each element:

**code/advanced_components/app/ts/host/steps/host_03.ts**

```
23 @Component({
24   selector: 'host-sample-app',
25   template: `
26   <div class="ui message" popup
27        message="Clicked the message">
28     <div class="header">
29       Learning Directives
30     </div>
31 
32     <p>
33       This should use our Popup diretive
34     </p>
35   </div>
36 
37   <i class="alarm icon" popup
38      message="Clicked the alarm icon"></i>
39   `
40 })
41 export class HostSampleApp3 {
42 }
```

Notice that we use the popup directive twice, and we pass a different message each time we use it. This means when we run the app, we're able to click either on the message or on the alarm icon, and we'll see different messages:

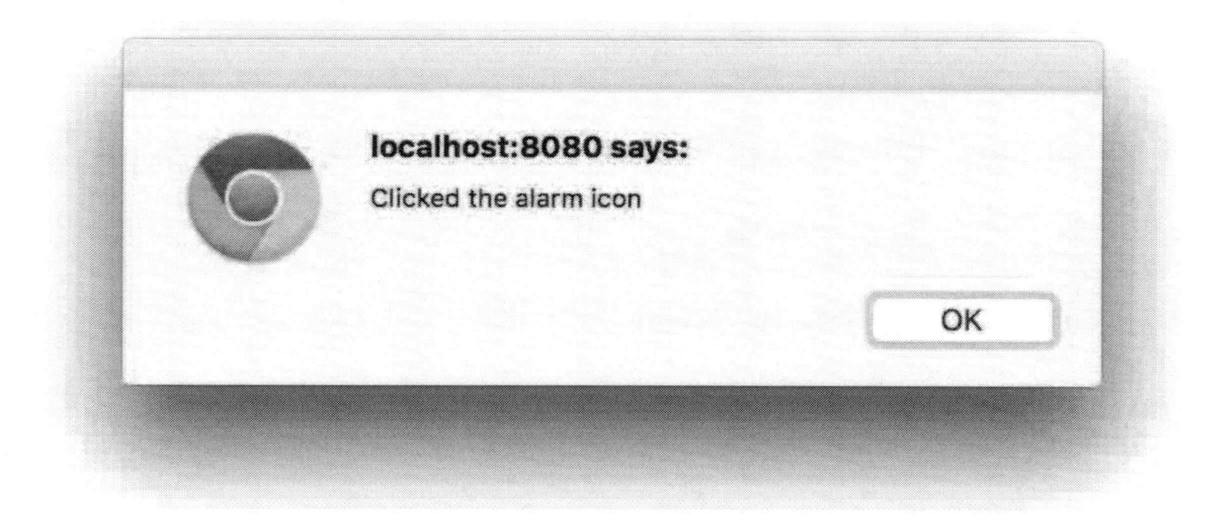


Popup 1

Popup 2

## Adding a Button using `exportAs`

Now let's say we have a new requirement: we want to trigger the alert manually by clicking a button. How could we trigger the popup message from **outside** the host element?

In order to achieve this, we need to make the **directive available from elsewhere in the template**.

As we discussed in previous chapters, the way to reference a component is by using **template variables**. We can reference directives the same way.

In order to give the templates a reference to a directive we use the `exportAt` attribute. This will allow the host element (or a child of the host element) to define a template variable that references the directive using the `#var="exportName"` syntax.

Let's add the `exportAs` attribute to our directive:

**code/advanced_components/app/ts/host/steps/host_04.ts**

```
4  @Directive({
5    selector: '[popup]',
6    inputs: ['message'],
7    exportAs: 'popup',
8    host: {
9      '(click)': 'displayMessage()'
10   }
11 })
12 class Popup {
13   message: String;
14
15   constructor(_elementRef: ElementRef) {
16     console.log(_elementRef);
17   }
18
19   displayMessage(): void {
20     alert(this.message);
21   }
22 }
```

And now we need to change the two elements to export the template variable:

**code/advanced_components/app/ts/host/steps/host_04.ts**

```
26   template: `
27   <div class="ui message" popup #popup1="popup"
28        message="Clicked the message">
29     <div class="header">
30       Learning Directives
31     </div>
32
33     <p>
34       This should use our Popup diretive
35     </p>
```

```
36    </div>
37
38    <i class="alarm icon" popup #p2="popup"
39       message="Clicked the alarm icon"></i>
```

See that we used the template var `#p1` for the `div.message` and `#p2` for the `icon`.

Now let's add two buttons, one to trigger each popup:

**code/advanced_components/app/ts/host/steps/host_04.ts**

```
40    <div style="margin-top: 20px;">
41      <button (click)="popup1.displayMessage()" class="ui button">
42        Display popup for message element
43      </button>
44
45      <button (click)="p2.displayMessage()" class="ui button">
46        Display popup for alarm icon
47      </button>
48    </div>
```

Now reload the page and click each of the buttons and each message will appear as expected.

## Creating a Message Pane with Content Projection

Sometimes when we are creating components we want to pass inner markup as an argument to the component. This technique is called *content projection*. The idea is that it lets us specify a bit of markup that will be expanded into a bigger template.

Angular 1 digged deep in the dictionary and called this *transclusion*.

Let's create a new directive that will render a nicely styled message like this:

**Learning Directives**

This should use our Popup diretive

**Popup 1**

Our goal is to write markup like this:

```
<div message header="My Message">
  This is the content of the message
</div>
```

Which will render into the more complicated HTML like:

```
<div class="ui message">
  <div class="header">
    My Message
  </div>

  <p>
    This is the content of the message
  </p>
</div>
```

We have two challenges here: we need to change the host element `<div>` to add the `ui` and `message` CSS classes, and we need to add the div's contents to a specific place in our markup.

## Changing the `host` CSS

To add attributes to the host element, we use the same attribute we used to add events to the host element: the `host` attribute. But now, instead of using the `(event)` notation, we define attribute names and attribute values. In our case using:

```
host: { 'class': 'ui message' }
```

Modified the host element, adding those to classes to the `class` attribute.

## Using `ng-content`

Our next challenge is to include the original host element children in a specific part of a view. To do that, we use the `ng-content` directive.

Since this directive needs a template, let's use a component instead and write the following code:

**code/advanced_components/app/ts/content-projection/content-projection.ts**

```
 7 @Component({
 8   selector: '[message]',
 9   inputs: ['header'],
10   host: {
11     'class': 'ui message'
12   },
13   template: `
14     <div class="header">
15       {{ header }}
16     </div>
17     <p>
18       <ng-content></ng-content>
19     </p>
20   `
21 })
22 export class Message {
23   header: string;
24 
25   ngOnInit(): void {
26     console.log('header', this.header);
27   }
28 }
```

A few highlights:

- We use the `inputs` attribute to indicate we want to receive a `header` attribute, set on the host element
- We set the host element's `class` attribute to `ui message` using the `host` attribute of our component
- We use `<ng-content></ng-content>` to project the host element's children into a specific location of our template

When we open the app in the browser and inspect the message div, we see it worked exactly like we planned:

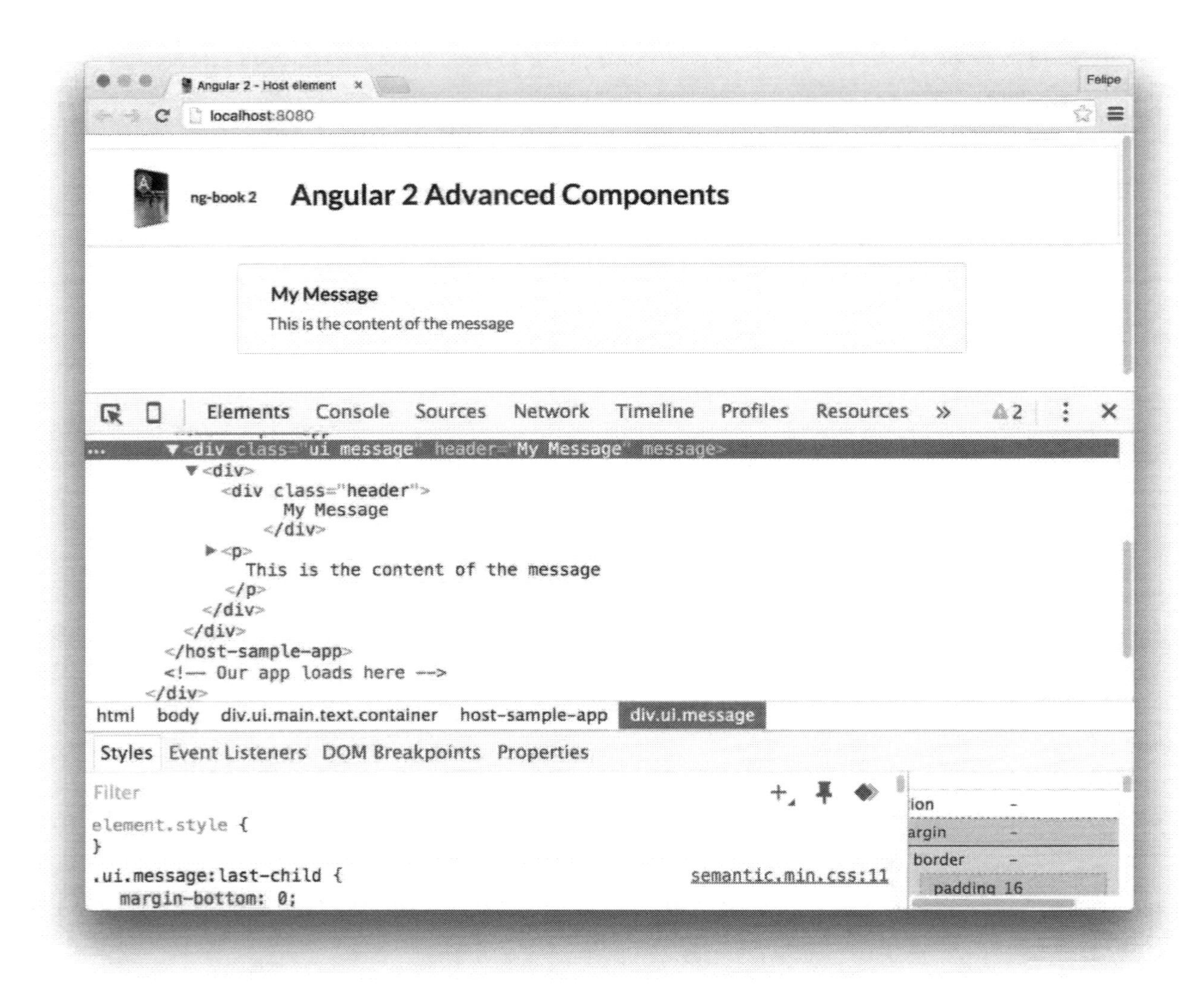


**projected content**

# Querying Neighbor Directives - Writing Tabs

It's great when you can create a component that fully encapsulates its own behavior.

However, as a component grows in features, it might make sense to split it up into several smaller components that work together.

A great example of components that work together is a tab pane that has multiple tabs. The tab panel or *tab set*, as it's usually called, is composed of multiple *tabs*. In this scenario we have a parent component (the tabset) and multiple child components (the tabs). The tabset and the tabs don't make sense separately, but putting all of the logic in one component is cumbersome. So in this example, we're going to cover how to make separate components that work together.

Let's start writing those components in a way that we'll be able to use the following markup:

```
1 <tabset>
2   <tab title="Tab 1">Tab 1</tab>
3   <tab title="Tab 2">Tab 2</tab>
4   ...
5 </tabset>
```

We're going to use Semantic UI Tab component[112] to render the tabs.

## Tab Component

Let's start by writing the Tab component:

**code/advanced_components/app/ts/tabs/tabs.ts**

```
11 @Component({
12   selector: 'tab',
13   inputs: ['title'],
14   template: `
15   <div class="ui bottom attached tab segment"
16        [class.active]="active">
17
18     <ng-content></ng-content>
19
20   </div>
21   `
22 })
23 class Tab {
24   title: string;
25   active: boolean = false;
26   name: string;
27 }
```

There are not many new concepts here. We're declaring a component that will use the `tab` selector, and it will allow a `title` input to be set.

Then we're rendering a `<div>` and using the content projection concept we learned on the previous section to inline the contents of the `<tab>` directive inside the **div**.

Next we declare 3 properties on our components: *title*, *active* and *name*. One thing to notice is the `@Input('title')` annotation we added to the `title` property. This annotation is a way to ask Angular to automatically bind the value of the *input* `title` into the *property* `title`.

[112]http://semantic-ui.com/modules/tab.html#/examples

## Tabset Component

Now let's move on to the Tabset component that will be used to wrap the tabs:

**code/advanced_components/app/ts/tabs/tabs.ts**

```
29 @Component({
30   selector: 'tabset',
31   template: `
32   <div class="ui top attached tabular menu">
33     <a *ngFor="let tab of tabs"
34        class="item"
35        [class.active]="tab.active"
36        (click)="setActive(tab)">
37 
38       {{ tab.title }}
39 
40     </a>
41   </div>
42   <ng-content></ng-content>
43   `
44 })
45 class Tabset implements AfterContentInit {
46   @ContentChildren(Tab) tabs: QueryList<Tab>;
47 
48   constructor() {
49   }
50 
51   ngAfterContentInit() {
52     this.tabs.toArray()[0].active = true;
53   }
54 
55   setActive(tab: Tab) {
56     this.tabs.toArray().forEach((t) => t.active = false);
57     tab.active = true;
58   }
59 }
```

Let's break down the implementation so we can learn about the new concepts it introduces.

### Tabset @Component Annotation

The @Component section doesn't have many new ideas. We're using the <tabset> tab as our selector.

The template itself uses `ngFor` to iterate through the tabs and if the tab has the *active* flag set to true, it will add the *active* CSS class to the `<a>` element that renders the tab.

We also specify that we are rendering the tabs themselves after the initial *div*, right where **ng-content** is.

#### `Tabset` class

Now let's turn our attention to the `Tabset` class. The first new idea we see here is that the `Tabset` class is implementing `AfterContentInit`. This *lifecycle hook* will tell Angular to call a method of our class (`ngAfterContentInit`) once the contents of the child directives has been initialized.

#### `Tabset` `ContentChildren` and `QueryList`

Next thing we do is declare the `tabs` property that will hold every `Tab` component we declare inside the **tabset**. Notice that instead of declaring this list as an array of `Tabs`, we use the class `QueryList`, passing a generic of `Tab`. Why is this?

`QueryList` is a class provided by Angular and when we use `QueryList` with a `ContentChildren` Angular populates this with the **components that match the query** and then **keeps the items up to date** if the state of the application changes.

However, `QueryList` requires a `ContentChildren` to populate it, so let's take a look at that now.

On the `tabs` instance variable, we add the `@ContentChildren(Tab)` annotation. This annotation will tell Angular to inject all the direct child directives (of the `Tab` type) into the `tabs` parameter. We then assign it to the `tabs` property of our component. With this **we now have access to all the child `Tab` components**.

### Initializing the `Tabset`

When this component is initialized, we want to make the first tab active. To do this we use the `ngAfterContentInit` function (that is described by the `AfterContentInit` hook). Notice that we use `this.tabs.toArray()` to cast the Angular's `QueryList` into a native TypeScript array.

#### `Tabset setActive`

Finally we define a `setActive` method. This method is used when we click a tab on our template e.g. using `(click)="setActive(tab)"`. This function will iterate through all the tabs, setting their `active` properties to false. Then we set the tab we clicked active.

## Using the `Tabset`

Now the next step is to code the application component that makes use of both of the components we created. Here's how we do it:

**code/advanced_components/app/ts/tabs/tabs.ts**

```
@Component({
  selector: 'tabs-sample-app',
  template: `
  <tabset>
    <tab title="First tab">
      Lorem ipsum dolor sit amet, consectetur adipisicing elit.
      Quibusdam magni quia ut harum facilis, ullam deleniti porro
      dignissimos quasi at molestiae sapiente natus, neque voluptatum
      ad consequuntur cupiditate nemo sunt.
    </tab>
    <tab *ngFor="let tab of tabs" [title]="tab.title">
      {{ tab.content }}
    </tab>
  </tabset>
  `
})
export class TabsSampleApp {
  tabs: any;

  constructor() {
    this.tabs = [
      { title: 'About', content: 'This is the About tab' },
      { title: 'Blog', content: 'This is our blog' },
      { title: 'Contact us', content: 'Contact us here' },
    ];
  }
}
```

We're declaring that we're using **tabs-sample-app** as our component's selector and using the **Tabset** and **Tab** components.

On the template we then create a **tabset** and we add first a static tab (First tab) and we add a few more tabs from the **tabs** property of the component controller class, to illustrate how we can render tabs dynamically.

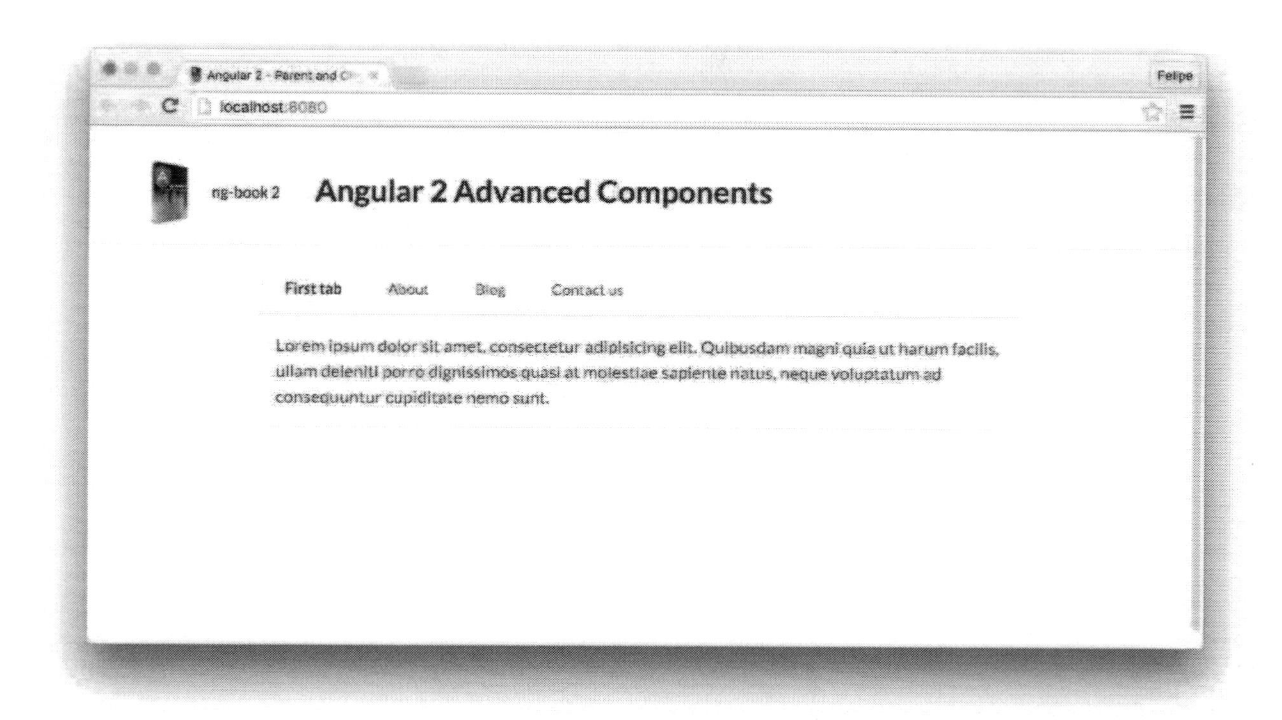


**Tabset application**

# Lifecycle Hooks

Lifecycle hooks are the way Angular allows you to add code that runs before or after each step of the directive lifecycle.

The list of hooks Angular offers are:

- `OnInit`
- `OnDestroy`
- `DoCheck`
- `OnChanges`
- `AfterContentInit`
- `AfterContentChecked`
- `AfterViewInit`
- `AfterViewChecked`

Using these hooks each follow a similar pattern:

In order to be notified about those events you

1. declare that your directive class `implements` the interface and then
2. declare the `ng` method of the hook (e.g. `ngOnInit`)

Every method name is `ng` plus the name of the hook. For example, for `OnInit` we declare the method `ngOnInit`, for `AfterContentInit` we declare `ngAfterContentInit` and so on.

When Angular knows that a component implements these functions, it will invoke them at the appropriate time.

Let's take a look at each hook individually and when we would use each of them.

### `OnInit` and `OnDestroy`

The `OnInit` hook is called when your directive properties have been initialized, and before any of the child directive properties are initialized.

Similarly, the `OnDestroy` hook is called when the directive instance is destroyed. This is typically used if we need to do some cleanup every time our directive is destroyed.

In order to illustrate let's write a component that implements both `OnInit` and `OnDestroy`:

**code/advanced_components/app/ts/lifecycle-hooks/lifecycle_01.ts**

```
 9 @Component({
10   selector: 'on-init',
11   template: `
12   <div class="ui label">
13     <i class="cubes icon"></i> Init/Destroy
14   </div>
15   `
16 })
17 class OnInitCmp implements OnInit, OnDestroy {
18   ngOnInit(): void {
19     console.log('On init');
20   }
21 
22   ngOnDestroy(): void {
23     console.log('On destroy');
24   }
25 }
```

For this component, we're just logging *On init* and *On destroy* to the console when the hooks are called.

Now in order to test those hooks let's use our component in our app component using `ngFor` to conditionally display it based on a boolean property. Let's also add a button that allows us to toggle that flag. This way, when the flag is false, our component will be *removed* from the page, causing the `OnDestroy` hook to be called. Similarly when the flag is toggled to true, the `OnInit` hook will be called.

Here's how our app component will look:

**code/advanced_components/app/ts/lifecycle-hooks/lifecycle_01.ts**

```
27  @Component({
28    selector: 'lifecycle-sample-app',
29    template: `
30    <h4 class="ui horizontal divider header">
31      OnInit and OnDestroy
32    </h4>
33
34    <button class="ui primary button" (click)="toggle()">
35      Toggle
36    </button>
37    <on-init *ngIf="display"></on-init>
38    `
39  })
40  export class LifecycleSampleApp1 {
41    display: boolean;
42
43    constructor() {
44      this.display = true;
45    }
46
47    toggle(): void {
48      this.display = !this.display;
49    }
50  }
```

When we first run the application, we can see that the `OnInit` hook was called when the component was first instantiated:

**Initial state of our component**

When I click the **Toggle** button for the first time, the component is destroyed and the hook is called as expected:

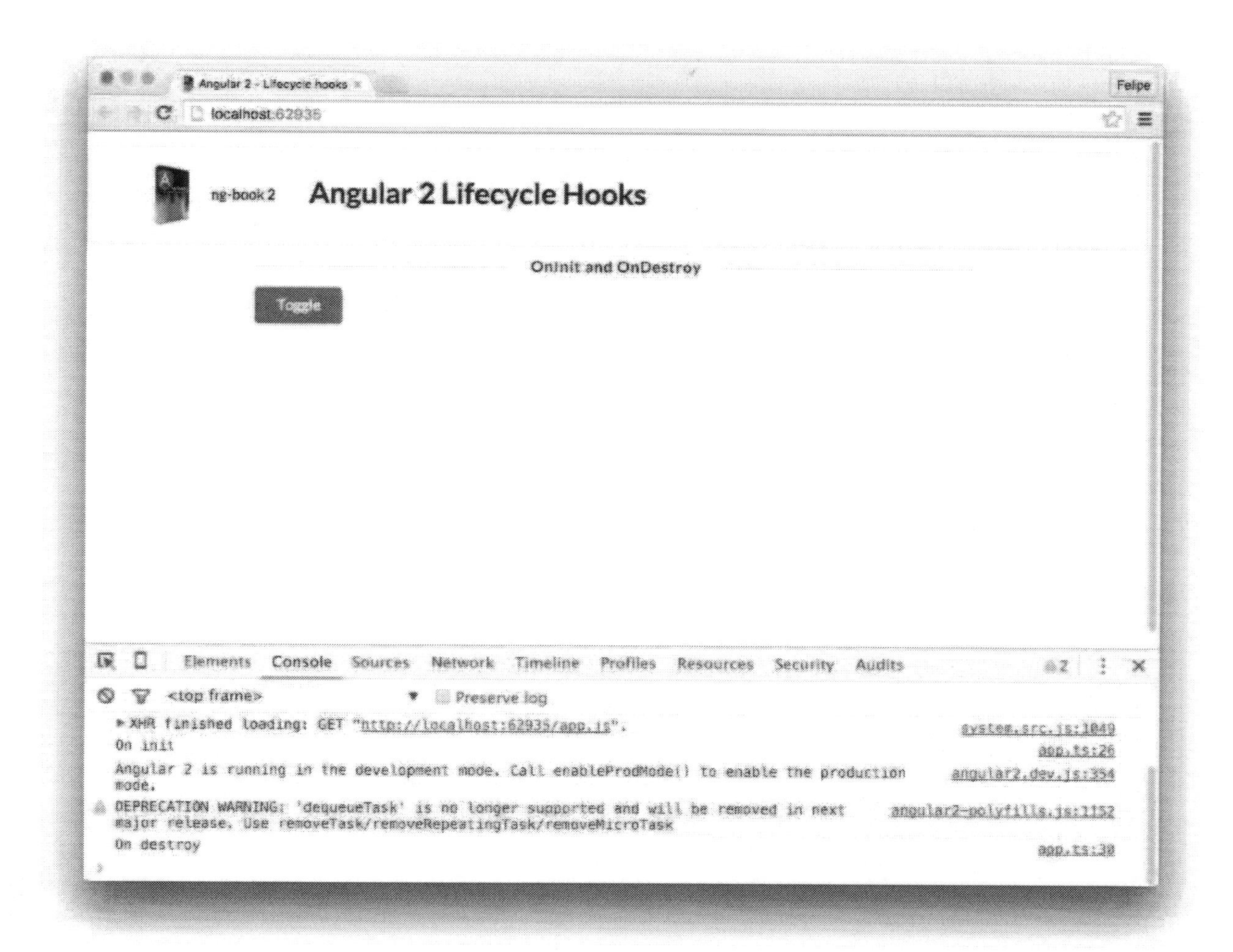


**OnDestroy hook**

And if we click it another time:

OnDestroy hook

### OnChanges

The `OnChanges` hook is called after one or more of our component properties have been changed. The `ngOnChanges` method receives a parameter which tells which properties have changed.

To understand this better, let's write a comment block component that has two inputs: *name* and *comment*:

**code/advanced_components/app/ts/lifecycle-hooks/lifecycle_02.ts**

```
30 @Component({
31   selector: 'on-change',
32   template: `
33   <div class="ui comments">
34     <div class="comment">
35       <a class="avatar">
36         <img src="app/images/avatars/matt.jpg">
37       </a>
38       <div class="content">
39         <a class="author">{{name}}</a>
40         <div class="text">
41           {{comment}}
42         </div>
43       </div>
44     </div>
45   </div>
46   `
47 })
48 class OnChangeCmp implements OnChanges {
49   @Input('name') name: string;
50   @Input('comment') comment: string;
51 
52   ngOnChanges(changes: {[propName: string]: SimpleChange}): void {
53     console.log('Changes', changes);
54   }
55 }
```

The important thing about this component is that it implements the `OnChanges` interface, and it declares the `ngOnChanges` method with this signature:

**code/advanced_components/app/ts/lifecycle-hooks/lifecycle_02.ts**

```
52   ngOnChanges(changes: {[propName: string]: SimpleChange}): void {
53     console.log('Changes', changes);
54   }
```

This method will be triggered whenever the values of either the *name* or *comment* properties change. When that happens, we receive an object that maps changed fields to `SimpleChange` objects.

Each `SimpleChange` instance has two fields: `currentValue` and `previousValue`. If both name and comment properties change for our component, we expect the value of `changes` in our method to be something like:

```
1  {
2    name: {
3      currentValue: 'new name value',
4      previousValue: 'old name value'
5    },
6    comment: {
7      currentValue: 'new comment value',
8      previousValue: 'old comment value'
9    }
10 }
```

Now, let's change the app component to use our component and also add a little form where we can play with the name and comment properties of our component:

**code/advanced_components/app/ts/lifecycle-hooks/lifecycle_02.ts**

```
57 @Component({
58   selector: 'lifecycle-sample-app',
59   template: `
60   <h4 class="ui horizontal divider header">
61     OnInit and OnDestroy
62   </h4>
63 
64   <button class="ui primary button" (click)="toggle()">
65     Toggle
66   </button>
67   <on-init *ngIf="display"></on-init>
68 
69   <h4 class="ui horizontal divider header">
70     OnChange
71   </h4>
72 
73   <div class="ui form">
74     <div class="field">
75       <label>Name</label>
76       <input type="text" #namefld value="{{name}}"
77              (keyup)="setValues(namefld, commentfld)">
78     </div>
79 
80     <div class="field">
81       <label>Comment</label>
82       <textarea (keyup)="setValues(namefld, commentfld)"
83                 rows="2" #commentfld>{{comment}}</textarea>
```

```
84      </div>
85    </div>
86
87    <on-change [name]="name" [comment]="comment"></on-change>
88    `
89  })
90  export class LifecycleSampleApp2 {
91    display: boolean;
92    name: string;
93    comment: string;
94
95    constructor() {
96      this.display = true;
97      this.name = 'Felipe Coury';
98      this.comment = 'I am learning so much!';
99    }
100
101   setValues(namefld, commentfld): void {
102     this.name = namefld.value;
103     this.comment = commentfld.value;
104   }
105
106   toggle(): void {
107     this.display = !this.display;
108   }
109 }
```

The important pieces that we added here where the template areas where we declare a new form with name and comment fields:

**code/advanced_components/app/ts/lifecycle-hooks/lifecycle_02.ts**

```
73    <div class="ui form">
74      <div class="field">
75        <label>Name</label>
76        <input type="text" #namefld value="{{name}}"
77               (keyup)="setValues(namefld, commentfld)">
78      </div>
79
80      <div class="field">
81        <label>Comment</label>
82        <textarea (keyup)="setValues(namefld, commentfld)"
83                  rows="2" #commentfld>{{comment}}</textarea>
```

```
84      </div>
85    </div>
```

Here, when the *keyup* event is fired for either the name or comment fields, we are calling setValues with the template vars namefld and commentfld that represent the input and textarea.

This method just takes the value from those fields and updates the name and comment properties accordingly:

**code/advanced_components/app/ts/lifecycle-hooks/lifecycle_02.ts**

```
101   setValues(namefld, commentfld): void {
102     this.name = namefld.value;
103     this.comment = commentfld.value;
104   }
```

So now, the first time we open the app, we can see that our `OnChanges` hook is called:

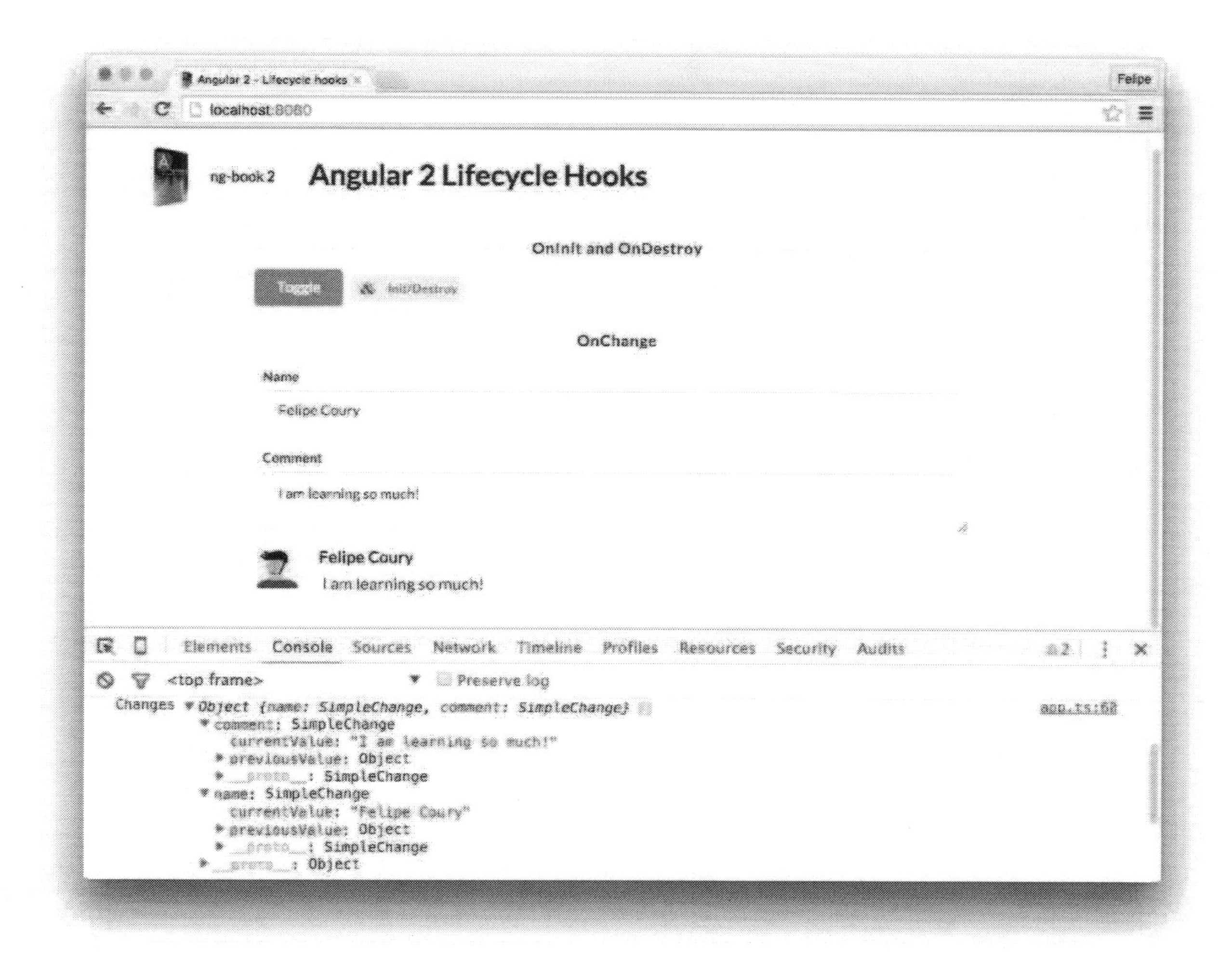


**OnChanges**

This happens when the initial values are set, on the constructor of the `LifecycleSampleApp` component.

Now if we play with the name, we can see that the hook is called repeatedly. In the case below, we pasted the name *Nate Murray* on top of the previous name, and the values for the changes are displayed as expected:

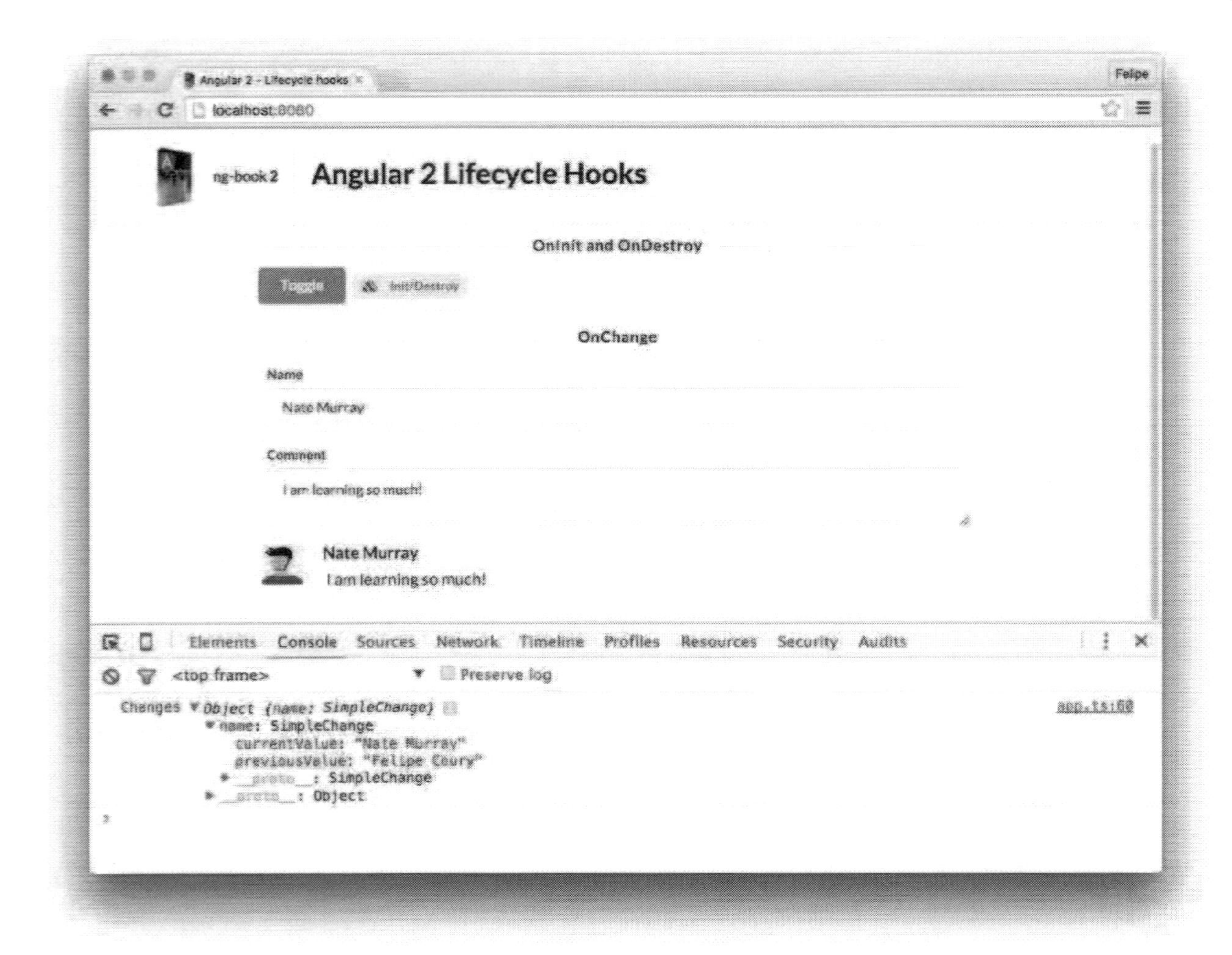


OnChanges

### DoCheck

The default notification system implemented by `OnChanges` is triggered every time the Angular change detection mechanism notices there was a change on any of the directive properties.

However, there may be times when the overhead added by this change notification may be too much, specially if performance is a concern.

There may be times when we just want to do something in case an item was removed or added, or if only a particular property changed, for instance.

If we run into one of these scenarios, we can use the `DoCheck` hook.

It's important to note that the `OnChanges` hook gets overriden by `DoCheck` so if we implement both, `OnChanges` will be ignored.

### Checking for changes

In order to evaluate what changed, Angular provides *differs*. **Differs will evaluate a given property of your directive to determine *what* changed**.

There are two types of built-in differs: *iterable differs* and *key-value differs*.

### Iterable differs

Iterable differs should be used when we have a list-like structure and we're only interested on knowing things that were added or removed from that list.

### Key-value differs

Key-value differs should be used for dictionary-like structures, and work at the key level. This differ will identify changes when a new key is added, when a key removed and when the value of a key changed.

### Rendering a comment with `do-check-item`

To illustrate these concepts, let's build a component that renders a stream of comments, like below:

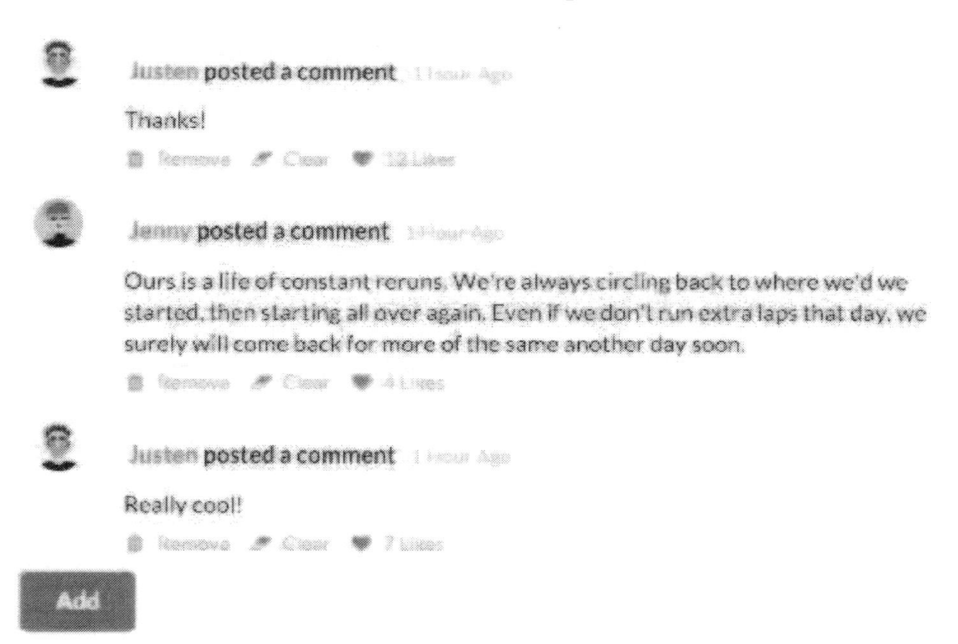


**DoCheck example**

First, let's write a component that will render one individual comment:

**code/advanced_components/app/ts/lifecycle-hooks/lifecycle_03.ts**

```
61 @Component({
62   selector: 'do-check-item',
63   outputs: ['onRemove'],
64   template: `
65   <div class="ui feed">
66     <div class="event">
67       <div class="label" *ngIf="comment.author">
68         <img src="/app/images/avatars/{{comment.author.toLowerCase()}}.jpg">
69       </div>
70       <div class="content">
71         <div class="summary">
72           <a class="user">
73             {{comment.author}}
74           </a> posted a comment
75           <div class="date">
76             1 Hour Ago
77           </div>
78         </div>
79         <div class="extra text">
80           {{comment.comment}}
81         </div>
82         <div class="meta">
83           <a class="trash" (click)="remove()">
84             <i class="trash icon"></i> Remove
85           </a>
86           <a class="trash" (click)="clear()">
87             <i class="eraser icon"></i> Clear
88           </a>
89           <a class="like" (click)="like()">
90             <i class="like icon"></i> {{comment.likes}} Likes
91           </a>
92         </div>
93       </div>
94     </div>
95   </div>
96   `
97 })
```

Here we are declaring the component metadata. Our component will receive the comment that should be rendered and it will emit an event with the remove button icon clicked.

Moving on to the component implementation:

**code/advanced_components/app/ts/lifecycle-hooks/lifecycle_03.ts**

```
98  class DoCheckItem implements DoCheck {
99    @Input('comment') comment: any;
100   onRemove: EventEmitter<any>;
101   differ: any;
```

On the class declaration we indicate we're implementing the `DoCheck` interface. We then declare the input property `comment`, and the output event `onRemove`. We also declare a `differ` property.

**code/advanced_components/app/ts/lifecycle-hooks/lifecycle_03.ts**

```
103   constructor(differs: KeyValueDiffers) {
104     this.differ = differs.find([]).create(null);
105     this.onRemove = new EventEmitter();
106   }
```

On the constructor we're receiving a `KeyValueDiffers` instance on the `differs` variable. We then use this variable to create an instance of the key value differ using this syntax `differs.find([]).create(null)`. We're also initializing our event emitter `onRemove`.

Next, let's implement the `ngDoCheck` method, required by the interface:

**code/advanced_components/app/ts/lifecycle-hooks/lifecycle_03.ts**

```
108   ngDoCheck(): void {
109     var changes = this.differ.diff(this.comment);
110 
111     if (changes) {
112       changes.forEachAddedItem(r => this.logChange('added', r));
113       changes.forEachRemovedItem(r => this.logChange('removed', r));
114       changes.forEachChangedItem(r => this.logChange('changed', r));
115     }
116   }
```

This is how you check for changes, if you're using a key-value differ. You call the `diff` method, providing the property you want to check. In our case, we want to know if there were changes to the `comment` property.

When no changes are detected, the returned value will be `null`. Now, if there are changes, we can call three different iterable methods on the differ:

- `forEachAddedItem`, for *keys* that were added

- `forEachRemovedItem`, for *keys* that were removed
- `forEachChangedItem`, for *keys* that were changed

Each method will call the provided callback with a *record*. For the key-value differ, this record will be an instance of the `KVChangeRecord` class.

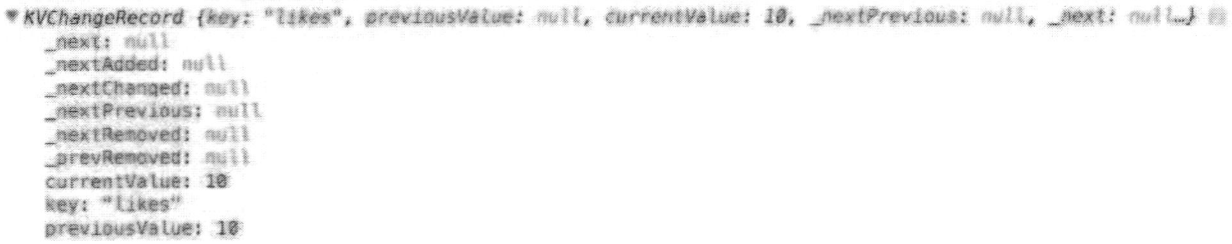

```
▼ KVChangeRecord {key: "likes", previousValue: null, currentValue: 10, _nextPrevious: null, _next: null…}
    _next: null
    _nextAdded: null
    _nextChanged: null
    _nextPrevious: null
    _nextRemoved: null
    _prevRemoved: null
    currentValue: 10
    key: "likes"
    previousValue: 10
```

**Example of a KVChangeRecord instance**

The important fields for understanding what changed are *key*, *previousValue* and *currentValue*.

Next, let's write a method that will log to the console a nice sentence about what changed:

**code/advanced_components/app/ts/lifecycle-hooks/lifecycle_03.ts**

```
118   logChange(action, r) {
119     if (action === 'changed') {
120       console.log(r.key, action, 'from', r.previousValue, 'to', r.currentValue);
121     }
122     if (action === 'added') {
123       console.log(action, r.key, 'with', r.currentValue);
124     }
125     if (action === 'removed') {
126       console.log(action, r.key, '(was ' + r.previousValue + ')');
127     }
128   }
```

Finally, let's write the methods that will help us change things on our component, to trigger our `DoCheck` hook:

**code/advanced_components/app/ts/lifecycle-hooks/lifecycle_03.ts**

```
130   remove(): void {
131     this.onRemove.emit(this.comment);
132   }
133 
134   clear(): void {
135     delete this.comment.comment;
136   }
137 
138   like(): void {
139     this.comment.likes += 1;
140   }
```

The `remove()` method will emit the event indicating that the user asked for this comment to be removed, the `clear()` method will remove the comment text from the comment object, and the `like()` method will increase to the like counter for the comment.

### Rendering a list of comments with `do-check`

Now that we have written a component for one individual comment, let's write a second component that will be responsible for rendering the list of comments:

**code/advanced_components/app/ts/lifecycle-hooks/lifecycle_03.ts**

```
143 @Component({
144   selector: 'do-check',
145   template: `
146   <do-check-item [comment]="comment"
147     *ngFor="let comment of comments" (onRemove)="removeComment($event)">
148   </do-check-item>
149 
150   <button class="ui primary button" (click)="addComment()">
151     Add
152   </button>
153   `
154 })
```

The component metadata is pretty straightforward: we're using the component we created above, and then using `ngFor` to iterate through a list of comments, renderding them. We also have a button that will allow the user to add more comments to the list.

Now let's implement our comment list class `DoCheckCmp`:

**code/advanced_components/app/ts/lifecycle-hooks/lifecycle_03.ts**

```
155 class DoCheckCmp implements DoCheck {
156   comments: any[];
157   iterable: boolean;
158   authors: string[];
159   texts: string[];
160   differ: any;
```

Here we declare the variables we'll use: `comments`, `iterable`, `authors`, and `texts`.

**code/advanced_components/app/ts/lifecycle-hooks/lifecycle_03.ts**

```
162   constructor(differs: IterableDiffers) {
163     this.differ = differs.find([]).create(null);
164     this.comments = [];
165 
166     this.authors = ['Elliot', 'Helen', 'Jenny', 'Joe', 'Justen', 'Matt'];
167     this.texts = [
168       "Ours is a life of constant reruns. We're always circling back to where we\
169 'd we started, then starting all over again. Even if we don't run extra laps tha\
170 t day, we surely will come back for more of the same another day soon.",
171       'Really cool!',
172       'Thanks!'
173     ];
174 
175     this.addComment();
176   }
```

For this component, we'll be using an iterable differ. We can see that the class we're using to create the differ is now `IterableDiffers`. However, the way we create a differ remains the same.

On the constructor we also initialize a list of authors and a list of comment texts to be used when adding new comments.

Finally, we call the `addComment()` method so we don't initialize the app with an empty list of comments.

The next three methods are used to add a new comment:

**code/advanced_components/app/ts/lifecycle-hooks/lifecycle_03.ts**

```
176   getRandomInt(max: number): number {
177     return Math.floor(Math.random() * (max + 1));
178   }
179
180   getRandomItem(array: string[]): string {
181     let pos: number = this.getRandomInt(array.length - 1);
182     return array[pos];
183   }
184
185   addComment(): void {
186     this.comments.push({
187       author: this.getRandomItem(this.authors),
188       comment: this.getRandomItem(this.texts),
189       likes: this.getRandomInt(20)
190     });
191   }
192
193   removeComment(comment) {
194     let pos = this.comments.indexOf(comment);
195     this.comments.splice(pos, 1);
196   }
```

We are declaring two methods that will return a random integer and a random item from an array, respectively.

Finally, the `addComment()` method will push a new comment to the list, with a random author, random text and a random number of likes.

Next, we have the `removeComment()` method, that will be used to remove one comment from the list:

**code/advanced_components/app/ts/lifecycle-hooks/lifecycle_03.ts**

```
193   removeComment(comment) {
194     let pos = this.comments.indexOf(comment);
195     this.comments.splice(pos, 1);
196   }
```

And finally we declare our change detection method `ngDoCheck()`:

**code/advanced_components/app/ts/lifecycle-hooks/lifecycle_03.ts**

```
198   ngDoCheck(): void {
199     var changes = this.differ.diff(this.comments);
200 
201     if (changes) {
202       changes.forEachAddedItem(r => console.log('Added', r.item));
203       changes.forEachRemovedItem(r => console.log('Removed', r.item));
204     }
205   }
```

The iterable differ behaves the same way as the key-value differ but it only provides methods for items that were added or removed.

When we run the app now, we get the list of comments with one comment:

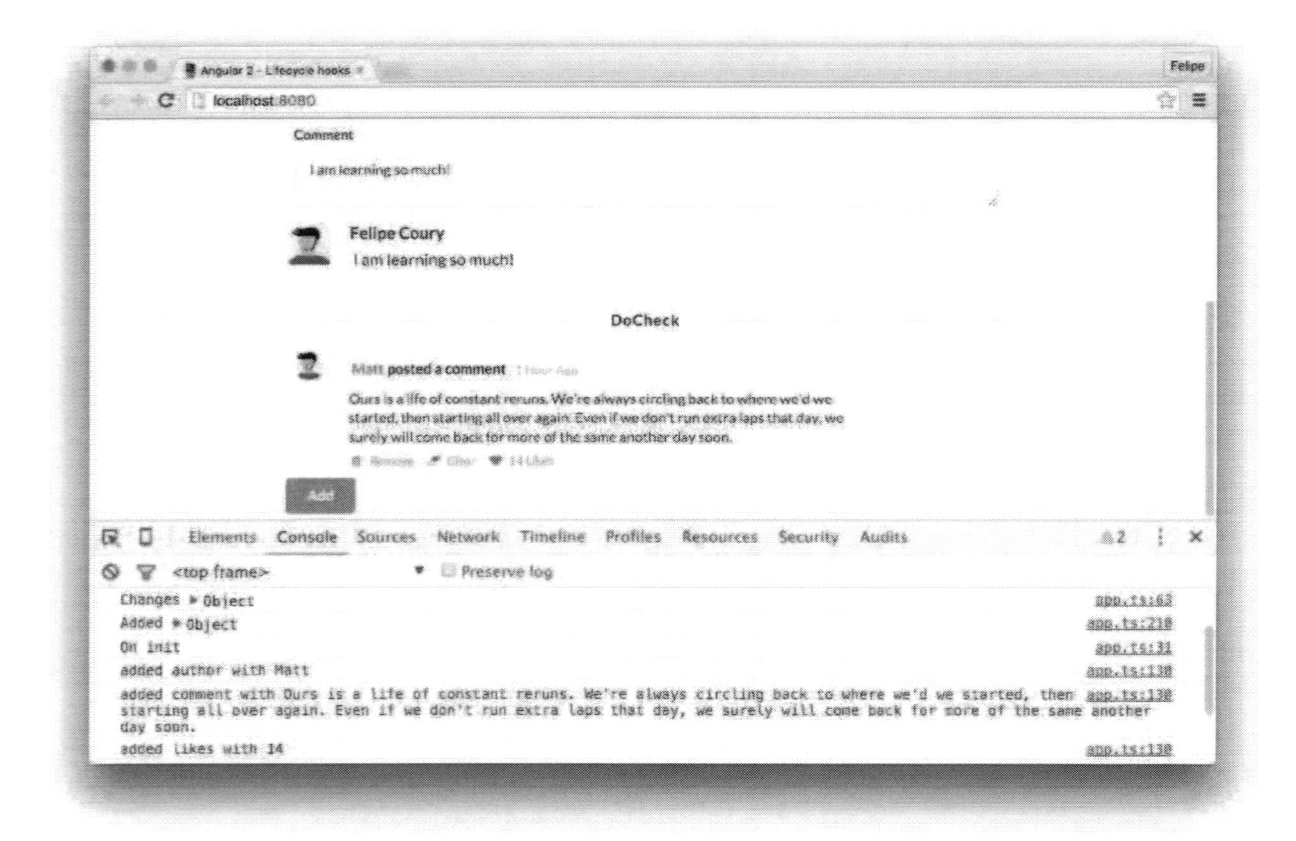


**Initial state**

We can also see that a few things were logged to the console, like:

```
1 added author with Matt
2 ...
3 added likes with 14
```

Let's see what happens when we add a new comment to the list by clicking the `Add` button:

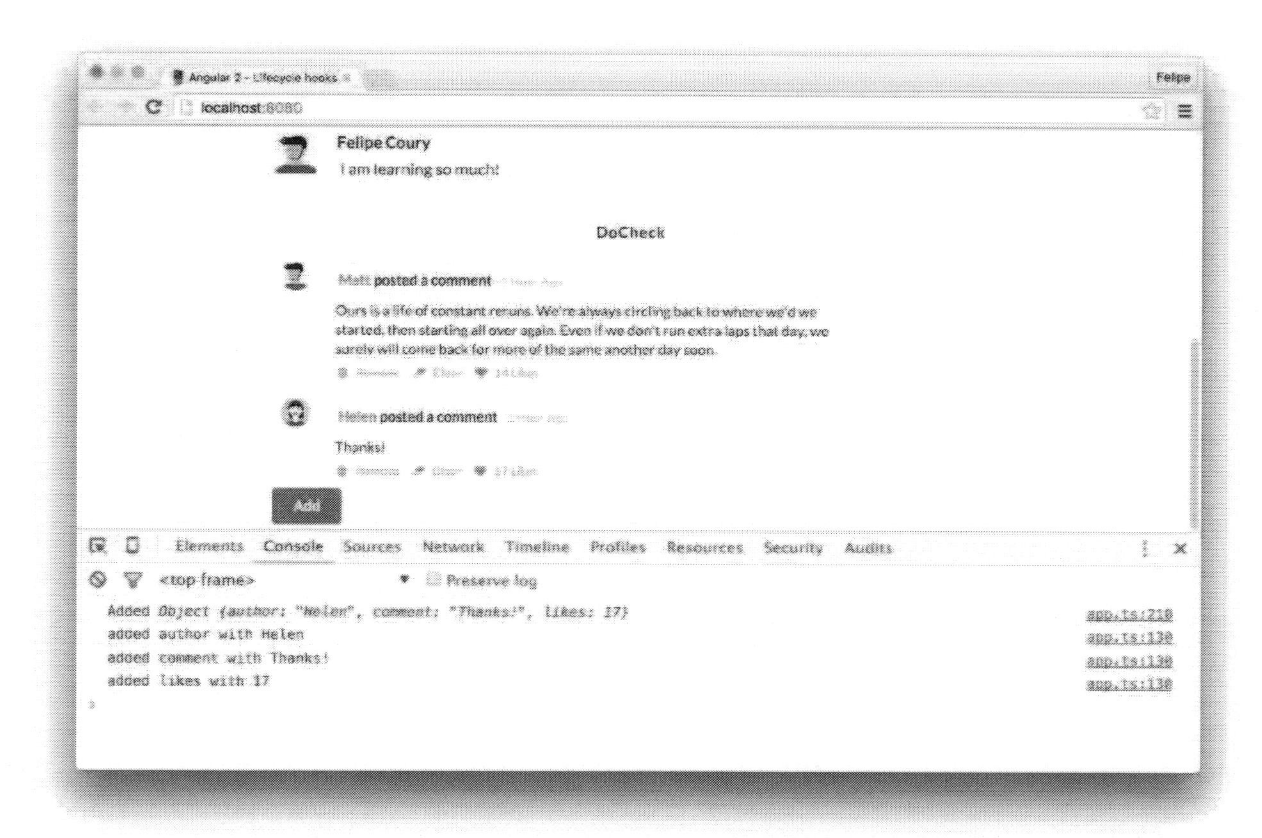


**Comment added**

We can see that the iterable differs identified that we added a new object to the list `{author: "Hellen", comment: "Thanks!", likes: 17}`.

We also got individual changes to the comment object logged, as detected by the key-value differ:

```
1 added author with Helen
2 added comment with Thanks!
3 added likes with 17
```

Now we can click the like button for this new comment:

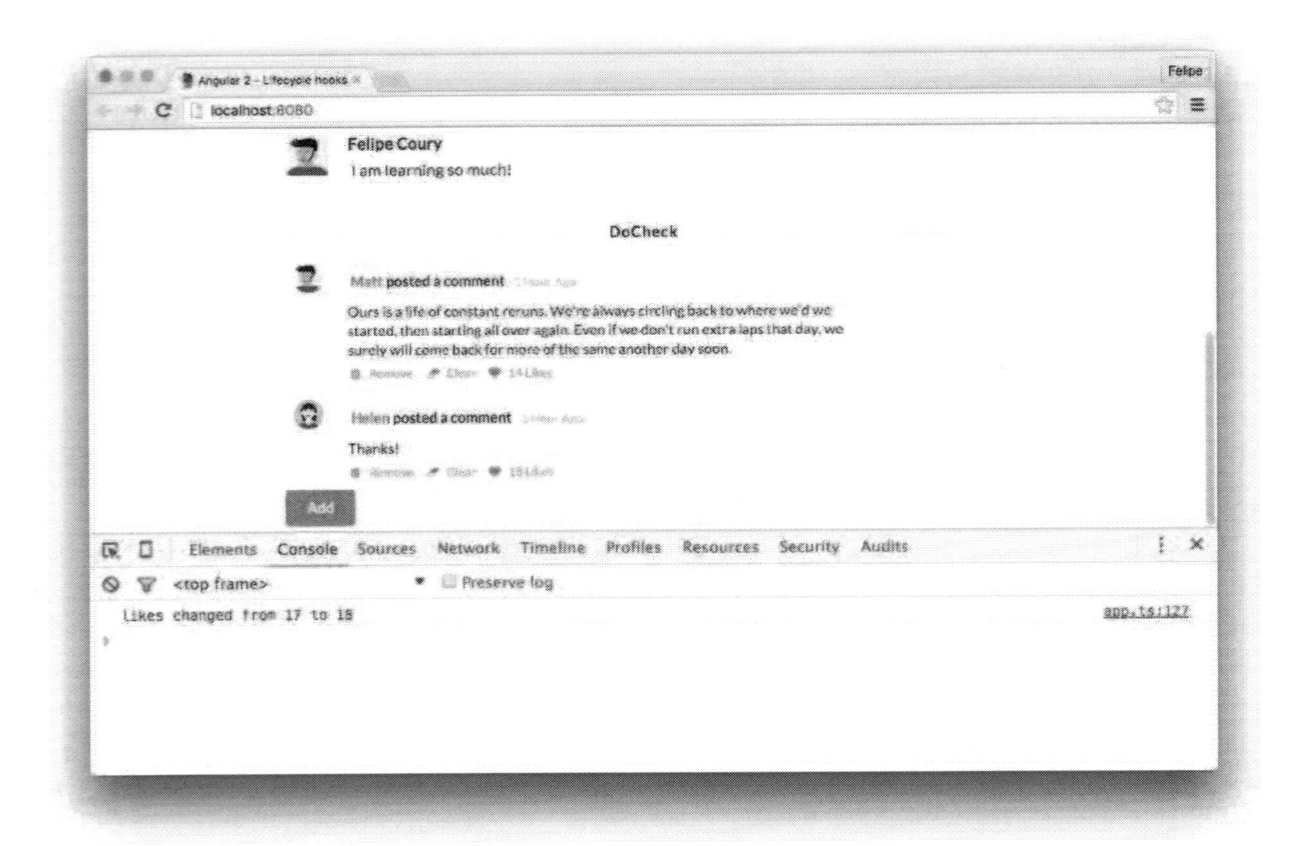


**Number of likes changed**

And now only the like change was detected.

If we click the *Clear* icon, it will remove the `comment` key from the comment object:

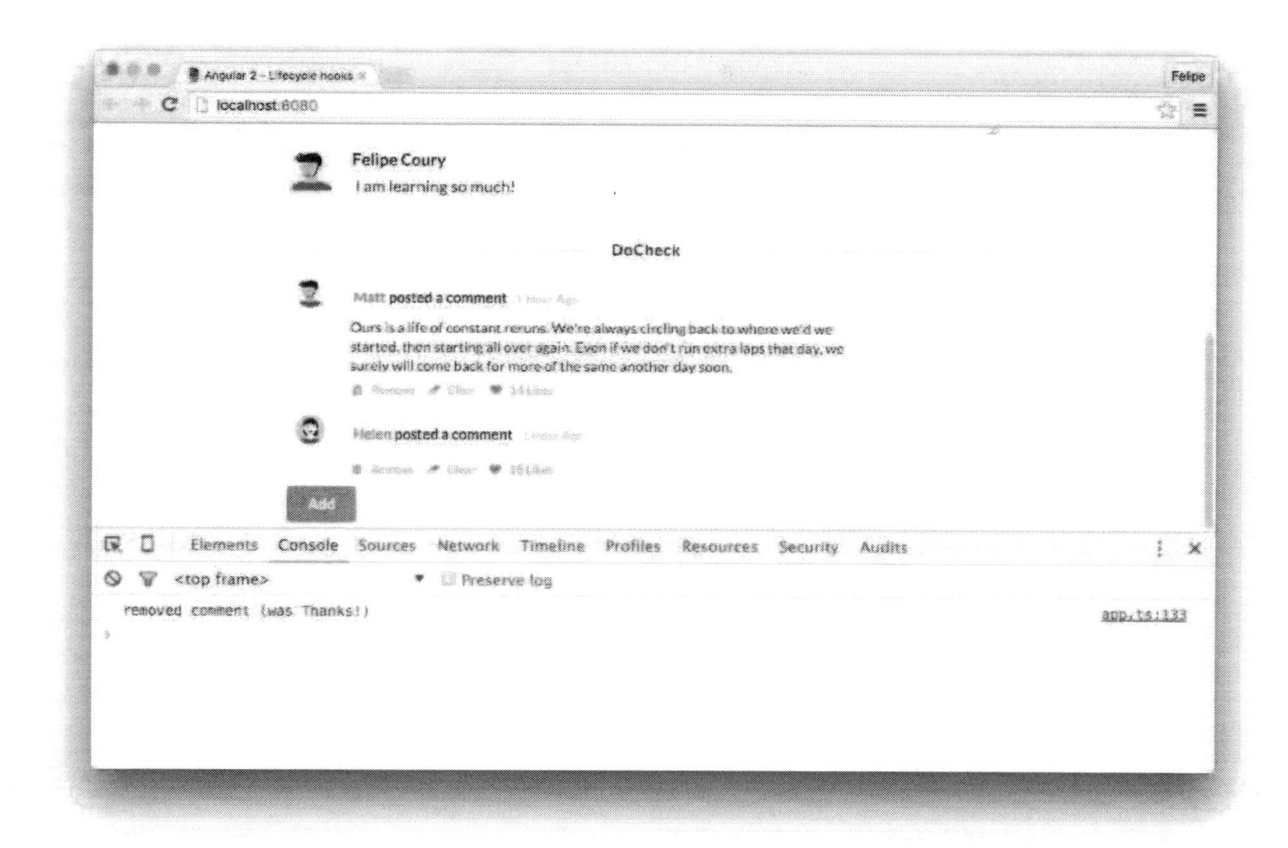


**Comment text cleared**

And the log confirms that we removed that key.

Finally, let's remove the last comment, by clicking the *Remove* icon:

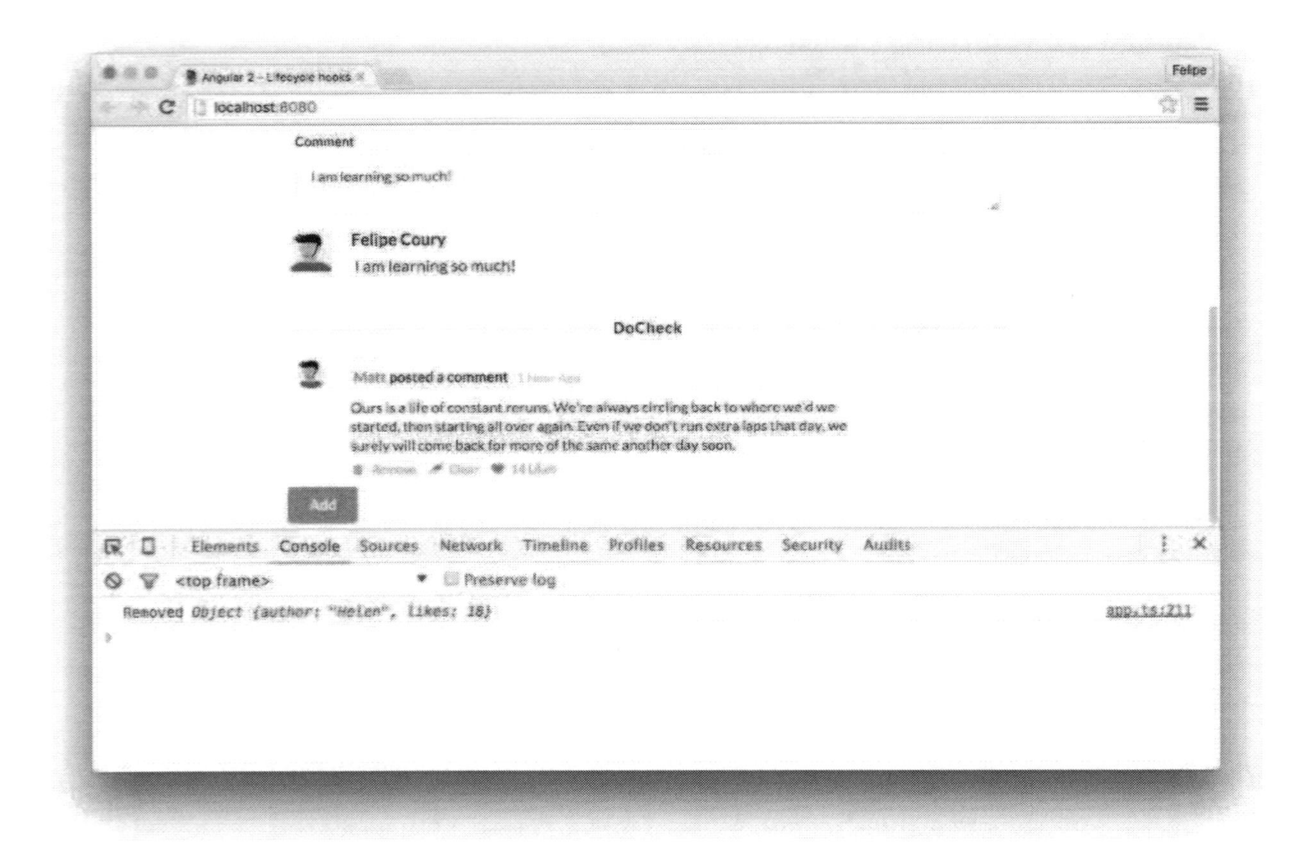


**Comment removed**

And as expected, we get a removed object log.

## AfterContentInit, AfterViewInit, AfterContentChecked and AfterViewChecked

The `AfterContentInit` hook is called after `OnInit`, right after the initialization of the content of the component or directive has finished.

The `AfterContentChecked` works similarly, but it's called after the directive check has finished. The check, in this context, is the change detection system check.

The other two hooks: `AfterViewInit` and `AfterViewChecked` are triggered right after the content ones above, right after the view has been fully initialized. Those two hooks are only applicable to components, and not to directives.

Also, the `AfterXXXInit` hooks are only called once during the directive lifecycle, while the `AfterXXXChecked` hooks are called after every change detection cycle.

To better understand this, let's write another component that logs to the console during each lifecycle hook. It will also have a counter that we can increment by clicking a button:

**code/advanced_components/app/ts/lifecycle-hooks/lifecycle_04.ts**

```
212 @Component({
213   selector: 'afters',
214   template: `
215   <div class="ui label">
216     <i class="list icon"></i> Counter: {{ counter }}
217   </div>
218 
219   <button class="ui primary button" (click)="inc()">
220     Increment
221   </button>
222   `
223 })
224 class AftersCmp implements OnInit, OnDestroy, DoCheck,
225                            OnChanges, AfterContentInit,
226                            AfterContentChecked, AfterViewInit,
227                            AfterViewChecked {
228   counter: number;
229 
230   constructor() {
231     console.log('AfterCmp --------- [constructor]');
232     this.counter = 1;
233   }
234   inc() {
235     console.log('AfterCmp --------- [counter]');
236     this.counter += 1;
237   }
238   ngOnInit() {
239     console.log('AfterCmp - OnInit');
240   }
241   ngOnDestroy() {
242     console.log('AfterCmp - OnDestroy');
243   }
244   ngDoCheck() {
245     console.log('AfterCmp - DoCheck');
246   }
247   ngOnChanges() {
248     console.log('AfterCmp - OnChanges');
249   }
250   ngAfterContentInit() {
251     console.log('AfterCmp - AfterContentInit');
252   }
```

```
253   ngAfterContentChecked() {
254     console.log('AfterCmp - AfterContentChecked');
255   }
256   ngAfterViewInit() {
257     console.log('AfterCmp - AfterViewInit');
258   }
259   ngAfterViewChecked() {
260     console.log('AfterCmp - AfterViewChecked');
261   }
262 }
```

Now let's add it to the app component, along with a `Toggle` button, like the one we used for the `OnDestroy` hook:

**code/advanced_components/app/ts/lifecycle-hooks/lifecycle_04.ts**

```
306   <afters *ngIf="displayAfters"></afters>
307   <button class="ui primary button" (click)="toggleAfters()">
308     Toggle
309   </button>
```

The final implementation for the app component now should look like this:

**code/advanced_components/app/ts/lifecycle-hooks/lifecycle_04.ts**

```
264 @Component({
265   selector: 'lifecycle-sample-app',
266   template: `
267   <h4 class="ui horizontal divider header">
268     OnInit and OnDestroy
269   </h4>
270 
271   <button class="ui primary button" (click)="toggle()">
272     Toggle
273   </button>
274   <on-init *ngIf="display"></on-init>
275 
276   <h4 class="ui horizontal divider header">
277     OnChange
278   </h4>
279 
280   <div class="ui form">
281     <div class="field">
```

```
282         <label>Name</label>
283         <input type="text" #namefld value="{{name}}"
284                (keyup)="setValues(namefld, commentfld)">
285       </div>
286 
287       <div class="field">
288         <label>Comment</label>
289         <textarea (keyup)="setValues(namefld, commentfld)"
290                   rows="2" #commentfld>{{comment}}</textarea>
291       </div>
292     </div>
293 
294     <on-change [name]="name" [comment]="comment"></on-change>
295 
296     <h4 class="ui horizontal divider header">
297       DoCheck
298     </h4>
299 
300     <do-check></do-check>
301 
302     <h4 class="ui horizontal divider header">
303       AfterContentInit, AfterViewInit, AfterContentChecked and AfterViewChecked
304     </h4>
305 
306     <afters *ngIf="displayAfters"></afters>
307     <button class="ui primary button" (click)="toggleAfters()">
308       Toggle
309     </button>
310     `
311   })
312   export class LifecycleSampleApp4 {
313     display: boolean;
314     displayAfters: boolean;
315     name: string;
316     comment: string;
317 
318     constructor() {
319       // OnInit and OnDestroy
320       this.display = true;
321 
322       // OnChange
323       this.name = 'Felipe Coury';
```

```
324     this.comment = 'I am learning so much!';
325
326     // AfterXXX
327     this.displayAfters = true;
328   }
329
330   setValues(namefld, commentfld) {
331     this.name = namefld.value;
332     this.comment = commentfld.value;
333   }
334
335   toggle(): void {
336     this.display = !this.display;
337   }
338
339   toggleAfters(): void {
340     this.displayAfters = !this.displayAfters;
341   }
342 }
```

When the application starts, we can see each hook is logged:

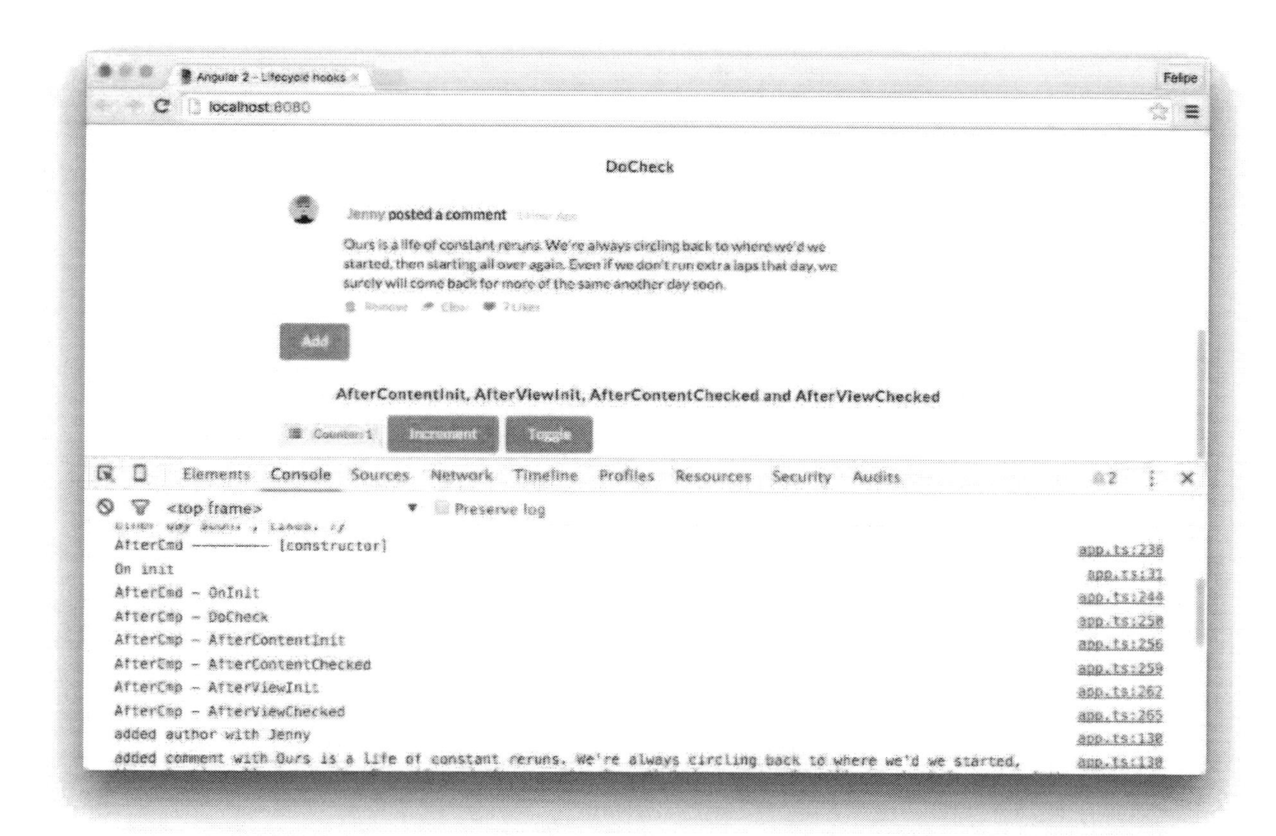


**App started**

Now let's clear the console and click the `Increment` button:

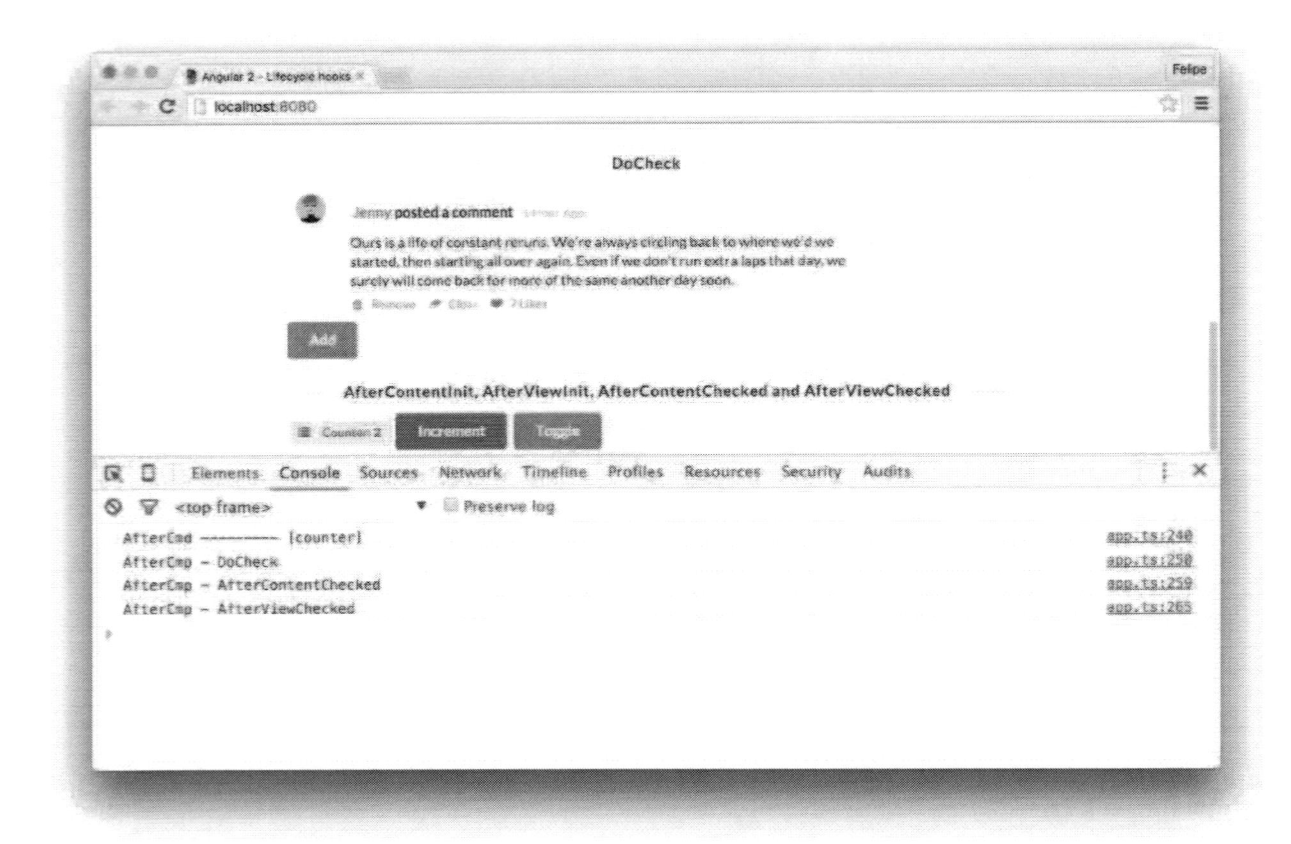


**After counter increment**

You can see that now only the `DoCheck`, `AfterContentCheck` and `AfterViewCheck` hooks were triggered.

Sure enough, if we click the `Toggle` button:

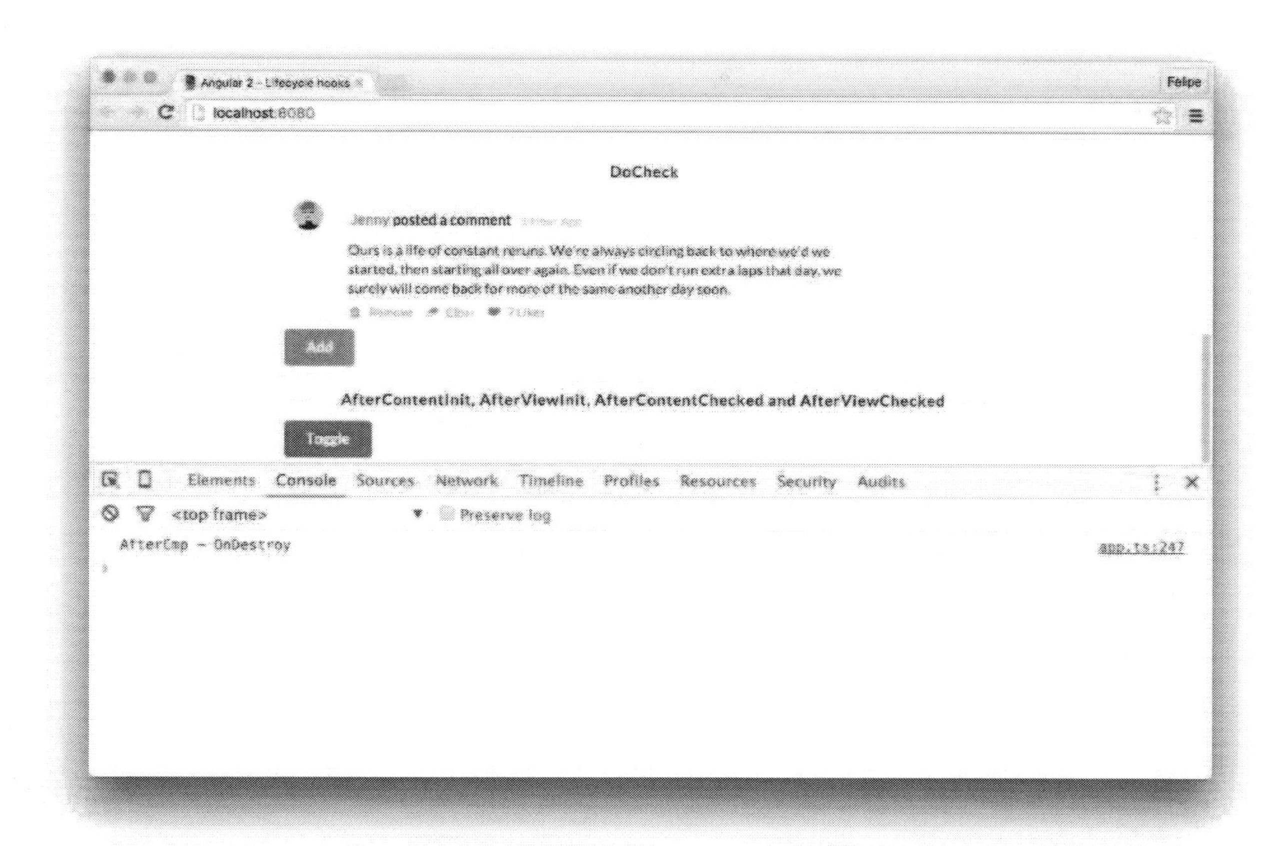


**App started**

And click it again:

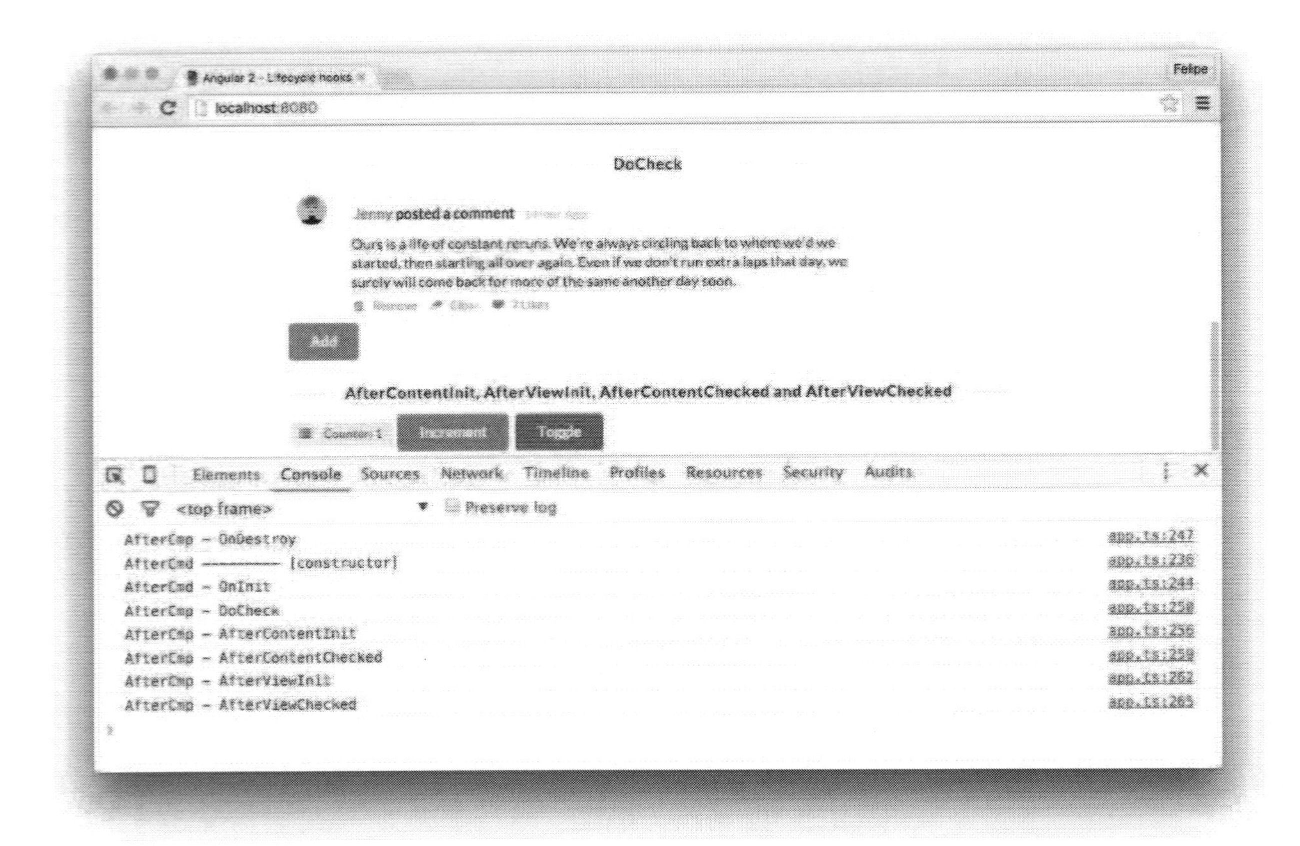


**App started**

All the hooks are triggered.

# Advanced Templates

Template elements are special elements used to create views that can be dynamically manipulated.

In order to make working with templates simpler, Angular provides some syntatic sugar to create templates, so we often don't create them by hand.

For instance, when we write:

```
1 <do-check-item
2   *ngFor="let comment of comments"
3   [comment]="comment"
4   (onRemove)="removeComment($event)">
5 </do-check-item>
```

This gets converted into:

```
<do-check-item
  template="ngFor let comment of comments; #i=index"
  [comment]="comment"
  (onRemove)="removeComment($event)">
</do-check-item>
```

Which then gets converted into:

```
<template
  ngFor
  [ngForOf]="comments"
  let-comment="$implicit"
  let-index="i">
  <do-check-item
    [comment]="comment"
    (onRemove)="removeComment($event)">
  </do-check-item>
</template>
```

It's important that we understand this underlying concept so we can build our own directives.

## Rewriting `ngIf` - `ngBookIf`

Let's create a directive that does exactly what `ngIf` does. Let's call it `ngBookIf`.

#### `ngBookIf @Directive`

We start by declaring the `@Directive` annotation for our class:

```
@Directive({
  selector: '[ngBookIf]',
  inputs: ['ngBookIf']
})
```

We're using `[ngBookIf]` as the selector because, as we learned above, when we use `*ngBookIf="condition"`, it will be converted to:

```
<template ngBookIf [ngBookIf]="condition">
```

Since `ngBookIf` is also an attribute we need to indicate that we're expecting to receive it as an input.

The job of this directive should be to add the directive template contents when the condition is true and remove it when it's false.

So when the condition is true, we will use a *view container*. The view container is used to attach one or more views to the directive.

We will use the view container to either:

- create a new view with our directive template embedded or
- clear the view container contents.

Before we do that, we need to inject the `ViewContainerRef` and the `TemplateRef`. They will be injected with the directive's view container and template.

Here's the code we'll need:

**code/advanced_components/app/ts/templates/if.ts**

```
13 class NgBookIf {
14   constructor(private viewContainer: ViewContainerRef,
15               private template: TemplateRef<any>) {}
```

Now that we have references to both the view container and the template, we will use a TypeScript property setter construct:

**code/advanced_components/app/ts/templates/if.ts**

```
14   constructor(private viewContainer: ViewContainerRef,
15               private template: TemplateRef<any>) {}
```

This method will be called every time we set a value on the `ngBookIf` property of our class. That is, this method will be called anytime the `condition` in `ngBookIf="condition"` changes.

Now we use the view container's `createEmbeddedView` method to attach the directive's template if the condition is true, or the `clear` method to remove everything from the view container.

### Using `ngBookIf`

In order to use our directive, we can write the following component:

**code/advanced_components/app/ts/templates/if.ts**

```
27 @Component({
28   selector: 'template-sample-app',
29   template: `
30   <button class="ui primary button" (click)="toggle()">
31     Toggle
32   </button>
33 
34   <div *ngBookIf="display">
35     The message is displayed
36   </div>
37 
38   `
39 })
40 export class IfTemplateSampleApp {
41   display: boolean;
42 
43   constructor() {
44     this.display = true;
45   }
46 
47   toggle() {
48     this.display = !this.display;
49   }
50 }
```

When we run the application, we can see that the directive works as expected: when we click the **Toggle** button the message *This message is displayed* is toggled on and off the page.

## Rewriting `ngFor` - `ngBookRepeat`

Now let's write a simplified version of the `ngFor` directive that Angular provides to handle repetition of templates for a given collection.

### `ngBookRepeat` template deconstruction

This directive will be used with the `*ngBookRepeat="let var of collection"` notation.

Like we did for the previous directive, we need to declare the selector as being `[ngBookRepeat]`. However the input parameter in this case won't be `ngBookRepeat` only.

If we look back at how Angular converts the `*something="let var in collection"` notation, we can see that the final form of the element is the equivalent of:

```
1 <template something [somethingOf]="collection" let-var="$implicit">
2   <!-- ... -->
3 </template>
```

As we can see, the attribute that's being passed isn't `something` but `somethingOf` instead. That's where our directive receives the collection we're iterating on.

In template that is generated, we're going to have a local view variable `#var`, that will receive the value from the `$implicit` local variable. That's the name of the local variable that Angular uses when "de-sugaring" the syntax into a `template`.

#### ngBookRepeat @Directive

Time to write the directive. First we have to write the directive annotation:

**code/advanced_components/app/ts/templates/for.ts**

```
15 @Directive({
16   selector: '[ngBookRepeat]',
17   inputs: ['ngBookRepeatOf']
18 })
```

#### ngBookRepeat class

Then we start writing the component class:

**code/advanced_components/app/ts/templates/for.ts**

```
19 class NgBookRepeat implements DoCheck {
20   private items: any;
21   private differ: IterableDiffer;
22   private views: Map<any, ViewRef> = new Map<any, ViewRef>();
23 
24 
25   constructor(private viewContainer: ViewContainerRef,
26               private template: TemplateRef<any>,
27               private changeDetector: ChangeDetectorRef,
28               private differs: IterableDiffers) {}
```

We are declaring some properties for our class:

- `items` holds the collection we're iterating on

- `differ` is an `IterableDiffer` (which we learned about in the Lifecycle Hooks section above) that will be used for change detection purposes
- `views` is a `Map` that will link a given item on the collection with the view that contains it

The constructor will receive the `viewContainer`, the template and an `IterableDiffers` instance (we discussed each of these things earlier in this chapter above).

Now, the next thing that's being injected is a change detector. We will have a deep dive in change detection in the next section. For now, let's say that this is the class that Angular creates to trigger the detection when properties of our directive change.

The next step is to write code that will trigger when we set the `ngBookRepeatOf` input:

**code/advanced_components/app/ts/templates/for.ts**

```
30    set ngBookRepeatOf(items) {
31      this.items = items;
32      if (this.items && !this.differ) {
33        this.differ = this.differs.find(items).create(this.changeDetector);
34      }
35    }
```

When we set this attribute, we're keeping the collection on the directive's `item` property and if the collection is valid and we don't have a differ yet, we create one.

To do that, we're creating an instance of `IterableDiffer` that reuses the directive's change detector (the one we injected in the constructor).

Now it's time to write the code that will react to a change on the collection. For this, we're going to use the **DoCheck** lifecycle hook by implementing the `ngDoCheck` method as follows:

**code/advanced_components/app/ts/templates/for.ts**

```
37    ngDoCheck(): void {
38      if (this.differ) {
39        let changes = this.differ.diff(this.items);
40        if (changes) {
41
42          changes.forEachAddedItem((change) => {
43            let view = this.viewContainer.createEmbeddedView(this.template,
44              {'$implicit': change.item});
45            this.views.set(change.item, view);
46          });
47          changes.forEachRemovedItem((change) => {
48            let view = this.views.get(change.item);
```

```
49            let idx = this.viewContainer.indexOf(view);
50            this.viewContainer.remove(idx);
51            this.views.delete(change.item);
52          });
53        }
54      }
55    }
```

Let's break this down a bit. First thing we do in this method is make sure we already instantiated the differ. If not, we do nothing.

Next, we ask the differ what changed. If there are changes, we first iterate through the itmes that were added using `changes.forEachAddedItem`. This method will receive a `CollectionChangeRecord` object for every element that was added.

Then for each element, we create a new embedded view using the view container's `createEmbeddedView` method.

```
1 let view = this.viewContainer.createEmbeddedView(this.template, {'$implicit': ch\
2 ange.item});
```

The second argument to `createEmbeddedView` is the *view context*. In this case, we're setting the `$implicit` local variable to `change.item`. This will allow us to reference the variable we declared back on the `*ngBookRepeat="let var of collection"` as `var` on that view. That is, the `var` in `let var` is the `$implicit` variable. We use `$implicit` because we don't know what name the user will assign to it when we're writing this component.

The final thing we need to do is to connect the item with the collection to its view. The reason behind this is that, if an item gets removed from the collection, we need to get rid of the correct view, as we do next.

Now for each item that was removed from the collection, we use the item-to-view map we keep to find the view. Then we ask the view container for the index of that view. We need that because the view container's `remove` method needs an index. Finally, we also remove the view from the item-to-view map.

## Trying out our directive

To test our new directive, let's write the following component:

**code/advanced_components/app/ts/templates/for.ts**

```
58 @Component({
59   selector: 'template-sample-app',
60   template: `
61   <ul>
62     <li *ngBookRepeat="let p of people">
63       {{ p.name }} is {{ p.age }}
64       <a href (click)="remove(p)">Remove</a>
65     </li>
66   </ul>
67 
68   <div class="ui form">
69     <div class="fields">
70       <div class="field">
71         <label>Name</label>
72         <input type="text" #name placeholder="Name">
73       </div>
74       <div class="field">
75         <label>Age</label>
76         <input type="text" #age placeholder="Age">
77       </div>
78     </div>
79   </div>
80   <div class="ui submit button"
81        (click)="add(name, age)">
82     Add
83   </div>
84   `
85 })
86 export class ForTemplateSampleApp {
87   people: any[];
88 
89   constructor() {
90     this.people = [
91       {name: 'Joe', age: 10},
92       {name: 'Patrick', age: 21},
93       {name: 'Melissa', age: 12},
94       {name: 'Kate', age: 19}
95     ];
96   }
97 
98   remove(p) {
```

```
99      let idx: number = this.people.indexOf(p);
100     this.people.splice(idx, 1);
101     return false;
102   }
103
104   add(name, age) {
105     this.people.push({name: name.value, age: age.value});
106     name.value = '';
107     age.value = '';
108   }
109 }
```

We're using our directive to iterate through a list of people:

**code/advanced_components/app/ts/templates/for.ts**

```
61   <ul>
62     <li *ngBookRepeat="let p of people">
63       {{ p.name }} is {{ p.age }}
64       <a href (click)="remove(p)">Remove</a>
65     </li>
66   </ul>
```

When we click **Remove** we remove the item from the collection, triggering the change detection.

We also provide a form that allows adding items to the collection:

**code/advanced_components/app/ts/templates/for.ts**

```
68   <div class="ui form">
69     <div class="fields">
70       <div class="field">
71         <label>Name</label>
72         <input type="text" #name placeholder="Name">
73       </div>
74       <div class="field">
75         <label>Age</label>
76         <input type="text" #age placeholder="Age">
77       </div>
78     </div>
79   </div>
80   <div class="ui submit button"
81        (click)="add(name, age)">
82     Add
83   </div>
```

# Change Detection

As a user interacts with our app, data (state) changes and our app needs to respond accordingly.

One of the big problems any modern JavaScript framework needs to solve is how to figure out when changes have happened and re-render components accordingly.

In order to make the view react to changes to components state, Angular uses *change detection.*

What are the things that can trigger changes in a component's state? The most obvious thing is user interaction. For instance, if we have a component:

```
@Component({
  selector: 'my-component',
  template: `
  Name: {{name}}
  <button (click)="changeName()">Change!</button>
  `
})
class MyComponent {
  name: string;
  constructor() {
    this.name = 'Felipe';
  }

  changeName() {
    this.name = 'Nate';
  }
}
```

We can see that when the user *clicks* on the **Change!** button, the component's *name* property will change.

Another source of change could be, for instance, a HTTP request:

```
@Component({
  selector: 'my-component',
  template: `
  Name: {{name}}
  `
})
class MyComponent {
  name: string;
  constructor(private http: Http) {
```

```
10       this.http.get('/names/1')
11         .map(res => res.json())
12         .subscribe(data => this.name = data.name);
13     }
14   }
```

And finally, we could have a timer that would trigger the change:

```
1   @Component({
2     selector: 'my-component',
3     template: `
4     Name: {{name}}
5     `
6   })
7   class MyComponent {
8     name: string;
9     constructor() {
10      setTimeout(() => this.name = 'Felipe', 2000);
11    }
12  }
```

But how does Angular become aware of these changes?

The first thing to know is that each component gets a change detector.

Like we've seen before, a typical application will have a number of components that will interact with each other, creating a dependency tree like below:

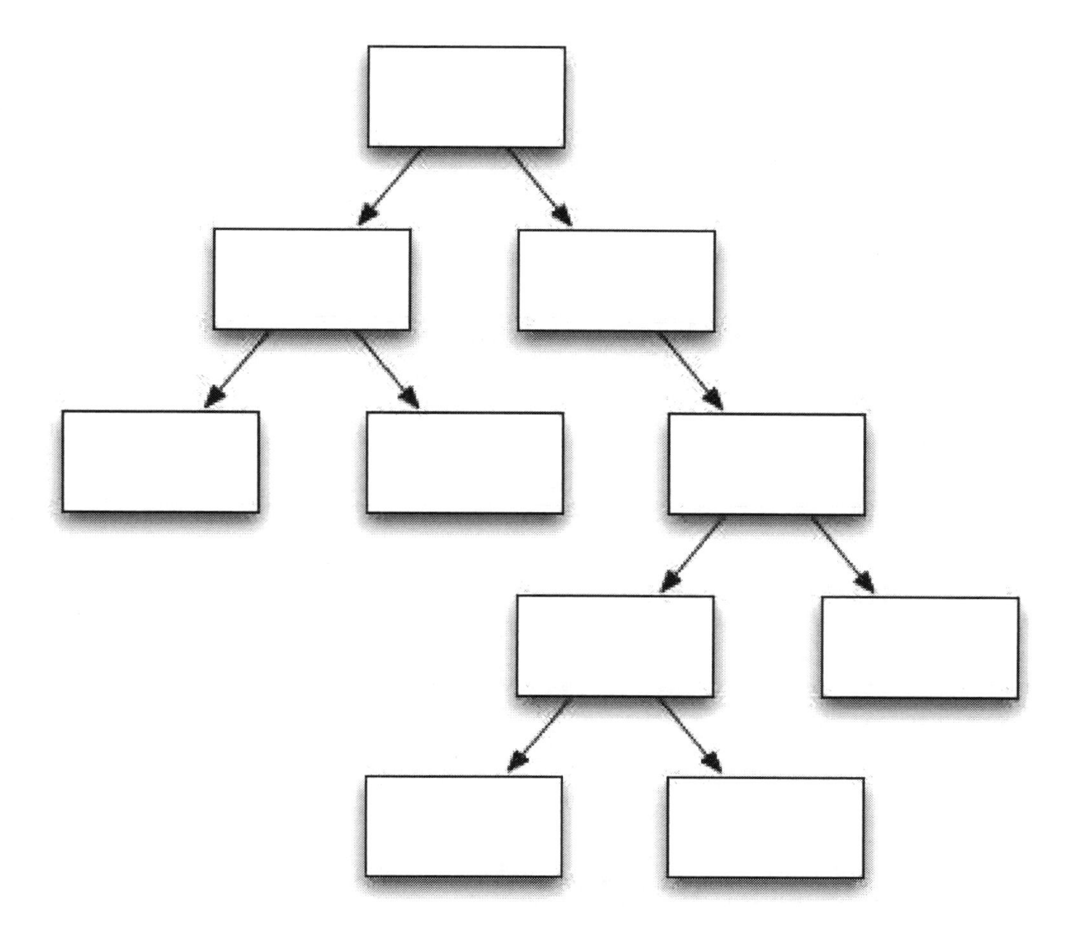

**Component tree**

For each component on our tree, a change detector is created and so we end up with a tree of change detectors:

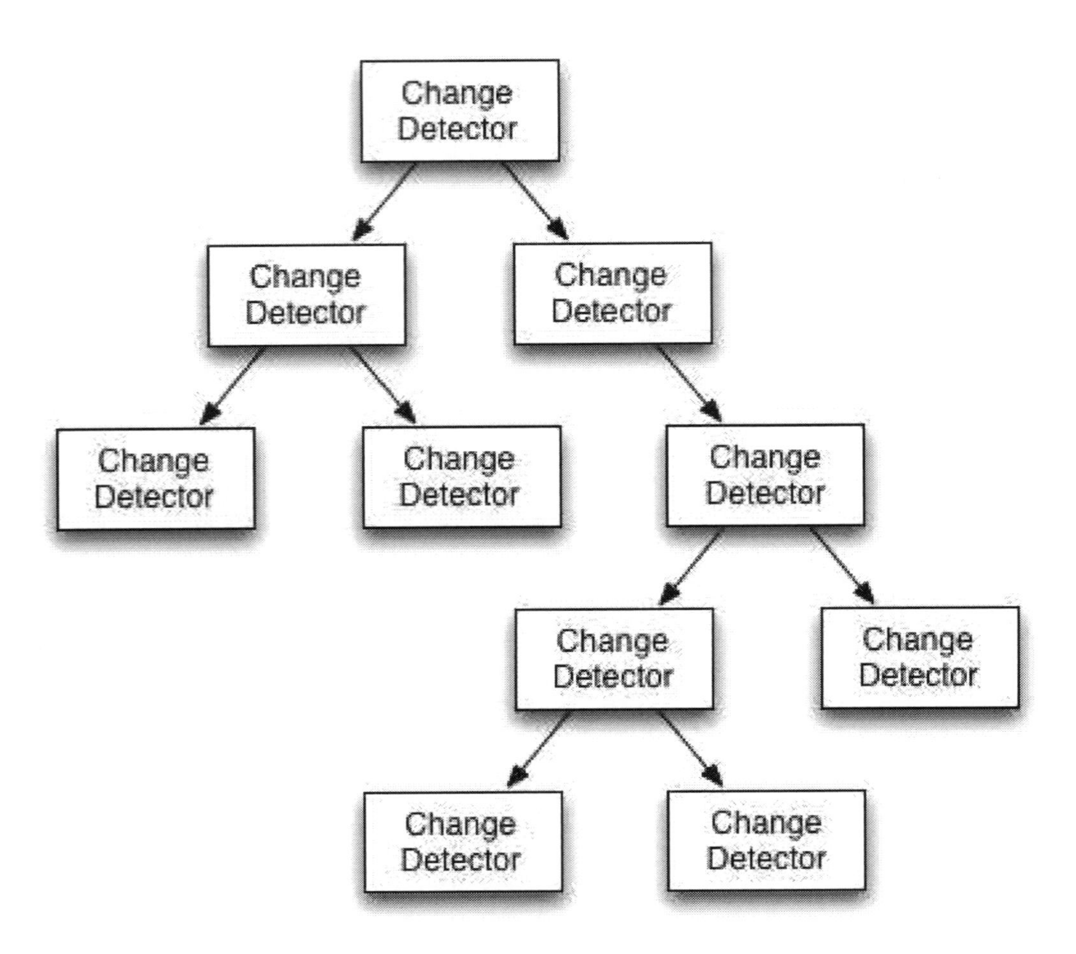


**Change detector tree**

When one of the the components change, no matter where it is in the tree, a change detection pass is triggered for the whole tree. This happens because Angular scans for changes from the top component node, all the way to the bottom leaves of the tree.

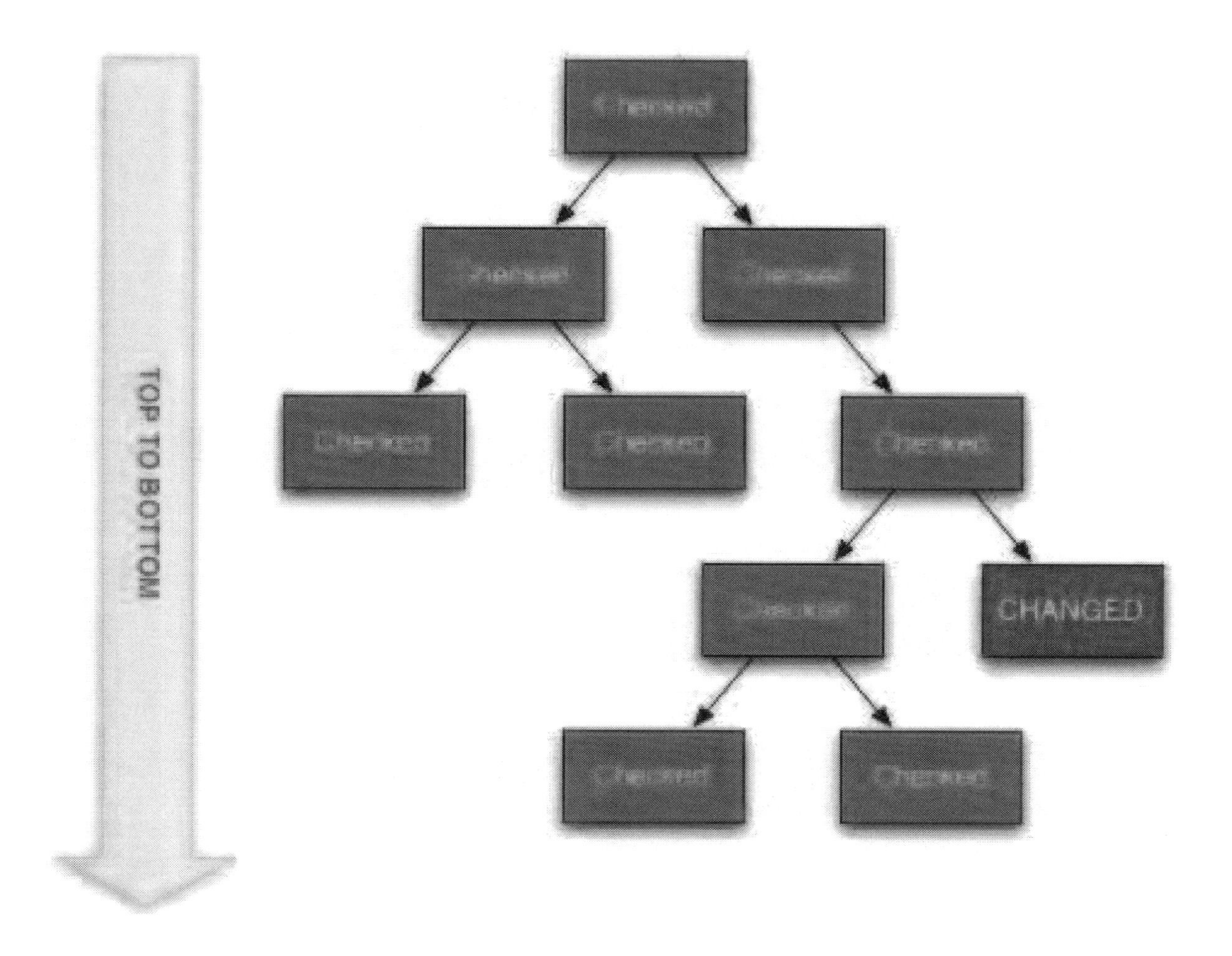


**Default change detection**

In our diagram above, the component in blue changed, but as we can see, it triggered checks for the whole component tree. Objects that were checked are indicated in red (note that the component itself was also checked).

It is natural to think that this check may be a very expensive operation. However, due to a number of optimizations (that make Angular code eligible for further optimization by the JavaScript engine), it's actually surprisingly fast.

## Customizing Change Detection

There are times that the built-in or default change detection mechanism may be overkill. One example is if you're using immutable objects or if your application architecture relies on observables. In these cases, Angular provides mechanisms for configuring the change detection system so that you get very fast performance.

The first way to change the change detector behavior is by telling a component that it should only be checked if one of its *input values* change.

To recap, an input value is an attribute your component receives from the outside world. For instance, in this code:

```
1  class Person {
2    constructor(public name: string, public age: string) {}
3  }
4
5  @Component({
6    selector: 'mycomp',
7    inputs: ['person'],
8    template: `
9      <div>
10       <span class="name">{person.name}</span>
11       is {person.age} years old.
12     </div>
13   `
14 })
15 class MyComp {
16 }
```

We have `person` as an input attribute. Now, if we want to make this component change only when its input attribute changes, we just need to change the change detection strategy, by setting its `changeDetection` attribute to `ChangeDetectionStrategy.OnPush`.

By the way, if you're curious, the default value for `changeDetection` is `ChangeDetectionStrategy.Default`.

Let's write a small experiment with two components. The first one will use the default change detection behavior and the other will use the `OnPush` strategy:

**code/advanced_components/app/ts/change-detection/onpush.ts**

```
1  import {
2    Component,
3    Input,
4    ChangeDetectionStrategy,
5  } from '@angular/core';
6
7
8  class Profile {
```

```
 9    constructor(private first: string, private last: string) {}
10
11    lastChanged() {
12      return new Date();
13    }
14  }
```

So we start with some imports and we declare a `Person` class that will be used as the input in both of our components. Notice that we also created a method called `lastChange()` on the `Profile` class. It will help us determine when the change detection is triggered. When a given component is marked as needing to be checked, this method will be called, since it's present on the template. So this method will reliably indicate the last time the component was checked for changes.

Next, we declare the `DefaultCmp` that will use the default change detection strategy:

**code/advanced_components/app/ts/change-detection/onpush.ts**

```
16  @Component({
17    selector: 'default',
18    template: `
19    <h4 class="ui horizontal divider header">
20      Default Strategy
21    </h4>
22
23    <form class="ui form">
24      <div class="field">
25        <label>First Name</label>
26        <input
27          type="text"
28          [(ngModel)]="profile.first"
29          name="first"
30          placeholder="First Name">
31      </div>
32      <div class="field">
33        <label>Last Name</label>
34        <input
35          type="text"
36          [(ngModel)]="profile.last"
37          name="last"
38          placeholder="Last Name">
39      </div>
40    </form>
41    <div>
```

```
42      {{profile.lastChanged() | date:'medium'}}
43    </div>
44    `
45  })
46  export class DefaultCmp {
47    @Input() profile: Profile;
48  }
```

And a second component using OnPush strategy:

**code/advanced_components/app/ts/change-detection/onpush.ts**

```
50  @Component({
51    selector: 'on-push',
52    changeDetection: ChangeDetectionStrategy.OnPush,
53    template: `
54    <h4 class="ui horizontal divider header">
55      OnPush Strategy
56    </h4>
57
58    <form class="ui form">
59      <div class="field">
60        <label>First Name</label>
61        <input
62          type="text"
63          [(ngModel)]="profile.first"
64          name="first"
65          placeholder="First Name">
66      </div>
67      <div class="field">
68        <label>Last Name</label>
69        <input
70          type="text"
71          [(ngModel)]="profile.last"
72          name="last"
73          placeholder="Last Name">
74      </div>
75    </form>
76    <div>
77      {{profile.lastChanged() | date:'medium'}}
78    </div>
79    `
80  })
```

```
81 export class OnPushCmp {
82   @Input() profile: Profile;
83 }
```

As we can see, both components use the same template. The only thing that is different is the header.

Finally, let's add the component that will render both components side by side:

**code/advanced_components/app/ts/change-detection/onpush.ts**

```
85 @Component({
86   selector: 'change-detection-sample-app',
87   template: `
88   <div class="ui page grid">
89     <div class="two column row">
90       <div class="column area">
91         <default [profile]="profile1"></default>
92       </div>
93       <div class="column area">
94         <on-push [profile]="profile2"></on-push>
95       </div>
96     </div>
97   </div>
98   `
99 })
100 export class OnPushChangeDetectionSampleApp {
101   profile1: Profile = new Profile('Felipe', 'Coury');
102   profile2: Profile = new Profile('Nate', 'Murray');
103 }
```

When we run this application, we should see both components rendered like below:

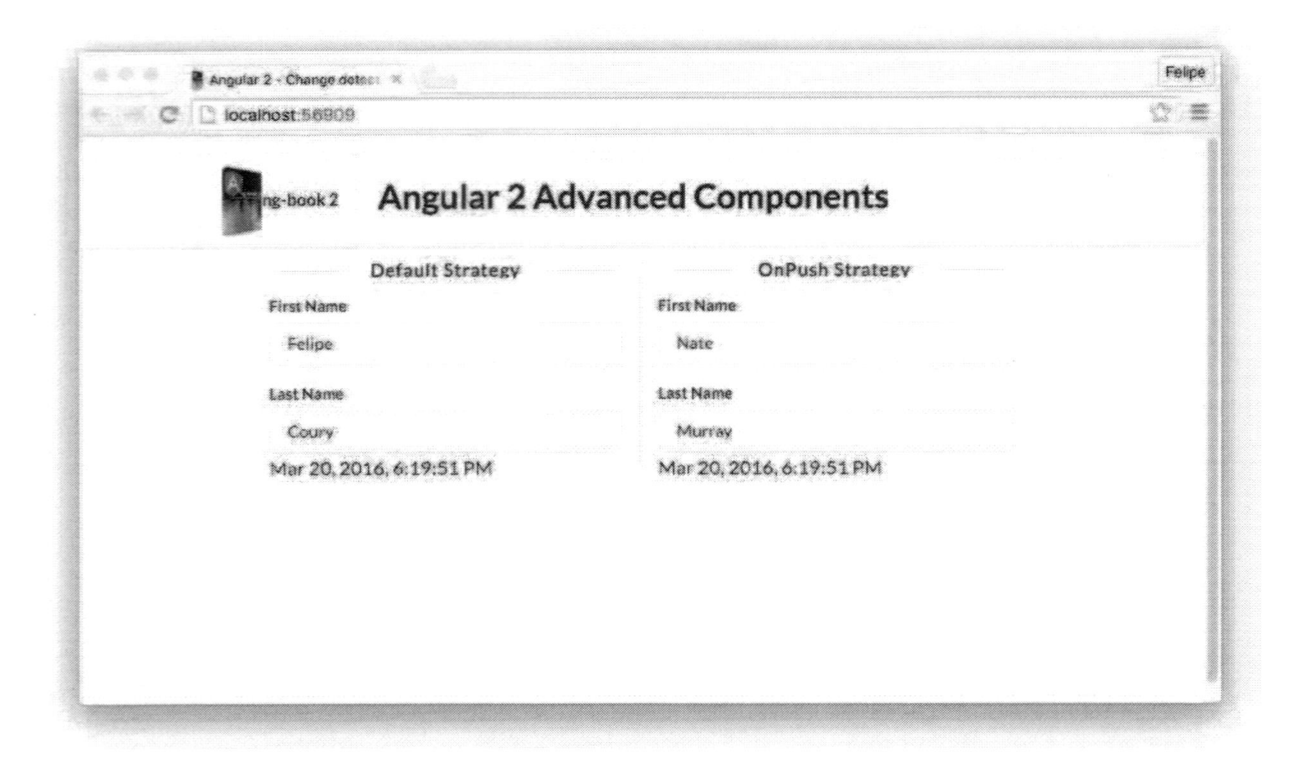


**Default vs. OnPush strategies**

When we change something on the component on the left, with the default strategy, we notice that the timestamp for the component on the right doesn't change:

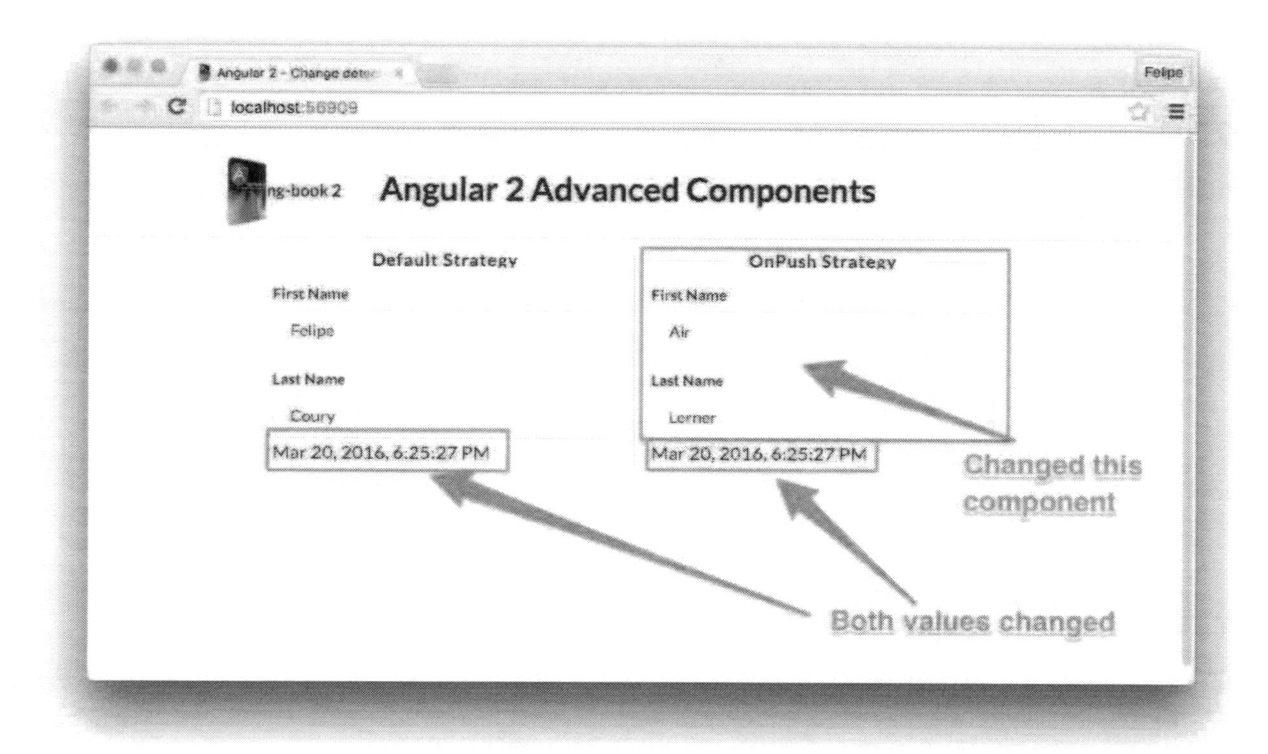


**OnPush changed, default got checked**

To understand why this happened, let's check this new tree of components:

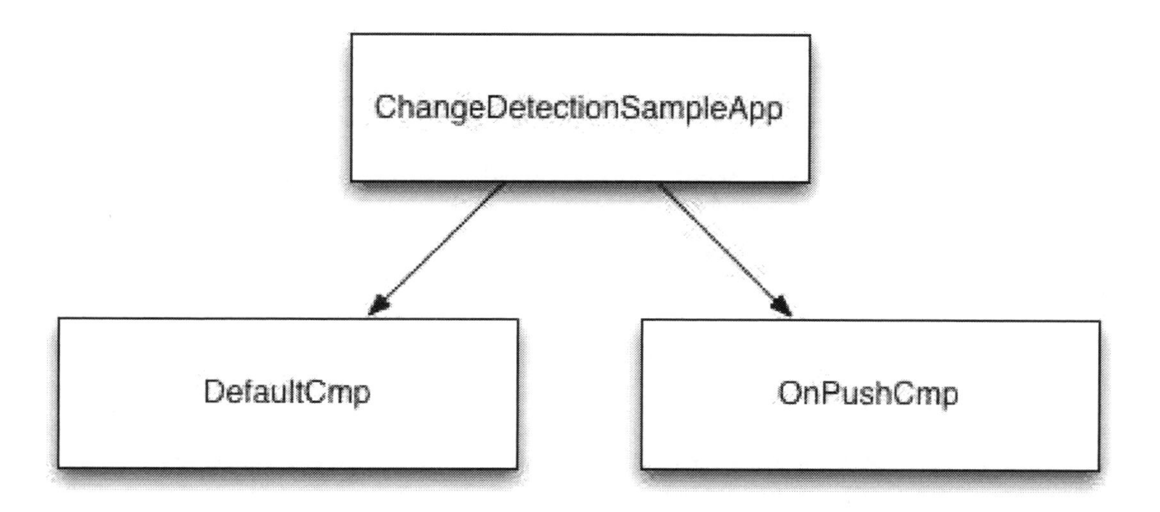


**Tree of components**

Angular checks for changes from the top to the bottom, so it queried first `ChangeDetectionSampleApp`, then `DefaultCmp` and finally `OnPushCmp`. When it inferred that `OnPushCmp` changed, it updates all the components of the tree, from top to bottom, making the `DefaultCmp` to be rendered again.

Now when we change the value of the component on the right:

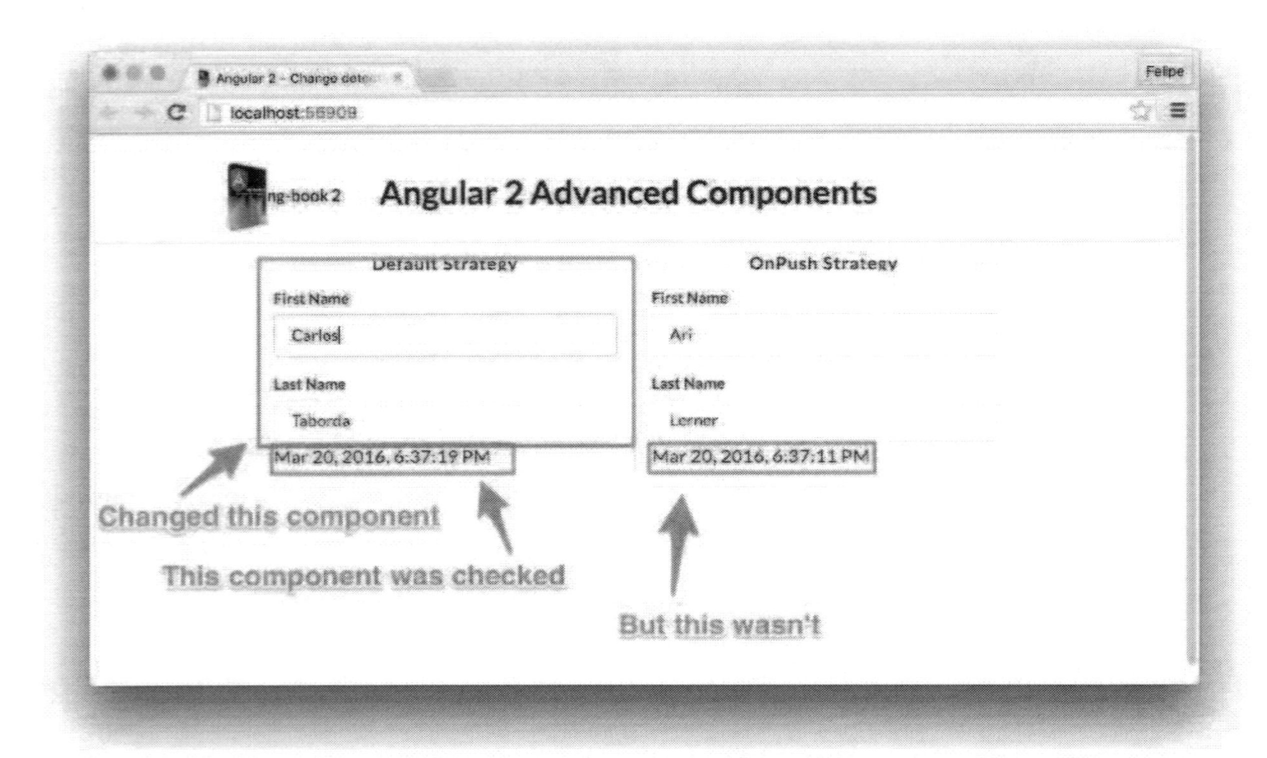


**Default changed, OnPush didn't get checked**

So now the change detection engine kicked in, the `DefaultCmp` component was checked but `OnPushCmp` wasn't. This happened because when we set the `OnPush` strategy for this component, it made the change detection kick in for this component *only* when one of its input attributes change. Changing other components of the tree doesn't trigger this component's change detector.

## Zones

Under the hood, Angular uses a library called Zones to automatically detect changes and trigger the change detection mechanism. Zones will automatically tell Angular that something changed under the most common scenarios:

- when a DOM Event occurs (like *click*, *change*, etc.)
- when an HTTP request is resolved
- when a Timer is trigger (*setTimeout* or *setInterval*)

However, there are scenarios where Zones won't be able to automatically identify that something changed. That's another scenario where the **OnPush** strategy can be very useful.

A few examples of things that is out of the Zones control, would be:

- using a third party library that runs asynchronously
- immutable data
- Observables

these are perfect candidates for using **OnPush** along with a technique to manually hint Angular that something changed.

## Observables and OnPush

Let's write a component that receives an **Observable** as a parameter. Every time we receive a value from this observable, we will increment a counter that is a property of the component.

If we were using the regular change detection strategy, any time we incremented the counter, we would get change detection triggered by Angular. However, we will have this component use the **OnPush** strategy and, instead of letting the change detector kick in for each increment, we'll only kick it when the number is a multiple of 5 or when the observable completes.

In order to do that, let's write our component:

**code/advanced_components/app/ts/change-detection/observables.ts**

```
1  import {
2    Component,
3    Input,
4    ChangeDetectorRef,
5    ChangeDetectionStrategy
6  } from '@angular/core';
7
8  import { Observable } from 'rxjs/Rx';
9
10 @Component({
11   selector: 'observable',
12   changeDetection: ChangeDetectionStrategy.OnPush,
13   template: `
14   <div>
15     <div>Total items: {{counter}}</div>
16   </div>
17   `
18 })
19 export class ObservableCmp {
20   @Input() items: Observable<number>;
```

```
21   counter = 0;
22 
23   constructor(private changeDetector: ChangeDetectorRef) {
24   }
25 
26   ngOnInit() {
27     this.items.subscribe((v) => {
28       console.log('got value', v);
29       this.counter++;
30       if (this.counter % 5 == 0) {
31         this.changeDetector.markForCheck();
32       }
33     },
34     null,
35     () => {
36       this.changeDetector.markForCheck();
37     });
38   }
39 }
```

Let's break down the code a bit so we can make sure we understand. First, we're declaring the component to take `items` as the input attribute and to use the `OnPush` detection strategy:

**code/advanced_components/app/ts/change-detection/observables.ts**

```
10 @Component({
11   selector: 'observable',
12   changeDetection: ChangeDetectionStrategy.OnPush,
13   template: `
14   <div>
15     <div>Total items: {{counter}}</div>
16   </div>
17   `
18 })
```

Next, we're storing our input attribute on the `items` property of the component class, and setting another property, called `counter`, to `0`.

**code/advanced_components/app/ts/change-detection/observables.ts**

```
19 export class ObservableCmp {
20   @Input() items: Observable<number>;
21   counter = 0;
```

Then we use the constructor to get hold of the component's change detector:

**code/advanced_components/app/ts/change-detection/observables.ts**

```
23   constructor(private changeDetector: ChangeDetectorRef) {
24   }
```

Then, during the component initialization, on the `ngOnInit` hook:

**code/advanced_components/app/ts/change-detection/observables.ts**

```
26   ngOnInit() {
27     this.items.subscribe((v) => {
28       console.log('got value', v);
29       this.counter++;
30       if (this.counter % 5 == 0) {
31         this.changeDetector.markForCheck();
32       }
33     },
34     null,
35     () => {
36       this.changeDetector.markForCheck();
37     });
38   }
```

We're subscribing to the Observable. The `subscribe` method takes three callbacks: **onNext**, **onError** and **onCompleted**.

Our onNext callback will print out the value we got, then increment the counter. Finally, if the current counter value is a multiple of 5, we call the change detector's `markForCheck` method. That's the method we use whenever we want to tell Angular that a change has been made, so the change detector should kick in.

Then for the **onError** callback, we're using `null`, indicating we don't want to handle this scenario.

Finally, for the **onComplete** callback, we're also triggering the change detector, so the final counter can be displayed.

Now, on to the application component code, that will create the subscriber:

**code/advanced_components/app/ts/change-detection/observables.ts**

```
41 @Component({
42   selector: 'change-detection-sample-app',
43   template: `
44   <observable [items]="itemObservable"></observable>
45   `
46 })
47 export class ObservableChangeDetectionSampleApp {
48   itemObservable: Observable<number>;
49 
50   constructor() {
51     this.itemObservable = Observable.timer(100, 100).take(101);
52   }
53 }
```

The important line here is the following:

```
1 this.itemObservable = Observable.timer(100, 100).take(101);
```

This line creates the Observable we're passing to the component on the `items` input attribute. We're passing two parameters to the `timer` method: the first is the number of milliseconds to wait before producing the first value and the second is the milliseconds to wait between values. So this observable will generate sequential values every 100 values forever.

Since we don't want the observable to run forever, we use the `take` method, to take only the first 101 values.

When we run this code, we'll see that the counter will only be updated for each 5 values obtained from the observer and also when the observable completes, generating a final value of 101:

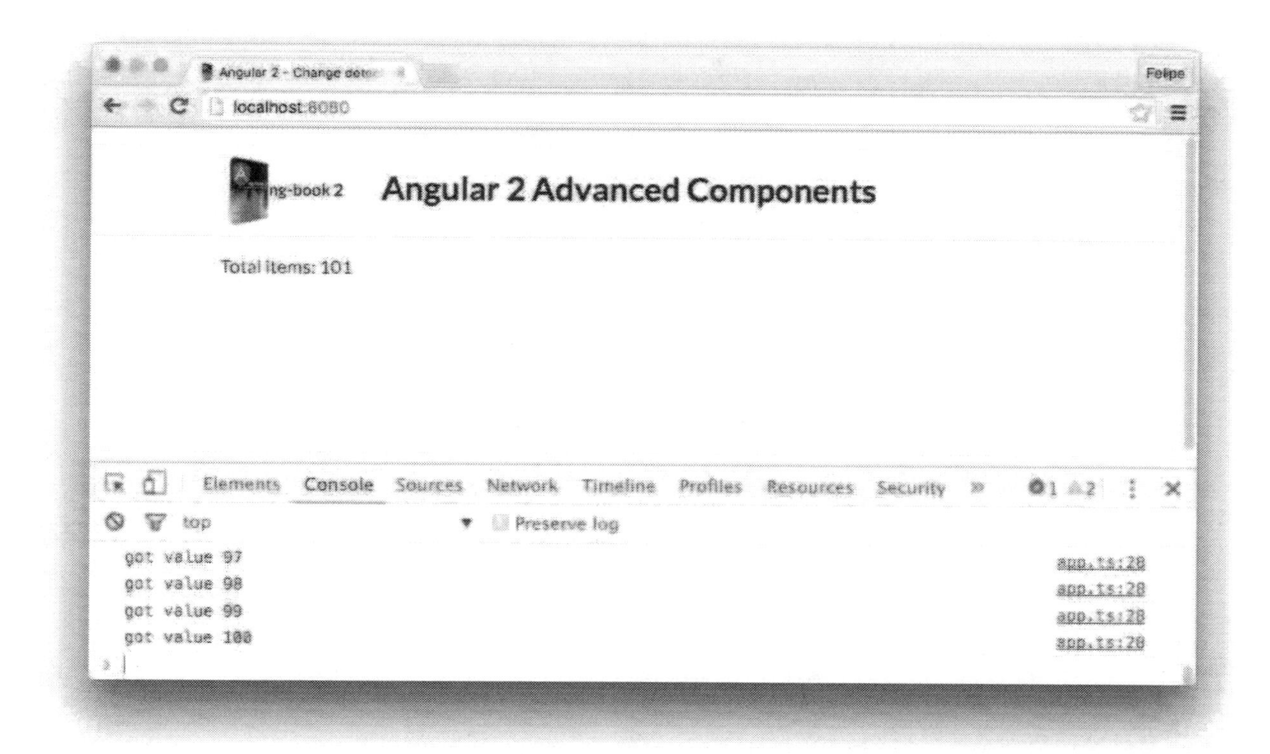


Manually triggering change detection

# Summary

Angular 2 provides us with many tools we can use for writing advanced components. Using the techniques in this chapter you will be able to write nearly any component functionality you wish.

However, there's one important concept that you'll use in many advanced components that we haven't talked about yet: Dependency Injection.

With dependency injection we can hook our components into many other parts of the system. In the next chapter we'll talk about what DI is, how you can use it in your apps, and common patterns for injecting services.

# Testing

After spending hours, days, months on a web app you're finally ready to release it to the world. Plenty of hard work and time has been poured into it and now it's time for it to pay off... and then boom: a blocking bug shows up that prevents anyone from signing up.

## Test driven?

Testing can help reveal bugs before they appear, instill confidence in your web application, and makes it easy to onboard new developers into the application. There is little doubt about the power of testing amongst the world of software development. However, there is debate about how to go about it.

Is it better to write the tests first and then write the implementation to make those tests pass or would it be better to validate that code that we've already written is correct? It's pretty odd to think that this is a source of contention across the development community, but there is a debate that can get pretty heated as to which is the *right* way to handle testing.

In our experience, particularly when coming from a prototype-heavy background, we focus on building test-able code. Although your experience may differ, we have found that while we are prototyping applications, testing individual pieces of code that are likely to change can double or triple the amount of work it takes to keep them up. In contrast, we focus on building our applications in small components, keeping large amounts of functionality broken into several methods which allows us to test the functionality of a part of the larger picture. This is what we mean when we say *testable* code.

An alternative methodology to prototyping (and then testing after) is called "Red-Green-Refactor". The idea is that you **write your tests first** and they fail (red) because you haven't written any code yet. Only after you have failing tests do you go on to write your implementation code until it all passes (green).

Of course, the decision of *what* to test is up to you and your team, however we'll focus on *how* to test your applications in this chapter.

## End-to-end vs. Unit Testing

There are two major ways to test you applications: *end-to-end testing* or *unit testing*.

If you take a top-down approach on testing you write tests that see the application as a "black box" and you interact with the application like a user would and evaluate if the app seems to work from the "outside". This top-down technique of testing is called *End to End testing*.

In the Angular world, the tool that is mostly used is called Protractor[113]. Protractor is a tool that opens a browser and interacts with the application, collecting results, to check whether the testing expectations were met.

The second testing approach commonly used is to isolate each part of the application and test it in isolation. This form of testing is called *Unit Testing*.

In Unit Testing we write tests that provide a given input to a given aspect of that unit and evaluate the output to make sure it matches our expectations.

In this chapter we're going to be covering how to **unit test** your Angular apps.

# Testing Tools

In order to test our apps, we'll use two tools: Jasmine and Karma.

## Jasmine

Jasmine[114] is a behavior-driven development framework for testing JavaScript code.

Using Jasmine, you can set expectations about what your code should do when invoked.

For instance, let's assume we have a `sum` function on a `Calculator` object. We want to make sure that adding 1 and 1 results in 2. We could express that test (also called a _spec), by writing the following code:

```
1 describe('Calculator', () => {
2   it('sums 1 and 1 to 2', () => {
3     var calc = new Calculator();
4     expect(calc.sum(1, 1)).toEqual(2);
5   });
6 });
```

One of the nice things about Jasmine is how readable the tests are. You can see here that we expect the `calc.sub` operation to equal 2.

We organize our tests with `describe` blocks and `it` blocks.

[113]https://angular.github.io/protractor/#/
[114]http://jasmine.github.io/2.4/introduction.html

Normally we use `describe` for each logical unit we're testing and inside that each we use one `it` for each expectation you want to assert. However, this isn't a hard and fast rule. You'll often see an `it` block contain several expectations.

On the `Calculator` example above we have a very simple object. For that reason, we used one describe block for the whole class and one `it` block for each method.

This is not the case most of the times. For example, methods that produce different outcomes depending on the input will probably have more than one `it` block associated. On those cases, it's perfectly fine to have nested describes: one for the object and one for each method, and then different assertions inside individual `it` blocks.

We'll be looking at a lot of `describe` and `it` blocks throughout this chapter, so don't worry if it isn't clear when to use one vs. the other. We'll be showing lots of examples.

For more information about Jasmine and all its syntax, check out the Jasmine documentation page[115].

### Karma

With Jasmine we can describe our tests and their expectations. Now, in order to actually run the tests we need to have a browser environment.

That's where Karma comes in. Karma allows us to run JavaScript code within a browser like Chrome or Firefox, or on a **headless** browser (or a browser that doesn't expose a user interface) like PhantomJS.

## Writing Unit Tests

Our main focus on this section will be to understand how we write unit tests against different parts of our Angular apps.

We're going to learn to test **Services**, **Components**, **HTTP requests** and more. Along the way we're also going to see a couple of different techniques to make our code more testable.

## Angular Unit testing framework

Angular provides its own set of classes that build upon the Jasmine framework to help writing unit testing for the framework.

The main testing framework can be found on the `@angular/core/testing` package. (Although, for testing components we'll use the `@angular/compiler/testing` package and `@angular/platform-browser/testing` for some other helpers. But more on that later.)

[115]http://jasmine.github.io/2.4/introduction.html

If this is your first time testing Angular I want to prepare you for something: When you write tests for Angular, there is a bit of setup.

For instance, when we have dependencies to inject, we often manually configure them. When we want to test a component, we have to use testing-helpers to initialize them. And when we want to test routing, there are quite a few dependencies we need to structure.

If it feels like there is a lot of setup, don't worry: you'll get the hang of it and find that the setup doesn't change that much from project to project. Besides, we'll walk you through each step in this chapter.

As always, you can find all of the sample code for this chapter in the code download. Looking over the code directly in your favorite editor can provide a good overview of the details we cover in this chapter. We'd encourage you to keep the code open as you go through this chapter.

## Setting Up Testing

Earlier in the Routing Chapter we created an application for searching for music. In this chapter, let's write tests for that application.

Karma requires a configuration in order to run. So the first thing we need to do to setup Karma is to create a `karma.conf.js` file.

Let's `karma.conf.js` file on the root path of our project, like so:

**code/routes/music/karma.conf.js**

```
1  // Karma configuration
2  var path = require('path');
3  var cwd = process.cwd();
4
5  module.exports = function(config) {
6    config.set({
7      // base path that will be used to resolve all patterns (eg. files, exclude)
8      basePath: '',
9
10     // frameworks to use
11     // available frameworks: https://npmjs.org/browse/keyword/karma-adapter
12     frameworks: ['jasmine'],
13
14     // list of files / patterns to load in the browser
15     files: [
16       { pattern: 'test.bundle.js', watched: false }
17     ],
```

```
18
19      // list of files to exclude
20      exclude: [
21      ],
22
23      // preprocess matching files before serving them to the browser
24      // available preprocessors: https://npmjs.org/browse/keyword/karma-preproces\
25  sor
26      preprocessors: {
27        'test.bundle.js': ['webpack', 'sourcemap']
28      },
29
30      webpack: {
31        devtool: 'inline-source-map',
32        resolve: {
33          root: [path.resolve(cwd)],
34          modulesDirectories: ['node_modules', 'app', 'app/ts', 'test', '.'],
35          extensions: ['', '.ts', '.js', '.css'],
36          alias: {
37            'app': 'app'
38          }
39        },
40        module: {
41          loaders: [
42            { test: /\.ts$/, loader: 'ts-loader', exclude: [/node_modules/]}
43          ]
44        },
45        stats: {
46          colors: true,
47          reasons: true
48        },
49        watch: true,
50        debug: true
51      },
52
53      webpackServer: {
54        noInfo: true
55      },
56
57
58      // test results reporter to use
59      // possible values: 'dots', 'progress'
```

```
60      // available reporters: https://npmjs.org/browse/keyword/karma-reporter
61      reporters: ['spec'],
62
63
64      // web server port
65      port: 9876,
66
67
68      // enable / disable colors in the output (reporters and logs)
69      colors: true,
70
71
72      // level of logging
73      // possible values: config.LOG_DISABLE || config.LOG_ERROR || config.LOG_WAR\
74  N || config.LOG_INFO || config.LOG_DEBUG
75      logLevel: config.LOG_INFO,
76
77
78      // enable / disable watching file and executing tests whenever any file chan\
79  ges
80      autoWatch: true,
81
82
83      // start these browsers
84      // available browser launchers: https://npmjs.org/browse/keyword/karma-launc\
85  her
86      browsers: ['PhantomJS'],
87
88
89      // Continuous Integration mode
90      // if true, Karma captures browsers, runs the tests and exits
91      singleRun: false
92    })
93  }
```

Don't worry too much about this file's contents right now, just keep in mind a few things about it:

- sets PhantomJS as the target testing browser;
- uses Jasmine karma framework for testing;
- uses a WebPack bundle called `test.bundle.js` that basically wraps all our testing and app code;

The next step is to create a new `test` folder to hold our test files.

```
1  mkdir test
```

## Testing Services and HTTP

Services in Angular start out their life as plain classes. In one sense, this makes our services easy to test because we can sometimes test them directly without using Angular.

With Karma configuration done, let's start testing the `SpotifyService` class. If we remember, this service works by interacting with the Spotify API to retrieve album, track and artist information.

Inside the `test` folder, let's create a `service` subfolder where all our service tests will go. Finally, let's create our first test file inside it, called `SpotifyService.spec.ts`.

Now we can start putting this test file together. The first thing we need to do is import the test helpers from the `@angular/core/testing` package:

**code/routes/music/test/services/SpotifyService.spec.ts**

```
1  import {
2    inject,
3    fakeAsync,
4    tick,
5    TestBed
6  } from '@angular/core/testing';
```

Next, we'll import a couple more classes:

**code/routes/music/test/services/SpotifyService.spec.ts**

```
7  import {MockBackend} from '@angular/http/testing';
8  import {
9    Http,
10   ConnectionBackend,
11   BaseRequestOptions,
12   Response,
13   ResponseOptions
14 } from '@angular/http';
```

Since our service uses HTTP requests, we'll import the `MockBackend` class from `@angular/http/testing` package. This class will help us set expectations and verify HTTP requests.

The last thing we need to import is the class we're testing:

**code/routes/music/test/services/SpotifyService.spec.ts**

```
16  import {SpotifyService} from '../../app/ts/services/SpotifyService';
```

## HTTP Considerations

We could start writing our tests right now, but during each test execution we would be calling out and hitting the Spotify server. This is far from ideal for two reasons:

1. HTTP requests are relatively slow and as our test suite grows, we'd notice it takes longer and longer to run all of the tests.
2. Spotify's API has a quota, and if our whole team is running the tests, we might use up our API call resources needlessly
3. If we are offline or if Spotify is down or inaccessible our tests would start breaking, even though our code might technically be correct

This is a good hint when writing unit tests: isolate everything that you don't control before testing.

In our case, this piece is the Spotify service. The solution is that we will replace the HTTP request with something that would behave like it, but will **not hit the real Spotify server**.

Doing this in the testing world is called *mocking* a dependency. They are sometimes also called *stubbing* a dependency.

You can read more about the difference between Mocks and Stubs in this article Mocks are not Stubs[116]

Let's pretend we're writing code that depends on a given `Car` class.

This class has a bunch of methods: you can `start` a car instance, `stop` it, `park` it and `getSpeed` of that car.

Let's see how we could use stubs and mocks to write tests that depend on this class.

## Stubs

**Stubs** are objects we create on the fly, with a subset of the behaviors our dependency has.

Let's write a test that just interacts with the `start` method of the class.

You could create a *stub* of that `Car` class on-the-fly and inject that into the class you're testing:

[116]http://martinfowler.com/articles/mocksArentStubs.html

```
describe('Speedtrap', function() {
  it('tickets a car at more than 60mph', function() {
    var stubCar = { getSpeed: function() { return 61; } };
    var speedTrap = new SpeedTrap(stubCar);
    speedTrap.ticketCount = 0;
    speedTrap.checkSpeed();
    expect(speedTrap.ticketCount).toEqual(1);
  });
});
```

This would be a typical case for using a stub and we'd probably only use it locally to that test.

## Mocks

**Mocks** in our case will be a more complete representation of objects, that overrides parts or all of the behavior of the dependency. Mocks can, and most of the time will be reused by more than one test across our suite.

They will also be used sometimes to assert that given methods were called the way they were supposed to be called.

One example of a mock version of our `Car` class would be:

```
class MockCar {
  startCallCount: number = 0;

  start() {
    this.startCallCount++;
  }
}
```

And it would be used to write another test like this:

```
describe('CarRemote', function() {
  it('starts the car when the start key is held', function() {
    var car = new MockCar();
    var remote = new CarRemote();
    remote.holdButton('start');
    expect(car.startCallCount).toEqual(1);
  });
});
```

The biggest difference between a mock and a stub is that:

- a stub provides a subset of functionality with "manual" behavior overrides whereas
- a mock generally sets expectations and verifies that certain methods were called

### Http MockBackend

Now that we have this background in mind, let's go back to writing our service test code.

Interacting with the live Spotify service every time we run our tests is a poor idea but thankfully Angular provides us with a way to create fake HTTP calls with `MockBackend`.

This class can be injected into a `Http` instance and gives us control of how we want the HTTP interaction to act. We can interfere and assert in a variety of different ways: we can manually set a response, simulate an HTTP error, and add expectations, like asserting the URL being requested matches what we want, if the provided request parameters are correct and a lot more.

So the idea here is that we're going to provide our code with a "fake" `Http` library. This "fake" library will appear to our code to be the real `Http` library: all of the methods will match, it will return responses and so on. However, we're not *actually* going to make the requests.

In fact, beyond not making the requests, our `MockBackend` will actually allow us to setup *expectations* and watch for behaviors we expect.

### `TestBed.configureTestingModule` and Providers

When we test our Angular apps we need to make sure we configure the top-level `NgModule` that we will use for this test. When we do this, we can configure providers, declare components, and import other modules: just like you would when using `NgModules` generally.

Sometimes when testing Angular code, we *manually setup injections*. This is good because it gives us more control over what we're actually testing.

So in the case of testing `Http` requests, we don't want to inject the "real" `Http` class, but instead we want to inject something that looks like `Http`, but really intercepts the requests and returns the responses we configure.

To do that, we create a version of the `Http` class that uses `MockBackend` internally.

To do this, we use the `TestBed.configureTestingModule` in the `beforeEach` hook. This hook takes a callback function that will be called before each test is run, giving us a great opportunity to configure alternative class implementations.

**code/routes/music/test/services/SpotifyService.spec.ts**

```
18 describe('SpotifyService', () => {
19   beforeEach(() => {
20     TestBed.configureTestingModule({
21       providers: [
22         BaseRequestOptions,
23         MockBackend,
24         SpotifyService,
25         { provide: Http,
```

```
26          useFactory: (backend: ConnectionBackend,
27                       defaultOptions: BaseRequestOptions) => {
28                         return new Http(backend, defaultOptions);
29                       }, deps: [MockBackend, BaseRequestOptions] },
30        ]
31      });
32    });
```

Notice that `TestBed.configureTestingModule` accepts an **array of providers** in the `providers` key to be used by the test injector.

`BaseRequestOptions` and `SpotifyService` are just the default implementation of those classes. But the last provider is a little more complicated :

**code/routes/music/test/services/SpotifyService.spec.ts**

```
25        { provide: Http,
26          useFactory: (backend: ConnectionBackend,
27                       defaultOptions: BaseRequestOptions) => {
28                         return new Http(backend, defaultOptions);
29                       }, deps: [MockBackend, BaseRequestOptions] },
30        ]
```

This code uses `provide` with `useFactory` to create a version of the `Http` class, using a factory (that's what `useFactory` does).

That factory has a signature that expects `ConnectionBackend` and a `BaseRequestOption` instances. The second key on that object is `deps: [MockBackend, BaseRequestOptions]`. That indicates that we'll be using `MockBackend` as the first parameter of the factory and `BaseRequestOptions` (the default implementation) as the second.

Finally, we return our customized `Http` class with the `MockBackend` as a result of that function.

What benefit do we get from this? Well now every time (in our test) that our code requests `Http` as an injection, it will instead receive our customized `Http` instance.

This is a powerful idea that we'll use a lot in testing: use dependency injection to customize dependencies and isolate the functionality you're trying to test.

## Testing `getTrack`

Now, when writing tests for the service, we want to verify that we're calling the correct URL.

If you haven't looked at the Routing chapter music example in a while, you can find the code for this example here

Let's write a test for the `getTrack` method:

**code/routes/music/app/ts/services/SpotifyService.ts**

```
39   getTrack(id: string): Observable<any[]> {
40     return this.query(`/tracks/${id}`);
41   }
```

If you remember how that method works, it uses the `query` method, that builds the URL based on the parameters it receives:

**code/routes/music/app/ts/services/SpotifyService.ts**

```
19   query(URL: string, params?: Array<string>): Observable<any[]> {
20     let queryURL: string = `${SpotifyService.BASE_URL}${URL}`;
21     if (params) {
22       queryURL = `${queryURL}?${params.join('&')}`;
23     }
24 
25     return this.http.request(queryURL).map((res: any) => res.json());
26   }
```

Since we're passing `/tracks/${id}` we assume that when calling `getTrack('TRACK_ID')` the expected URL will be `https://api.spotify.com/v1/tracks/TRACK_ID`.

Here is how we write the test for this:

```
 1 describe('getTrack', () => {
 2   it('retrieves using the track ID',
 3     inject([SpotifyService, MockBackend], fakeAsync((spotifyService, mockBackend\
 4 ) => {
 5       var res;
 6       mockBackend.connections.subscribe(c => {
 7         expect(c.request.url).toBe('https://api.spotify.com/v1/tracks/TRACK_ID');
 8         let response = new ResponseOptions({body: '{"name": "felipe"}'});
 9         c.mockRespond(new Response(response));
10       });
11       spotifyService.getTrack('TRACK_ID').subscribe((_res) => {
12         res = _res;
13       });
14       tick();
15       expect(res.name).toBe('felipe');
16     }))
17   );
18 });
```

This seems like a lot to grasp at first, so let's break it down a bit:

Every time we write tests with dependencies, we need to ask Angular injector to provide us with the instances of those classes. To do that we use:

```
1  inject([Class1, ..., ClassN], (instance1, ..., instanceN) => {
2    ... testing code ...
3  })
```

When you are testing code that returns either a Promise or an RxJS Observable, you can use `fakeAsync` helper to test that code as if it were synchronous. This way every Promises are fulfilled and Observables are notified immediately after you call `tick()`.

So in this code:

```
1  inject([SpotifyService, MockBackend], fakeAsync((spotifyService, mockBackend) =>\
2   {
3    ...
4  }));
```

We're getting two variables: `spotifyService` and `mockBackend`. The first one has a concrete instance of the `SpotifyService` and the second is an instance `MockBackend` class. Notice that the arguments to the inner function (`spotifyService, mockBackend`) are injections of the classes specified in the first argument array of the `inject` function (`SpotifyService` and `MockBackend`).

We're also running inside `fakeAsync` which means that async code will be run synchronously when `tick()` is called.

Now that we've setup the injections and context for our test, we can start writing our "actual" test. We start by declaring a `res` variable that will eventually get the HTTP call response. Next we subscribe to `mockBackend.connections`:

```
1  var res;
2  mockBackend.connections.subscribe(c => { ... });
```

Here we're saying that whenever a new connection comes in to `mockBackend` we want to be notified (e.g. call this function).

We want to verify that the `SpotifyService` is calling out to the correct URL given the track id `TRACK_ID`. So what we do is specify an *expectation* that the URL is as we would expect. We can get the URL from the connection `c` via `c.request.url`. So we setup an expectation that `c.request.url` should be the string `'https://api.spotify.com/v1/tracks/TRACK_ID'`:

```
1 expect(c.request.url).toBe('https://api.spotify.com/v1/tracks/TRACK_ID');
```

When our test is run, if the request URL doesn't match, then the test will fail.

Now that we've received our request and verified that it is correct, we need to craft a response. We do this by creating a new `ResponseOptions` instance. Here we specify that it will return the JSON string: `{"name": "felipe"}` as the body of the response.

```
1 let response = new ResponseOptions({body: '{"name": "felipe"}'});
```

Finally, we tell the connection to replace the response with a `Response` object that wraps the `ResponseOptions` instance we created:

```
1 c.mockRespond(new Response(response));
```

An interesting thing to note here is that your callback function in `subscribe` can be as sophisticated as you wish it to be. You could have conditional logic based on the URL, query parameters, or anything you can read from the request object etc.

This allows us to write tests for nearly every possible scenario our code might encounter.

We have now everything setup to call the `getTrack` method with `TRACK_ID` as a parameter and tracking the response in our `res` variable:

```
1 spotifyService.getTrack('TRACK_ID').subscribe((_res) => {
2   res = _res;
3 });
```

If we ended our test here, we would be waiting for the HTTP call to be made and the response to be fulfilled before the callback function would be triggered. It would also happen on a different execution path and we'd have to orchestrate our code to sync things up. Thankfully using `fakeAsync` takes that problem away. All we need to do is call `tick()` and, like magic, our async code will be executed:

```
1 tick();
```

We now perform one final check just to make sure our response we setup is the one we received:

```
1 expect(res.name).toBe('felipe');
```

If you think about it, the code for all the methods of this service are *very* similar. So let's extract the snippet we use to setup the URL expectation into a function called expectURL:

**code/routes/music/test/services/SpotifyService.spec.ts**

```
35    function expectURL(backend: MockBackend, url: string) {
36      backend.connections.subscribe(c => {
37        expect(c.request.url).toBe(url);
38        let response = new ResponseOptions({body: '{"name": "felipe"}'});
39        c.mockRespond(new Response(response));
40      });
41    }
```

Following the same lines, it should be very simple to create similar tests for `getArtist` and `getAlbum` methods:

**code/routes/music/test/services/SpotifyService.spec.ts**

```
57    describe('getArtist', () => {
58      it('retrieves using the artist ID',
59        inject([SpotifyService, MockBackend], fakeAsync((svc, backend) => {
60          var res;
61          expectURL(backend, 'https://api.spotify.com/v1/artists/ARTIST_ID');
62          svc.getArtist('ARTIST_ID').subscribe((_res) => {
63            res = _res;
64          });
65          tick();
66          expect(res.name).toBe('felipe');
67        }))
68      );
69    });
70
71    describe('getAlbum', () => {
72      it('retrieves using the album ID',
73        inject([SpotifyService, MockBackend], fakeAsync((svc, backend) => {
74          var res;
75          expectURL(backend, 'https://api.spotify.com/v1/albums/ALBUM_ID');
76          svc.getAlbum('ALBUM_ID').subscribe((_res) => {
77            res = _res;
78          });
79          tick();
80          expect(res.name).toBe('felipe');
81        }))
82      );
83    });
```

Now `searchTrack` is slightly different: instead of calling `query`, this method uses the `search` method:

**code/routes/music/app/ts/services/SpotifyService.ts**

```
35   searchTrack(query: string): Observable<any[]> {
36     return this.search(query, 'track');
37   }
```

And then `search` calls `query` with `/search` as the first argument and an Array containing `q=<query>` and `type=track` as the second argument:

**code/routes/music/app/ts/services/SpotifyService.ts**

```
28   search(query: string, type: string): Observable<any[]> {
29     return this.query(`/search`, [
30       `q=${query}`,
31       `type=${type}`
32     ]);
33   }
```

Finally, `query` will transform the parameters into a URL *path* with a *QueryString*. So now, the URL we expect to call ends with **/search?q=<query>&type=track**.

Let's now write the test for `searchTrack` that takes into consideration what we learned above:

**code/routes/music/test/services/SpotifyService.spec.ts**

```
85   describe('searchTrack', () => {
86     it('searches type and term',
87       inject([SpotifyService, MockBackend], fakeAsync((svc, backend) => {
88         var res;
89         expectURL(backend, 'https://api.spotify.com/v1/search?q=TERM&type=track'\
90 );
91         svc.searchTrack("TERM").subscribe((_res) => {
92           res = _res;
93         });
94         tick();
95         expect(res.name).toBe('felipe');
96       }))
97     );
98   });
```

The test ended up also being very similar to the ones we wrote so far. Let's review what this test does:

- it hooks into the HTTP lifecycle, by adding a callback when a new HTTP connection is initiated
- it sets an expectation for the URL we expect the connection to use including the query type and the search term
- it calls the method we're testing, `searchTrack`
- it then tells Angular to complete all the pending async calls
- it finally asserts that we have the expected response

In essence, when testing services our goals should be:

1. Isolate all the dependencies by using stubs or mocks
2. In case of async calls, use `fakeAsync` and `tick` to make sure they are fulfilled
3. Call the service method you're testing
4. Assert that the returning value from the method matches what we expect

Now let's move on to the classes that usually consume the services: components.

## Testing Routing to Components

When testing components, we can either:

1. write tests that will interact with the component from the outside, passing attributes in and checking how the markup is affected or
2. test individual component methods and their output.

Those test strategies are known as **black box** and **white box** testing, respectively. During this section, we'll see a mix of both.

We'll begin by writing tests for the `ArtistComponent` class, which is one of the simpler components we have. This initial set of tests will test the component's internals, so it falls into the **white box** category of testing.

Before we jump into it, let's remember what `ArtistComponent` does:

The first thing we do on the class constructor is retrieve the **id** from the routeParams collection:

**code/routes/music/app/ts/components/ArtistComponent.ts**

```
31   constructor(private route: ActivatedRoute, private spotify: SpotifyService,
32               private location: Location) {
33     route.params.subscribe(params => { this.id = params['id']; });
34   }
```

And with that we have our first obstacle. How can we retrieve the ID of a route without an available running router?

## Creating a Router for Testing

Remember that when we write tests in Angular we manually configure many of the classes that are injected. Routing (and testing components) has a daunting number of dependencies that we need to inject. That said, once it's configured, it isn't something we change very much and it's very easy to use.

When we test write tests it's often convenient to use `beforeEach` with `TestBed.configureTestingModule` to set the dependencies that can be injected. In the case of testing our `ArtistComponent` we're going to create a custom function that will create and configure our router for testing:

**code/routes/music/test/components/ArtistComponent.spec.ts**

```
16 describe('ArtistComponent', () => {
17   beforeEach(() => {
18     configureMusicTests();
19   });
```

We define `configureMusicTests` in the helper file `MusicTestHelpers.ts`. Let's look at that now.

Here's the implementation of `configureMusicTests`. Don't worry, we'll explain each part:

**code/routes/music/test/MusicTestHelpers.ts**

```
66 export function configureMusicTests() {
67   const mockSpotifyService: MockSpotifyService = new MockSpotifyService();
68
69   TestBed.configureTestingModule({
70     imports: [
71       { // TODO RouterTestingModule.withRoutes coming soon
72         ngModule: RouterTestingModule,
73         providers: [provideRoutes(routerConfig)]
74       },
75       TestModule
```

```
76      ],
77      providers: [
78        mockSpotifyService.getProviders(),
79        {
80          provide: ActivatedRoute,
81          useFactory: (r: Router) => r.routerState.root, deps: [ Router ]
82        }
83      ]
84    });
85  }
```

We start by creating an instance of `MockSpotifyService` that we will use to mock the real implementation of `SpotifyService`.

Next we use a class called `TestBed` and call `configureTestingModule`. `TestBed` is a helper library that ships with Angular to help make testing easier.

In this case, `TestBed.configureTestingModule` is used to configure the `NgModule` used for testing. You can see that we provide an `NgModule` configuration as the argument which has:

- `imports` and
- `providers`

In our `imports` we're importing

- The `RouterTestingModule` and configuring it with our `routerConfig` - this configures the routes for testing
- The `TestModule` - which is the `NgModule` which declares all of the components we will test (see `MusicTestHelpers.ts` for the full details)

In `providers`

- We provide the `MockSpotifyService` (via `mockSpotifyService.getProviders()`)
- and the `ActivatedRoute`

Let's take a closer look at these starting with the `Router`.

#### `Router`

One thing we haven't talked about yet is what routes we want to use when testing. There are many different ways of doing this. First we'll look at what we're doing here:

**code/routes/music/test/MusicTestHelpers.ts**

```
30  @Component({
31    selector: 'blank-cmp',
32    template: ``
33  })
34  export class BlankCmp {
35  }
36
37  @Component({
38    selector: 'root-cmp',
39    template: `<router-outlet></router-outlet>`
40  })
41  export class RootCmp {
42  }
43
44  export const routerConfig: Routes = [
45    { path: '', component: BlankCmp },
46    { path: 'search', component: SearchComponent },
47    { path: 'artists/:id', component: ArtistComponent },
48    { path: 'tracks/:id', component: TrackComponent },
49    { path: 'albums/:id', component: AlbumComponent }
50  ];
```

Here instead of redirecting (like we do in the real router config) for the empty URL, we're just using `BlankCmp`.

Of course, if you want to use the same `RouterConfig` as in your top-level app then all you need to do is `export` it somewhere and `import` it here.

If you have a more complex scenario where you need to test lots of different route configurations, you could even accept a parameter to the `musicTestProviders` function where you use a new router configuration each time.

There are many possibilities here and you'll need to pick whichever fits best for your team. This configuration works for cases where your routes are relatively static and one configuration works for all of the tests.

Now that we have all of the dependencies, we create the `new Router` and call `r.initialNavigation()` on it.

#### `ActivatedRoute`

The `ActivatedRoute` service keeps track of the "current route". It requires the `Router` itself as a dependency so we put it in `deps` and inject it.

**MockSpotifyService**

Earlier we tested our `SpotifyService` by mocking out the HTTP library that backed it. Instead here, we're going to **mock out the whole service itself**. Let's look at how we can mock out this, or any, service.

## Mocking dependencies

If you look inside `music/test` you'll find a `mocks/spotify.ts` file. Let's take a look:

**code/routes/music/test/mocks/spotify.ts**

```
import {SpyObject} from './helper';
import {SpotifyService} from '../../app/ts/services/SpotifyService';

export class MockSpotifyService extends SpyObject {
  getAlbumSpy;
  getArtistSpy;
  getTrackSpy;
  searchTrackSpy;
  mockObservable;
  fakeResponse;
```

Here we're declaring the `MockSpotifyService` class, which will be a mocked version of the real `SpotifyService`. These instance variables will act as *spies*.

## Spies

A *spy* is a specific type of mock object that gives us two benefits:

1. we can simulate return values and
2. count how many times the method was called and with which parameters.

In order to use spies with Angular, we're using the internal `SpyObject` class (it's used by Angular to test itself).

You can either declare a class by creating a new `SpyObject` on the fly or you can make your mock class inherit from `SpyObject`, like we're doing in our code.

The great thing inheriting or using this class gives us is the `spy` method. The `spy` method lets us override a method and force a return value (as well as watch and ensure the method was called). We use `spy` on our class constructor:

**code/routes/music/test/mocks/spotify.ts**

```
12   constructor() {
13     super(SpotifyService);
14 
15     this.fakeResponse = null;
16     this.getAlbumSpy = this.spy('getAlbum').andReturn(this);
17     this.getArtistSpy = this.spy('getArtist').andReturn(this);
18     this.getTrackSpy = this.spy('getTrack').andReturn(this);
19     this.searchTrackSpy = this.spy('searchTrack').andReturn(this);
20   }
```

The first line of the constructor call's the `SpyObject` constructor, passing the concrete class we're mocking. Calling `super(...)` is optional, but when you do the mock class will inherit all the concrete class methods, so you can override just the pieces you're testing.

If you're curious about how `SpyObject` is implemented you can check it on the `angular/angular` repository, on the file `/modules/angular2/src/testing/testing_-internal.ts`[117]

After calling `super`, we're intializing the `fakeResponse` field, that we'll use later to `null`.

Next we declare spies that will replace the concrete class methods. Having a reference to them will be helpful to set expectations and simulate responses while writing our tests.

When we use the `SpotifyService` within the `ArtistComponent`, the real `getArtist` method returns an `Observable` and the method we're calling from our components is the subscribe method:

**code/routes/music/app/ts/components/ArtistComponent.ts**

```
36   ngOnInit(): void {
37     this.spotify
38       .getArtist(this.id)
39       .subscribe((res: any) => this.renderArtist(res));
40   }
```

However, in our mock service, we're going to do something tricky: instead of returning an observable from `getArtist`, we're returning `this`, the `MockSpotifyService` itself. That means the return value of `this.spotify.getArtist(this.id)` above will be the `MockSpotifyService`.

There's one problem with doing this though: our `ArtistComponent` was expecting to call `subscribe` on an Observable. To account for this, we're going to define `subscribe` on our `MockSpotifyService`:

[117] https://github.com/angular/angular/blob/b0cebdba6b65c1e9e7eb5bf801ea42dc7c4a7f25/modules/angular2/src/testing/testing_internal.ts#L205

**code/routes/music/test/mocks/spotify.ts**

```
22    subscribe(callback) {
23      callback(this.fakeResponse);
24    }
```

Now when `subscribe` is called on our mock, we're immediately calling the callback, making the async call happen synchronously.

The other thing you'll notice is that we're calling the callback function with `this.fakeResponse`. This leads us to the next method:

**code/routes/music/test/mocks/spotify.ts**

```
26    setResponse(json: any): void {
27      this.fakeResponse = json;
28    }
```

This method doesn't replace anything on the concrete service, but is instead a helper method to allow the test code to set a given response (that would come from the service on the concrete class) and with that simulate different responses.

**code/routes/music/test/mocks/spotify.ts**

```
30    getProviders(): Array<any> {
31      return [{ provide: SpotifyService, useValue: this }];
32    }
```

This last method is a helper method to be used in `TestBed.configureTestingModule` `providers` like we'll see later when we get back to writing component tests.

Here's what our `MockSpotifyService` looks like altogether:

**code/routes/music/test/mocks/spotify.ts**

```
1   import {SpyObject} from './helper';
2   import {SpotifyService} from '../../app/ts/services/SpotifyService';
3
4   export class MockSpotifyService extends SpyObject {
5     getAlbumSpy;
6     getArtistSpy;
7     getTrackSpy;
8     searchTrackSpy;
9     mockObservable;
```

```
10    fakeResponse;
11
12    constructor() {
13      super(SpotifyService);
14
15      this.fakeResponse = null;
16      this.getAlbumSpy = this.spy('getAlbum').andReturn(this);
17      this.getArtistSpy = this.spy('getArtist').andReturn(this);
18      this.getTrackSpy = this.spy('getTrack').andReturn(this);
19      this.searchTrackSpy = this.spy('searchTrack').andReturn(this);
20    }
21
22    subscribe(callback) {
23      callback(this.fakeResponse);
24    }
25
26    setResponse(json: any): void {
27      this.fakeResponse = json;
28    }
29
30    getProviders(): Array<any> {
31      return [{ provide: SpotifyService, useValue: this }];
32    }
33  }
```

## Back to Testing Code

Now that we have all our dependencies under control, it is easier to write our tests. Let's write our test for our `ArtistComponent`.

As usual, we start with imports:

**code/routes/music/test/components/ArtistComponent.spec.ts**

```
1  import {
2    inject,
3    fakeAsync,
4  } from '@angular/core/testing';
5  import { Router } from '@angular/router';
6  import { Location } from '@angular/common';
7  import { MockSpotifyService } from '../mocks/spotify';
8  import { SpotifyService } from '../../app/ts/services/SpotifyService';
9  import {
```

```
10    advance,
11    createRoot,
12    RootCmp,
13    configureMusicTests
14  } from '../MusicTestHelpers';
```

Next, before we can start to describe our tests `configureMusicTests` to ensure we can access our `musicTestProviders` in each test:

**code/routes/music/test/components/ArtistComponent.spec.ts**

```
16  describe('ArtistComponent', () => {
17    beforeEach(() => {
18      configureMusicTests();
19    });
```

Next, we'll write a test for everything that happens during the initialization of the component. First, let's take a refresh look at what happens on initialization of our `ArtistComponent`:

**code/routes/music/app/ts/components/ArtistComponent.ts**

```
27  export class ArtistComponent implements OnInit {
28    id: string;
29    artist: Object;
30
31    constructor(private route: ActivatedRoute, private spotify: SpotifyService,
32                private location: Location) {
33      route.params.subscribe(params => { this.id = params['id']; });
34    }
35
36    ngOnInit(): void {
37      this.spotify
38        .getArtist(this.id)
39        .subscribe((res: any) => this.renderArtist(res));
40    }
```

Remember that during the creation of the component, we use `route.params` to retrieve the current route `id` param and store it on the `id` attribute of the class.

When the component is initialized `ngOnInit` is triggered by Angular (because we declared that this component `implements OnInit`. We then use the `SpotifyService` to retrieve the artist for the received `id`, and we subscribe to the returned `observable`. When the artist is finally retrieved, we call renderArtist, passing the artist data.

An important idea here is that we used dependency injection to get the `SpotifyService`, but remember, **we created a `MockSpotifyService`!**

So in order to test this behavior, let's:

1. Use our router to navigate to the `ArtistComponent`, which will initialize the component
2. Check our `MockSpotifyService` and ensure that the `ArtistComponent` did, indeed, try to get the artist with the appropriate id.

Here's the code for our test:

**code/routes/music/test/components/ArtistComponent.spec.ts**

```
21    describe('initialization', () => {
22      it('retrieves the artist', fakeAsync(
23        inject([Router, SpotifyService],
24               (router: Router,
25                mockSpotifyService: MockSpotifyService) => {
26          const fixture = createRoot(router, RootCmp);
27
28          router.navigateByUrl('/artists/2');
29          advance(fixture);
30
31          expect(mockSpotifyService.getArtistSpy).toHaveBeenCalledWith('2');
32        })));
33    });
```

Let's take it step by step.

### `fakeAsync` and `advance`

We start by wrapping the test in `fakeAsync`. Without getting too bogged down in the details, by using `fakeAsync` we're able to have more control over when change detection and asynchronous operations occur. A consequence of this is that we need to explicitly tell our components that they need to detect changes after we make changes in our tests.

Normally you don't need to worry about this when writing your apps, as zones tend to do the right thing, but during tests we manipulate the change detection process more carefully.

If you skip a few lines down you'll notice that we're using a function called `advance` that comes from our `MusicTestHelpers`. Let's take a look at that function:

**code/routes/music/test/MusicTestHelpers.ts**

```
52 export function advance(fixture: ComponentFixture<any>): void {
53   tick();
54   fixture.detectChanges();
55 }
```

So we see here that `advance` does two things:

1. It tells the component to detect changes and
2. Calls `tick()`

When we use `fakeAsync`, timers are actually synchronous and we use `tick()` to simulate the asynchronous passage of time.

Practically speaking, in our tests we'll call `advance` whenever we want Angular to "work it's magic". So for instance, whenever we navigate to a new route, update a form element, make an HTTP request etc. we'll call `advance` to give Angular a chance to do it's thing.

#### `inject`

In our test we need some dependencies. We use `inject` to get them. The `inject` function takes two arguments:

1. An array of *tokens* to inject
2. A function into which to provide the injections

And what classes will `inject` use? The providers we defined in `TestBed.configureTestingModule providers`.

Notice that we're injecting:

1. `Router`
2. `SpotifyService`

The `Router` that will be injected is the `Router` we configured in `musicTestProviders` above.

For `SpotifyService`, notice that we're requesting injection of the *token* `SpotifyService`, but we're receiving a `MockSpotifyService`. A little tricky, but hopefully it makes sense given what we've talked about so far.

## Testing `ArtistComponent`'s Initialization

Let's review the contents of our actual test:

**code/routes/music/test/components/ArtistComponent.spec.ts**

```
26        const fixture = createRoot(router, RootCmp);
27
28        router.navigateByUrl('/artists/2');
29        advance(fixture);
30
31        expect(mockSpotifyService.getArtistSpy).toHaveBeenCalledWith('2');
```

We start by creating an instance of our `RootCmp` by using `createRoot`. Let's look at the `createRoot` helper function:

**code/routes/music/test/MusicTestHelpers.ts**

```
57  export function createRoot(router: Router,
58                             componentType: any): ComponentFixture<any> {
59    const f = TestBed.createComponent(componentType);
60    advance(f);
61    (<any>router).initialNavigation();
62    advance(f);
63    return f;
64  }
```

Notice here that when we call `createRoot` we

1. Create an instance of the root component
2. `advance` it
3. Tell the router to setup it's `initialNavigation`
4. `advance` again
5. return the new root component.

This is something we'll do a lot when we want to test a component that depends on routing, so it's handy to have this helper function around.

Notice that we're using the `TestBed` library again to call `TestBed.createComponent`. This function creates a component of the appropriate type.

`RootCmp` is an empty component that we created in `MusicTestHelpers`. You definitely don't need to create an empty component for your root component, but I like to do it this way because it lets us test our child component (`ArtistComponent`) more-or-less in isolation. That is, we don't have to worry about the effects of the parent app component.

That said, maybe you *want* to make sure that the child component operates correctly in context. In that case instead of using `RootCmp` you'd probably want to use your app's normal parent component.

Next we use `router` to navigate to the url `/artists/2` and `advance`. When we navigate to that URL, `ArtistComponent` should be initialized, so we assert that the `getArtist` method of the `SpotifyService` was called with the proper value.

## Testing `ArtistComponent` Methods

Recall that the `ArtistComponent` has an `href` which calls the `back()` function.

**code/routes/music/app/ts/components/ArtistComponent.ts**

```
42   back(): void {
43     this.location.back();
44   }
```

Let's test that when the `back` method is called, the router will redirect the user back to the previous location.

The current location state is controlled by the `Location` service. When we need to send the user back to the previous location, we use the `Location`'s `back` method.

Here is how we test the `back` method:

**code/routes/music/test/components/ArtistComponent.spec.ts**

```
35   describe('back', () => {
36     it('returns to the previous location', fakeAsync(
37       inject([Router, Location],
38              (router: Router, location: Location) => {
39         const fixture = createRoot(router, RootCmp);
40         expect(location.path()).toEqual('/');
41
42         router.navigateByUrl('/artists/2');
43         advance(fixture);
44         expect(location.path()).toEqual('/artists/2');
45
46         const artist = fixture.debugElement.children[1].componentInstance;
47         artist.back();
48         advance(fixture);
49
50         expect(location.path()).toEqual('/');
51       })));
52   });
```

The initial structure is similar: we inject our dependencies and create a new component.

We have a new `expectation` - we assert that the `location.path()` is equal to what we expect it to be.

We also have another new idea: we're accessing the methods on the `ArtistComponent` itself. We get a reference to our `ArtistComponent` instance through the line

`fixture.debugElement.children[1].componentInstance.`

Now that we have the instance of the component, we're able to call methods on it directly, like `back()`.

After we call `back()` we `advance` and then verify that the `location.path()` is what we `expected` it to be.

## Testing `ArtistComponent` DOM Template Values

The last thing we need to test on `ArtistComponent` is the template that renders the artist.

**code/routes/music/app/ts/components/ArtistComponent.ts**

```
15    template: `
16    <div *ngIf="artist">
17      <h1>{{ artist.name }}</h1>
18
19      <p>
20        <img src="{{ artist.images[0].url }}">
21      </p>
22
23      <p><a href (click)="back()">Back</a></p>
24    </div>
25    `
```

Remember that the instance variable `artist` is set by the result of the `SpotifyService getArtist` call. Since we're mocking the `SpotifyService` with `MockSpotifyService`, the data we should have in our template should be whatever the `mockSpotifyService` returns. Let's look at how we do this:

**code/routes/music/test/components/ArtistComponent.spec.ts**

```
54    describe('renderArtist', () => {
55      it('renders album info', fakeAsync(
56        inject([Router, SpotifyService],
57               (router: Router,
58                mockSpotifyService: MockSpotifyService) => {
59          const fixture = createRoot(router, RootCmp);
60
61          let artist = {name: 'ARTIST NAME', images: [{url: 'IMAGE_1'}]};
62          mockSpotifyService.setResponse(artist);
63
64          router.navigateByUrl('/artists/2');
65          advance(fixture);
66
67          const compiled = fixture.debugElement.nativeElement;
68
69          expect(compiled.querySelector('h1').innerHTML).toContain('ARTIST NAME');
70          expect(compiled.querySelector('img').src).toContain('IMAGE_1');
71        })));
72    });
```

The first thing that's new here is that we're *manually setting the response* of the `mockSpotifyService` with `setResponse`.

The `artist` variable is a *fixture* that represents what we get from the Spotify API when we call the `artists` endpoint at `GET https://api.spotify.com/v1/artists/{id}`.

Here's what the real JSON looks like:

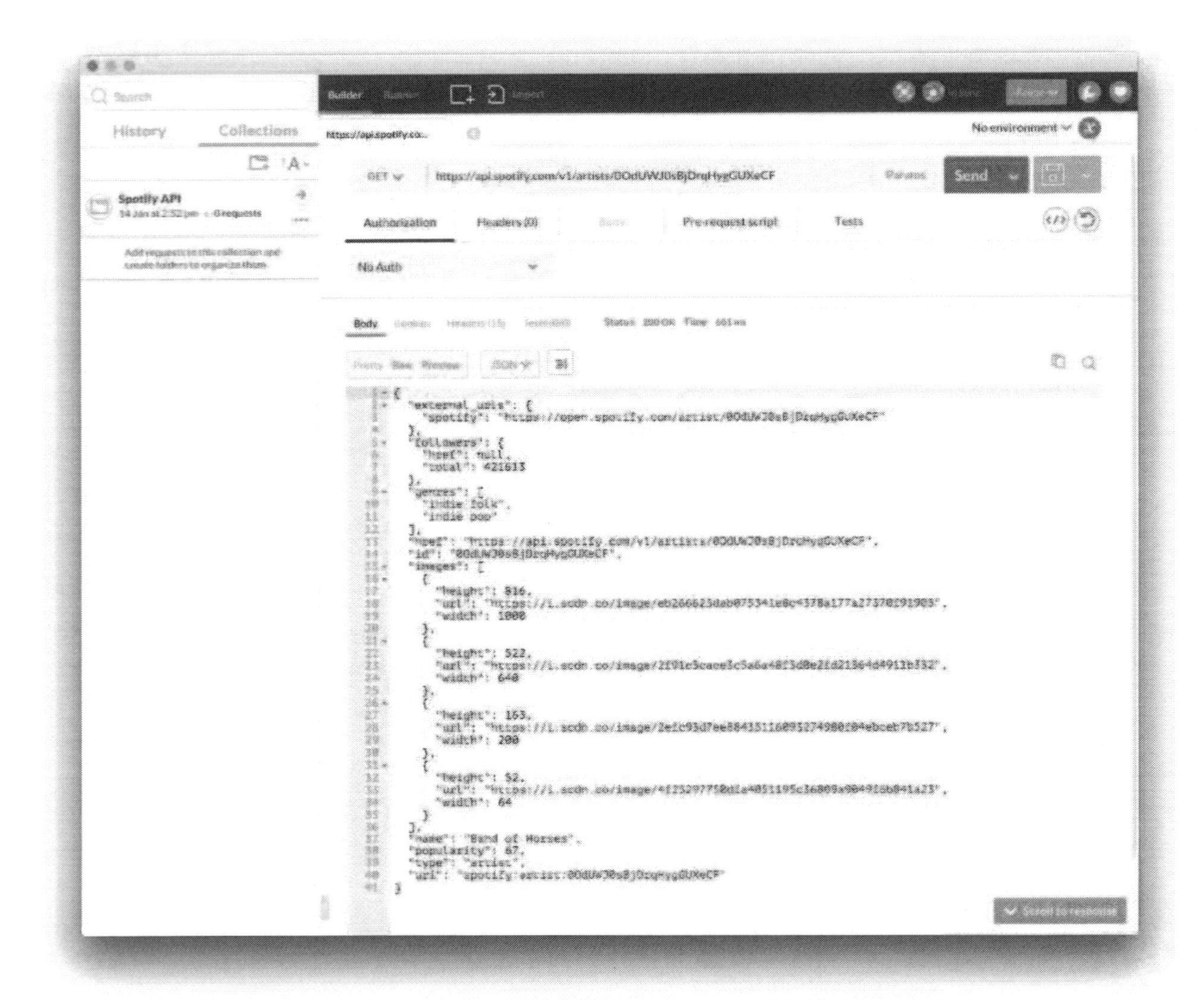

**Postman - Spotify Get Artist Endpoint**

However, for this test we need only the `name` and `images` properties.

When we call the `setResponse` method, that response will be used for the next call we make to any of the service methods. In this case, we want the method `getArtist` to return this response.

Next we navigate with the router and `advance`. Now that the view is rendered, we can use the DOM representation of the component's view to check if the artist was properly rendered.

We do that by getting the `nativeElement` property of the `DebugElement` with the line `fixture.debugElement.nativeElement`.

In our assertions, we expect to see H1 tag containing the artist's name, in our case the string `ARTIST NAME` (because of our `artist` fixture above).

To check those conditions, we use the `NativeElement`'s `querySelector` method. This method will return the first element that matches the provided CSS selector.

For the H1 we check that the text is indeed `ARTIST NAME` and for the image, we check its `src` property is `IMAGE 1`.

With this, we are done testing the `ArtistComponent` class.

## Testing Forms

To write form tests, let's use the `DemoFormNgModel` component we created back in the `Forms` chapter. This example is a good candidate because it uses a few features of Angular's forms:

- it uses a `FormBuilder`
- has validations
- handles events

Here's the full code for that class:

**code/forms/app/forms/demo_form_with_events.ts**

```
1  import { Component } from '@angular/core';
2  import {
3    FormBuilder,
4    FormGroup,
5    Validators,
6    AbstractControl
7  } from '@angular/forms';
8
9  @Component({
10   selector: 'demo-form-with-events',
11   template: `
12   <div class="ui raised segment">
13     <h2 class="ui header">Demo Form: with events</h2>
14     <form [formGroup]="myForm"
15           (ngSubmit)="onSubmit(myForm.value)"
16           class="ui form">
17
18       <div class="field"
19           [class.error]="!sku.valid && sku.touched">
20         <label for="skuInput">SKU</label>
21         <input type="text"
22                class="form-control"
23                id="skuInput"
```

```
24                placeholder="SKU"
25                [formControl]="sku">
26          <div *ngIf="!sku.valid"
27            class="ui error message">SKU is invalid</div>
28          <div *ngIf="sku.hasError('required')"
29            class="ui error message">SKU is required</div>
30       </div>
31
32       <div *ngIf="!myForm.valid"
33         class="ui error message">Form is invalid</div>
34
35       <button type="submit" class="ui button">Submit</button>
36     </form>
37   </div>
38   `
39 })
40 export class DemoFormWithEvents {
41   myForm: FormGroup;
42   sku: AbstractControl;
43
44   constructor(fb: FormBuilder) {
45     this.myForm = fb.group({
46       'sku':  ['', Validators.required]
47     });
48
49     this.sku = this.myForm.controls['sku'];
50
51     this.sku.valueChanges.subscribe(
52       (value: string) => {
53         console.log('sku changed to:', value);
54       }
55     );
56
57     this.myForm.valueChanges.subscribe(
58       (form: any) => {
59         console.log('form changed to:', form);
60       }
61     );
62
63   }
64
65   onSubmit(form: any): void {
```

```
66      console.log('you submitted value:', form.sku);
67    }
68  }
```

Just to recap, this code will have the following behavior:

- when no value is present for the SKU field, two validation error will be displayed: *SKU is invalid* and *SKU is required*
- when the value of the SKU field changes, we are logging a message to the console
- when the form changes, we are also logging to the console
- when the form is submitted, we log yet another final message to the console

It seems that one obvious external dependency we have is the console. As we learned before, we need to somehow mock all external dependencies.

## Creating a `ConsoleSpy`

This time, instead of using a `SpyObject` to create a mock, let's do something simpler, since all we're using from the `console` is the `log` method.

We will replace the original `console` instance, that is held on the `window.console` object and replace by an object we control: a `ConsoleSpy`.

**code/forms/test/util.ts**

```
14  export class ConsoleSpy {
15    public logs: string[] = [];
16    log(...args) {
17      this.logs.push(args.join(' '));
18    }
19    warn(...args) {
20      this.log(...args);
21    }
22  }
```

The `ConsoleSpy` is an object that will take whatever is `logged`, naively convert it to a string, and store it in an internal list of things that were logged.

To accept a variable number of arguments on our version of the `console.log` method, we are using ES6 and TypeScript's *Rest parameters*[118].

This operator, represented by an ellipsis, like `...theArgs` as our function argument. In a nutshell using it indicates that we're going to capture all the remaining arguments from that point on. If we had something like `(a, b, ...theArgs)` and called `func(1, 2, 3, 4, 5)`, `a` would be `1`, `b` would be `2` and `theArgs` would have `[3, 4, 5]`.

You can play with it yourself if you have a recent version of Node.js[119] installed:

```
1  $ node --harmony
2  > var test = (a, b, ...theArgs) => console.log('a=',a,'b=',b,'theArgs=',theArgs);
3  undefined
4  > test(1,2,3,4,5);
5  a= 1 b= 2 theArgs= [ 3, 4, 5 ]
```

So instead of writing it to the console itself, we'll be storing them on an array. If the code under test calls `console.log` three times:

```
1  console.log('First message', 'is', 123);
2  console.log('Second message');
3  console.log('Third message');
```

We expect the `_logs` field to have an array of `['First message is 123', 'Second message', 'Third message']`.

## Installing the `ConsoleSpy`

To use our spy in our test we start by declaring two variables: `originalConsole` will keep a reference to the original console instance, and `fakeConsole` that will hold the *mocked* version of the console. We also declare a few variables that will be helpful in testing our `input` and `form` elements.

**code/forms/test/forms/demo_form_with_events.spec.ts**

```
19  describe('DemoFormWithEvents', () => {
20    let originalConsole, fakeConsole;
21    let el, input, form;
```

And then we can install the fake console and specify our providers:

[118] https://developer.mozilla.org/en/docs/Web/JavaScript/Reference/Functions/rest_parameters
[119] https://nodejs.org/en/

**code/forms/test/forms/demo_form_with_events.spec.ts**

```
23   beforeEach(() => {
24     // replace the real window.console with our spy
25     fakeConsole = new ConsoleSpy();
26     originalConsole = window.console;
27     (<any>window).console = fakeConsole;
28 
29     TestBed.configureTestingModule({
30       imports: [ FormsModule, ReactiveFormsModule ],
31       declarations: [ DemoFormWithEvents ]
32     });
33   });
```

Back to the testing code, the next thing we need to do is replace the real console instance with ours, saving the original instance.

Finally, on the `afterAll` method, we restore the original console instance to make sure it doesn't *leak* into other tests.

**code/forms/test/forms/demo_form_with_events.spec.ts**

```
35   // restores the real console
36   afterAll(() => (<any>window).console = originalConsole);
```

## Configuring the Testing Module

Notice that in the `beforeEach` we call `TestBed.configureTestingModule` - remember that `configureTestingModule` sets up the root `NgModule` for our tests.

In this case we're importing the two forms modules and declaring the `DemoFormWithEvents` component.

Now that we have control of the console, let's begin testing our form.

## Testing The Form

Now we need to test the validation errors and the events of the form.

The first thing we need to do is to get the references to the SKU input field and to the form elements:

**code/forms/test/forms/demo_form_with_events_bad.spec.ts**

```
38    it('validates and triggers events', fakeAsync((tcb) => {
39      let fixture = TestBed.createComponent(DemoFormWithEvents);
40
41      let el = fixture.debugElement.nativeElement;
42      let input = fixture.debugElement.query(By.css('input')).nativeElement;
43      let form = fixture.debugElement.query(By.css('form')).nativeElement;
44      fixture.detectChanges();
```

The last line tells Angular to commit all the pending changes, similar to what we did in the routing section above. Next, we will set the SKU input value to the empty string:

**code/forms/test/forms/demo_form_with_events_bad.spec.ts**

```
46      input.value = '';
47      dispatchEvent(input, 'input');
48      fixture.detectChanges();
49      tick();
```

Here we use `dispatchEvent` to notify Angular that the input element changed, and then we trigger the change detection a second time. Finally we use `tick()` to make sure all asynchronous code triggered up to this point gets executed.

The reason we are using `fakeAsync` and `tick` on this test, is to assure the form events are triggered. If we used `async` and `inject` instead, we would finish the code before the events were triggered.

Now that we have changed the input value, let's make sure the validation is working. We ask the component element (using the `el` variable) for all child elements that are error messages and then making sure we have both error messages displayed:

**code/forms/test/forms/demo_form_with_events_bad.spec.ts**

```
52      let msgs = el.querySelectorAll('.ui.error.message');
53      expect(msgs[0].innerHTML).toContain('SKU is invalid');
54      expect(msgs[1].innerHTML).toContain('SKU is required');
```

Next, we will do something similar, but this time we set a value to the SKU field:

**code/forms/test/forms/demo_form_with_events_bad.spec.ts**

```
57      input.value = 'XYZ';
58      dispatchEvent(input, 'input');
59      fixture.detectChanges();
60      tick();
```

And make sure all the error messages are gone:

**code/forms/test/forms/demo_form_with_events_bad.spec.ts**

```
62      msgs = el.querySelectorAll('.ui.error.message');
63      expect(msgs.length).toEqual(0);
```

Finally, we will trigger the submit event of the form:

**code/forms/test/forms/demo_form_with_events_bad.spec.ts**

```
65      fixture.detectChanges();
66      dispatchEvent(form, 'submit');
67      tick();
```

And finally we make sure the event was kicked by checking that the message we log to the console when the form is submitted is there:

**code/forms/test/forms/demo_form_with_events_bad.spec.ts**

```
69      // checks for the form submitted message
70      expect(fakeConsole.logs).toContain('you submitted value: XYZ');
```

We could continue and add new verifications for the other two events our form triggers: the SKU change and the form change events. However, our test is growing quite long.

When we run our tests, we see it passes:

```
DemoFormWithEvents
  ✓ validates and trigger events
```

**DemoFormWithEvents test output**

This test works, but stylistically we have some code smells:

- a really long `it` condition (more than 5-10 lines)
- more than one or two `expects` per `it` condition
- the word **and** on the test description

## Refactoring Our Form Test

Let's fix that by first extracting the code that creates the component and gets the component element and also the elements for the input and for the form:

**code/forms/test/forms/demo_form_with_events.spec.ts**

```
38  function createComponent(): ComponentFixture<any> {
39    let fixture = TestBed.createComponent(DemoFormWithEvents);
40    el = fixture.debugElement.nativeElement;
41    input = fixture.debugElement.query(By.css('input')).nativeElement;
42    form = fixture.debugElement.query(By.css('form')).nativeElement;
43    fixture.detectChanges();
44
45    return fixture;
46  }
```

The `createComponent` code is pretty straightforward: Creates the component with `TestBed.createComponent`, retrieves all the elements we need and calls `detectChanges`.

Now the first thing we want to test is that given an empty SKU field, we should see two error messages:

**code/forms/test/forms/demo_form_with_events.spec.ts**

```
48  it('displays errors with no sku', fakeAsync( () => {
49    let fixture = createComponent();
50    input.value = '';
51    dispatchEvent(input, 'input');
52    fixture.detectChanges();
53
54    // no value on sku field, all error messages are displayed
55    let msgs = el.querySelectorAll('.ui.error.message');
56    expect(msgs[0].innerHTML).toContain('SKU is invalid');
57    expect(msgs[1].innerHTML).toContain('SKU is required');
58  }));
```

See how much cleaner this is? Our test is focused and tests only one thing. Great job!

This new structure makes adding the second test easy. This time we want to test that, once we add a value to the SKU field, the error messages are gone:

**code/forms/test/forms/demo_form_with_events.spec.ts**

```
60    it('displays no errors when sku has a value', fakeAsync( () => {
61      let fixture = createComponent();
62      input.value = 'XYZ';
63      dispatchEvent(input, 'input');
64      fixture.detectChanges();
65
66      let msgs = el.querySelectorAll('.ui.error.message');
67      expect(msgs.length).toEqual(0);
68    }));
```

One thing you may have noticed is that so far, our tests are not using `fakeAsync`, but `async` plus `inject` instead.

That's another bonus of this refactoring: we will only use `fakeAsync` and `tick()` when we want to check if something was added to the console, because that's all our form's event handlers do.

The next test will do exactly that - when the SKU value changes, we should have a message logged to the console:

**code/forms/test/forms/demo_form_with_events.spec.ts**

```
70    it('handles sku value changes', fakeAsync( () => {
71      let fixture = createComponent();
72      input.value = 'XYZ';
73      dispatchEvent(input, 'input');
74      tick();
75
76      expect(fakeConsole.logs).toContain('sku changed to: XYZ');
77    }));
```

We can write similar code for both the form change...

**code/forms/test/forms/demo_form_with_events.spec.ts**

```
79    it('handles form changes', fakeAsync(() => {
80      let fixture = createComponent();
81      input.value = 'XYZ';
82      dispatchEvent(input, 'input');
83      tick();
84
85      expect(fakeConsole.logs).toContain('form changed to: [object Object]');
86    }));
```

... and the form submission events:

**code/forms/test/forms/demo_form_with_events.spec.ts**

```
  it('handles form submission', fakeAsync((tcb) => {
    let fixture = createComponent();
    input.value = 'ABC';
    dispatchEvent(input, 'input');
    tick();

    fixture.detectChanges();
    dispatchEvent(form, 'submit');
    tick();

    expect(fakeConsole.logs).toContain('you submitted value: ABC');
  }));
```

When we run the tests now, we get a much nicer output:

```
DemoFormWithEvents
  ✓ displays errors with no sku
  ✓ displays no errors when sku has a value
  ✓ handles sku value changes
  ✓ handles form changes
  ✓ handles form submission
```

**DemoFormWithEvents test output after refactoring**

Another great benefit from this refactor can be seen when something goes wrong. Let's go back to the component code and change the message when the form gets submitted, in order to force one of our tests to fail:

```
onSubmit(form: any): void {
  console.log('you have submitted the value:', form.sku);
}
```

If we ran the previous version of the test, here's what would happen:

```
DemoFormWithEvents
  ✗ validates and trigger events
      Expected [ 'sku changed to: ', 'form changed to: [object Object]', 'sku changed to: XYZ', 'form cha
nged to: [object Object]', 'you have submitted the value: XYZ' ] to contain 'you submitted value: XYZ'.
          at /Users/fcoury/code/ng-book2/manuscript/code/forms/test.bundle.js:41894
          at run (/Users/fcoury/code/ng-book2/manuscript/code/forms/test.bundle.js:5942)
          at zoneBoundFn (/Users/fcoury/code/ng-book2/manuscript/code/forms/test.bundle.js:5915)
          at lib$es6$promise$$internal$$tryCatch (/Users/fcoury/code/ng-book2/manuscript/code/forms/test.
```

**DemoFormWithEvents error output before refactoring**

It's not immediately obvious what failed. We have to read the error code to realize it was the submission message that failed. We also can't be sure if that was the only thing that broke on the component code, since we may have other test conditions after the one that failed that never had a chance to be executed.

Now, compare that to the error we get from our refactored code:

```
DemoFormWithEvents
  ✓ displays errors with no sku
  ✓ displays no errors when sku has a value
  ✓ handles sku value changes
  ✓ handles form changes
  ✗ handles form submission
      Expected [ 'sku changed to: ABC', 'form changed to: [object Object]', 'you have submitted the
value: ABC' ] to contain 'you submitted value: ABC'.
          at /Users/fcoury/code/ng-book2/manuscript/code/forms/test.bundle.js:41673
          at run (/Users/fcoury/code/ng-book2/manuscript/code/forms/test.bundle.js:5942)
          at zoneBoundFn (/Users/fcoury/code/ng-book2/manuscript/code/forms/test.bundle.js:5915)
          at lib$es6$promise$$internal$$tryCatch (/Users/fcoury/code/ng-book2/manuscript/code/forms/
```

**DemoFormWithEvents error output after refactoring**

This version makes it pretty obvious that the only thing that failed was the form submission event.

# Testing HTTP requests

We could test the HTTP interaction in our apps using the same strategy as we used so far: write a mock version of the `Http` class, since it is an external dependency.

But since the vast majority of single page apps written using frameworks like Angular use HTTP interaction to talk to APIs, the Angular testing library already provides a built in alternative: `MockBackend`.

We have used this class before in this chapter when we were testing the `SpotifyService` class.

Let's dive a little deeper now and see some more testing scenarios and also some good practices. In order to do this, let's write tests for the examples from the *HTTP chapter*.

First, let's see how we test different HTTP methods, like POST or DELETE and how to test the correct HTTP headers are being sent.

Back on the HTTP chapter, we created this example that covered how to do those things using `Http`.

## Testing a POST

The first test we'll write is to make sure we're doing a proper POST request on the `makePost` method:

**code/http/app/ts/components/MoreHTTPRequests.ts**

```
30   makePost(): void {
31     this.loading = true;
32     this.http.post(
33       'http://jsonplaceholder.typicode.com/posts',
34       JSON.stringify({
35         body: 'bar',
36         title: 'foo',
37         userId: 1
38       }))
39       .subscribe((res: Response) => {
40         this.data = res.json();
41         this.loading = false;
42       });
43   }
```

When writing our test for this method, our goal is to test two things:

1. the request method (POST) is correct and that
2. the URL we're hitting is also correct.

Here's how we turn that into a test:

**code/http/test/MoreHTTPRequests.spec.ts**

```
37   it('performs a POST',
38      async(inject([MockBackend], (backend) => {
39        let fixture = TestBed.createComponent(MoreHTTPRequests);
40        let comp = fixture.debugElement.componentInstance;
41
42        backend.connections.subscribe(c => {
43          expect(c.request.url)
44            .toBe('http://jsonplaceholder.typicode.com/posts');
45          expect(c.request.method).toBe(RequestMethod.Post);
46          c.mockRespond(new Response(<any>{body: '{"response": "OK"}'}));
47        });
48
49        comp.makePost();
```

```
50        expect(comp.data).toEqual({'response': 'OK'});
51      }))
52    );
```

Notice how we have a `subscribe` call to `backend.connections`. This will trigger our code whenever a new HTTP connection is established, giving us an opportunity to peek into the request and also provide the response we want.

This place is where you can:

- add request assertions, like checking the correct URL or HTTP method was requested
- set a mocked response, to force your code to deal with different responses, given different test scenarios

Angular uses an `enum` called `RequestMethod` to identify HTTP methods. Here are the supported methods:

```
1 export enum RequestMethod {
2   Get,
3   Post,
4   Put,
5   Delete,
6   Options,
7   Head,
8   Patch
9 }
```

Finally, after the call `makePost()` we're doing another check to make sure that the mock response we set was the one that was assigned to our component.

Now that we understand how this work, adding a second test for a DELETE method is easy.

## Testing `DELETE`

Here's how the `makeDelete` method is implemented:

**code/http/app/ts/components/MoreHTTPRequests.ts**

```
45    makeDelete(): void {
46      this.loading = true;
47      this.http.delete('http://jsonplaceholder.typicode.com/posts/1')
48        .subscribe((res: Response) => {
49          this.data = res.json();
50          this.loading = false;
51        });
52    }
```

And this is the code we use to test it:

**code/http/test/MoreHTTPRequests.spec.ts**

```
54    it('performs a DELETE',
55       async(inject([MockBackend], (backend) => {
56         let fixture = TestBed.createComponent(MoreHTTPRequests);
57         let comp = fixture.debugElement.componentInstance;
58
59         backend.connections.subscribe(c => {
60           expect(c.request.url)
61             .toBe('http://jsonplaceholder.typicode.com/posts/1');
62           expect(c.request.method).toBe(RequestMethod.Delete);
63           c.mockRespond(new Response(<any>{body: '{"response": "OK"}'}));
64         });
65
66         comp.makeDelete();
67         expect(comp.data).toEqual({'response': 'OK'});
68       }))
69     );
```

Everything here is the same, except for the URL that changes a bit and the HTTP method, which is now `RequestMethod.Delete`.

## Testing HTTP Headers

The last method we have to test on this class is `makeHeaders`:

**code/http/app/ts/components/MoreHTTPRequests.ts**

```
54    makeHeaders(): void {
55      let headers: Headers = new Headers();
56      headers.append('X-API-TOKEN', 'ng-book');
57
58      let opts: RequestOptions = new RequestOptions();
59      opts.headers = headers;
60
61      this.http.get('http://jsonplaceholder.typicode.com/posts/1', opts)
62        .subscribe((res: Response) => {
63          this.data = res.json();
64        });
65    }
```

In this case, what our test should focus on is making sure the header `X-API-TOKEN` is being properly set to `ng-book`:

**code/http/test/MoreHTTPRequests.spec.ts**

```
71    it('sends correct headers',
72       async(inject([MockBackend], (backend) => {
73         let fixture = TestBed.createComponent(MoreHTTPRequests);
74         let comp = fixture.debugElement.componentInstance;
75
76         backend.connections.subscribe(c => {
77           expect(c.request.url)
78             .toBe('http://jsonplaceholder.typicode.com/posts/1');
79           expect(c.request.headers.has('X-API-TOKEN')).toBeTruthy();
80           expect(c.request.headers.get('X-API-TOKEN')).toEqual('ng-book');
81           c.mockRespond(new Response(<any>{body: '{"response": "OK"}'}));
82         });
83
84         comp.makeHeaders();
85         expect(comp.data).toEqual({'response': 'OK'});
86       }))
87     );
```

The connection's `request.headers` attribute returns a `Headers` class instance and we're using two methods to perform two different assertions:

- the `has` method to check whether a given header was set, ignoring it's value

- the `get` method, that returns the value that was set

If having the header set is sufficient, use `has`. Otherwise, if you need to inspect the set value, use `get`.

And with that we finish the tests of different methods and headers on Angular. Time to move to a more complex example, that will be closer to what you will encounter when coding real world applications.

## Testing `YouTubeService`

The other example we built back on the HTTP chapter was a YouTube video search. The HTTP interaction for that example takes place on a service called `YouTubeService`:

**code/http/app/ts/components/YouTubeSearchComponent.ts**

```
47  /**
48   * YouTubeService connects to the YouTube API
49   * See: * https://developers.google.com/youtube/v3/docs/search/list
50   */
51  @Injectable()
52  export class YouTubeService {
53    constructor(private http: Http,
54                @Inject(YOUTUBE_API_KEY) private apiKey: string,
55                @Inject(YOUTUBE_API_URL) private apiUrl: string) {
56    }
57
58    search(query: string): Observable<SearchResult[]> {
59      let params: string = [
60        `q=${query}`,
61        `key=${this.apiKey}`,
62        `part=snippet`,
63        `type=video`,
64        `maxResults=10`
65      ].join('&');
66      let queryUrl: string = `${this.apiUrl}?${params}`;
67      return this.http.get(queryUrl)
68        .map((response: Response) => {
69          return (<any>response.json()).items.map(item => {
70            // console.log("raw item", item); // uncomment if you want to debug
71            return new SearchResult({
72              id: item.id.videoId,
73              title: item.snippet.title,
```

```
74              description: item.snippet.description,
75              thumbnailUrl: item.snippet.thumbnails.high.url
76            });
77          });
78        });
79    }
80  }
```

It uses the YouTube API to search for videos and parse the results into a `SearchResult` instance:

**code/http/app/ts/components/YouTubeSearchComponent.ts**

```
30  class SearchResult {
31    id: string;
32    title: string;
33    description: string;
34    thumbnailUrl: string;
35    videoUrl: string;
36
37    constructor(obj?: any) {
38      this.id              = obj && obj.id              || null;
39      this.title           = obj && obj.title           || null;
40      this.description     = obj && obj.description     || null;
41      this.thumbnailUrl    = obj && obj.thumbnailUrl    || null;
42      this.videoUrl        = obj && obj.videoUrl        ||
43                               `https://www.youtube.com/watch?v=${this.id}`;
44    }
45  }
```

The important aspects of this service we need to test are that:

- given a JSON response, the service is able to parse the video id, title, description and thumbnail
- the URL we are requesting uses the provided search term
- the URL starts with what is set on the `YOUTUBE_API_URL` constant
- the API key used matches the `YOUTUBE_API_KEY` constant

With that in mind, let's start writing our test:

**code/http/test/YouTubeSearchComponentBefore.spec.ts**

```
21  describe('MoreHTTPRequests (before)', () => {
22    beforeEach(() => {
23      TestBed.configureTestingModule({
24        providers: [
25          YouTubeService,
26          BaseRequestOptions,
27          MockBackend,
28          { provide: YOUTUBE_API_KEY, useValue: 'YOUTUBE_API_KEY' },
29          { provide: YOUTUBE_API_URL, useValue: 'YOUTUBE_API_URL' },
30          { provide: Http,
31            useFactory: (backend: ConnectionBackend,
32                         defaultOptions: BaseRequestOptions) => {
33                           return new Http(backend, defaultOptions);
34                         }, deps: [MockBackend, BaseRequestOptions] }
35        ]
36      });
37
38    });
```

As we did for every test we wrote on this chapter, we start by declaring how we want to setup our dependencies: we're using the real `YouTubeService` instance, but setting fake values for `YOUTUBE_-API_KEY` and `YOUTUBE_API_URL` constants. We also setting up the `Http` class to use a `MockBackend`.

Now, let's begin to write our first test case:

**code/http/test/YouTubeSearchComponentBefore.spec.ts**

```
40    describe('search', () => {
41      it('parses YouTube response',
42        inject([YouTubeService, MockBackend], fakeAsync((service, backend) => {
43          let res;
44
45          backend.connections.subscribe(c => {
46            c.mockRespond(new Response(<any>{
47              body: `
48              {
49                "items": [
50                  {
51                    "id": { "videoId": "VIDEO_ID" },
52                    "snippet": {
53                      "title": "TITLE",
```

```
54                      "description": "DESCRIPTION",
55                      "thumbnails": {
56                        "high": { "url": "THUMBNAIL_URL" }
57                      }}}]}`
58            }));
59          });
60
61          service.search('hey').subscribe(_res => {
62            res = _res;
63          });
64          tick();
65
66          let video = res[0];
67          expect(video.id).toEqual('VIDEO_ID');
68          expect(video.title).toEqual('TITLE');
69          expect(video.description).toEqual('DESCRIPTION');
70          expect(video.thumbnailUrl).toEqual('THUMBNAIL_URL');
71        }))
72      )
73    });
```

Here we are telling `Http` to return a fake response that will match the relevant fields what we expect the YouTube API to respond when we call the real URL. We do that by using the `mockRespond` method of the connection.

**code/http/test/YouTubeSearchComponentBefore.spec.ts**

```
61          service.search('hey').subscribe(_res => {
62            res = _res;
63          });
64          tick();
```

Next, we're calling the method we're testing: `search`. We're calling it with the term *hey* and capturing the response on the `res` variable.

If you noticed before, we're using `fakeAsync` that requires us to manually sync asynchronous code by calling `tick()`. When we do that here, we expect that the search finished executing and our `res` variable to have a value.

Now is the time to evaluate that value:

**code/http/test/YouTubeSearchComponentBefore.spec.ts**

```
66          let video = res[0];
67          expect(video.id).toEqual('VIDEO_ID');
68          expect(video.title).toEqual('TITLE');
69          expect(video.description).toEqual('DESCRIPTION');
70          expect(video.thumbnailUrl).toEqual('THUMBNAIL_URL');
```

We are getting the first element from the list of responses. We know it's a `SearchResult`, so we're now checking that each attribute was set correctly, based on our provided response: the id, title, description and thumbnail URL should all match.

With this, we completed our first goal when writing this test. However, didn't we just say that having a huge `it` method and having too many `expects` are testing code smells?

We did, so before we continue let's refactor this code to make isolated assertions easier.

Add the following helper fuction inside our `describe('search', ...)`:

**code/http/test/YouTubeSearchComponentAfter.spec.ts**

```
55      function search(term: string, response: any, callback) {
56        return inject([YouTubeService, MockBackend],
57          fakeAsync((service, backend) => {
58            var req;
59            var res;
60
61            backend.connections.subscribe(c => {
62              req = c.request;
63              c.mockRespond(new Response(<any>{body: response}));
64            });
65
66            service.search(term).subscribe(_res => {
67              res = _res;
68            });
69            tick();
70
71            callback(req, res);
72          })
73        )
74      }
```

Let's see what this function does: it uses `inject` and `fakeAsync` to perform the same thing we were doing before, but in a configurable way. We take a *search term*, a *response* and a *callback function*.

We use those parameters to call the `search` method with the search term, set the fake response and call the callback function after the request is finished, providing the request and the response objects.

This way, all our test need to do is call the function and check one of the objects.

Let's break the test we had before into four tests, each testing one specific aspect of the response:

**code/http/test/YouTubeSearchComponentAfter.spec.ts**

```
76    it('parses YouTube video id', search('hey', response, (req, res) => {
77      let video = res[0];
78      expect(video.id).toEqual('VIDEO_ID');
79    }));
80
81    it('parses YouTube video title', search('hey', response, (req, res) => {
82      let video = res[0];
83      expect(video.title).toEqual('TITLE');
84    }));
85
86    it('parses YouTube video description', search('hey', response, (req, res) =>\
87  {
88      let video = res[0];
89      expect(video.description).toEqual('DESCRIPTION');
90    }));
91
92    it('parses YouTube video thumbnail', search('hey', response, (req, res) => {
93      let video = res[0];
94      expect(video.description).toEqual('DESCRIPTION');
95    }));
```

Doesn't it look good? Small, focused tests that test only one thing. Great!

Now it should be really easy to add tests for the remaining goals we had:

**code/http/test/YouTubeSearchComponentAfter.spec.ts**

```
96    it('sends the query', search('term', response, (req, res) => {
97      expect(req.url).toContain('q=term');
98    }));
99
100   it('sends the API key', search('term', response, (req, res) => {
101     expect(req.url).toContain('key=YOUTUBE_API_KEY');
102   }));
103
104   it('uses the provided YouTube URL', search('term', response, (req, res) => {
```

```
105        expect(req.url).toMatch(/^YOUTUBE_API_URL\?/);
106      }));
```

Feel free to add more tests as you see fit. For example, you could add a test for when you have more than one `item` on the response, with different attributes. See if you can find other aspects of the code you'd like to test.

# Conclusion

The Angular team has done a great job building testing right into Angular. It's easy to test all of the aspects of our application: from controllers, to services, forms and HTTP. Even testing asynchronous code that was a difficult to test is now a breeze.

# Converting an Angular 1 App to Angular 2

If you've been using Angular for a while, then you probably already have production Angular 1 apps. Angular 2 is great, but there's no way we can drop everything and rewrite our entire production apps in Angular 2. What we need is a way to *incrementally* upgrade our Angular 1 app. Thankfully, Angular 2 has a fantastic way to do that.

The interoperability of Angular 1 (ng1) and Angular 2 (ng2) works really well. In this chapter, we're going to talk about how to upgrade your ng1 app to ng2 by writing a *hybrid* app. A hybrid app is running ng1 and ng2 simultaneously (and we can exchange data between them).

## Peripheral Concepts

When we talk about interoperability between Angular 1 and Angular 2, there's a lot of peripheral concepts. For instance:

**Mapping Angular 1 Concepts to Angular 2**: At a high level, ng2 Components are ng1 directives. We also use Services in both. However, this chapter is about using both ng1 and ng2, so we're going to assume you have basic knowledge of both. If you haven't used ng2 much, checkout the chapter on How Angular Works before reading this chapter.

**Preparing ng1 apps for ng2**: Angular 1.5 provides a new `.component` method to make "component-directives". `.component` is a great way to start preparing your ng1 app for ng2. Furthermore, creating thin controllers (or banning them altogether[120]) is a great way to refactor your ng1 app such that it's easier to integrate with ng2.

Another way to prepare your ng1 app is to reduce or eliminate your use of two-way data-binding in favor of a one-way data flow. In-part, you'd do this by reducing `$scope` changes that pass data between directives and instead use services to pass your data around.

These ideas are important and warrant further exploration. However, we're not going to extensively cover best-practices for pre-upgrade refactoring in this chapter.

Instead, here's what we **are** going to talk about:

**Writing hybrid ng1/ng2 apps**: ng2 provides a way to bootstrap your ng1 app and then write ng2 components and services. You can write ng2 components that will mix with ng1 components and it "just works". Furthermore, the dependency injection system supports passing between ng1 and ng2 (both directions), so you can write services which will run in either ng1 or ng2.

[120] http://teropa.info/blog/2014/10/24/how-ive-improved-my-angular-apps-by-banning-ng-controller.html

The best part? Change detection runs within Zones, so you don't need to call `$scope.apply` or worry much about change-detection at all.

## What We're Building

In this chapter, we're going to be converting an app called "Interest" - it's a Pinterest-like clone. The idea is that you can save a "Pin" which is a link with an image. The Pins will be shown in a list and you can "fav" (or unfav) a pin.

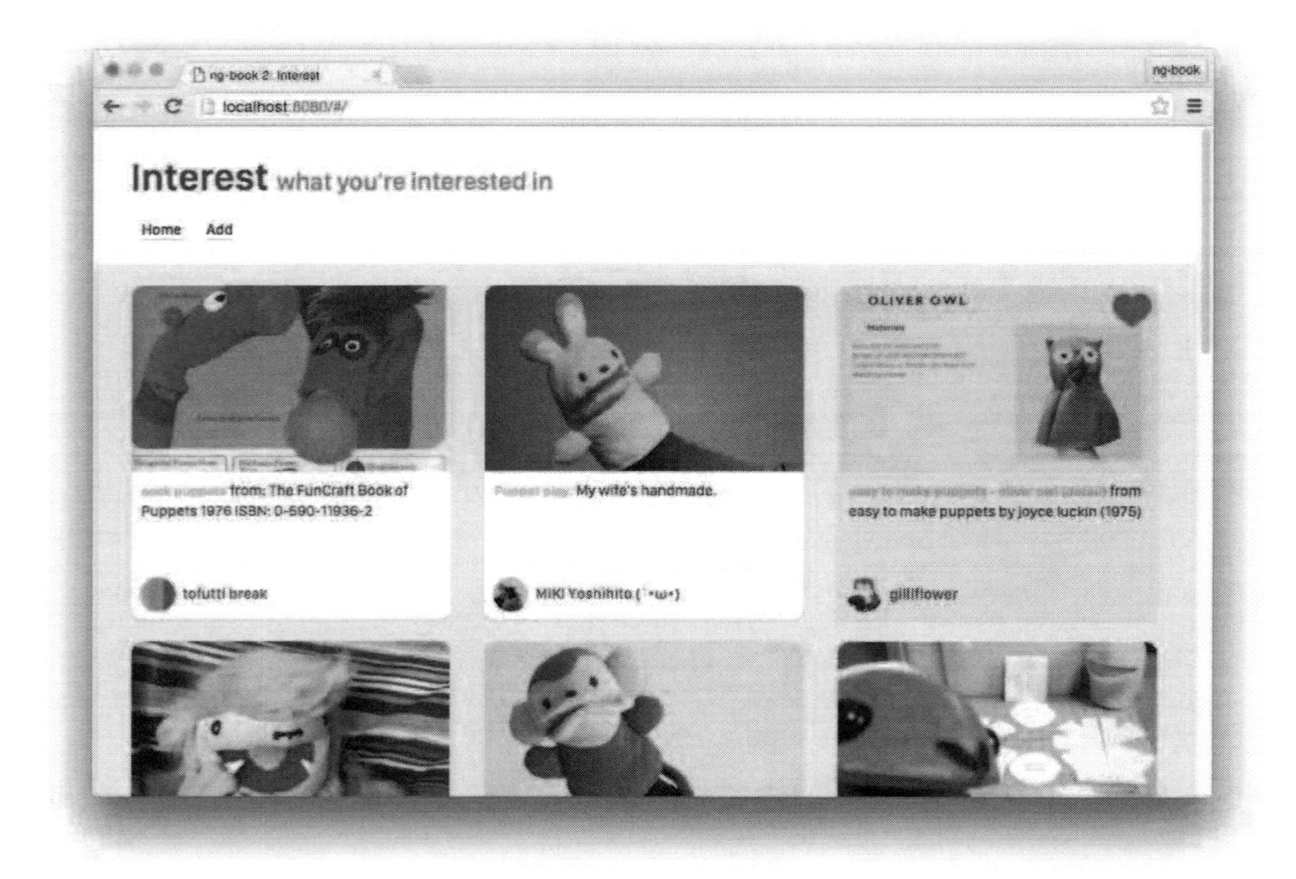


**Our completed Pinterest-like app**

You can find the completed code for both the ng1 version and the completed hybrid version in the sample code download under `code/conversion/ng1` and `code/conversion/hybrid`

Before we dive in, let's set the stage for interoperability between ng1 and ng2

# Mapping Angular 1 to Angular 2

From a high level, the five main parts of Angular 1 are:

- Directives
- Controllers
- Scopes
- Services
- Dependency Injection

Angular 2 changes this list significantly. You might have heard that at ngEurope 2014 Igor and Tobias from the Angular core team announced that they were killing off several "core" ideas in Angular 1 (video here[121]). Specifically, they announced that Angular 2 was killing off:

- `$scope` (& two-way binding by default)
- Directive Definition Objects
- Controllers
- `angular.module`

[121]https://www.youtube.com/watch?v=gNmWybAyBHI

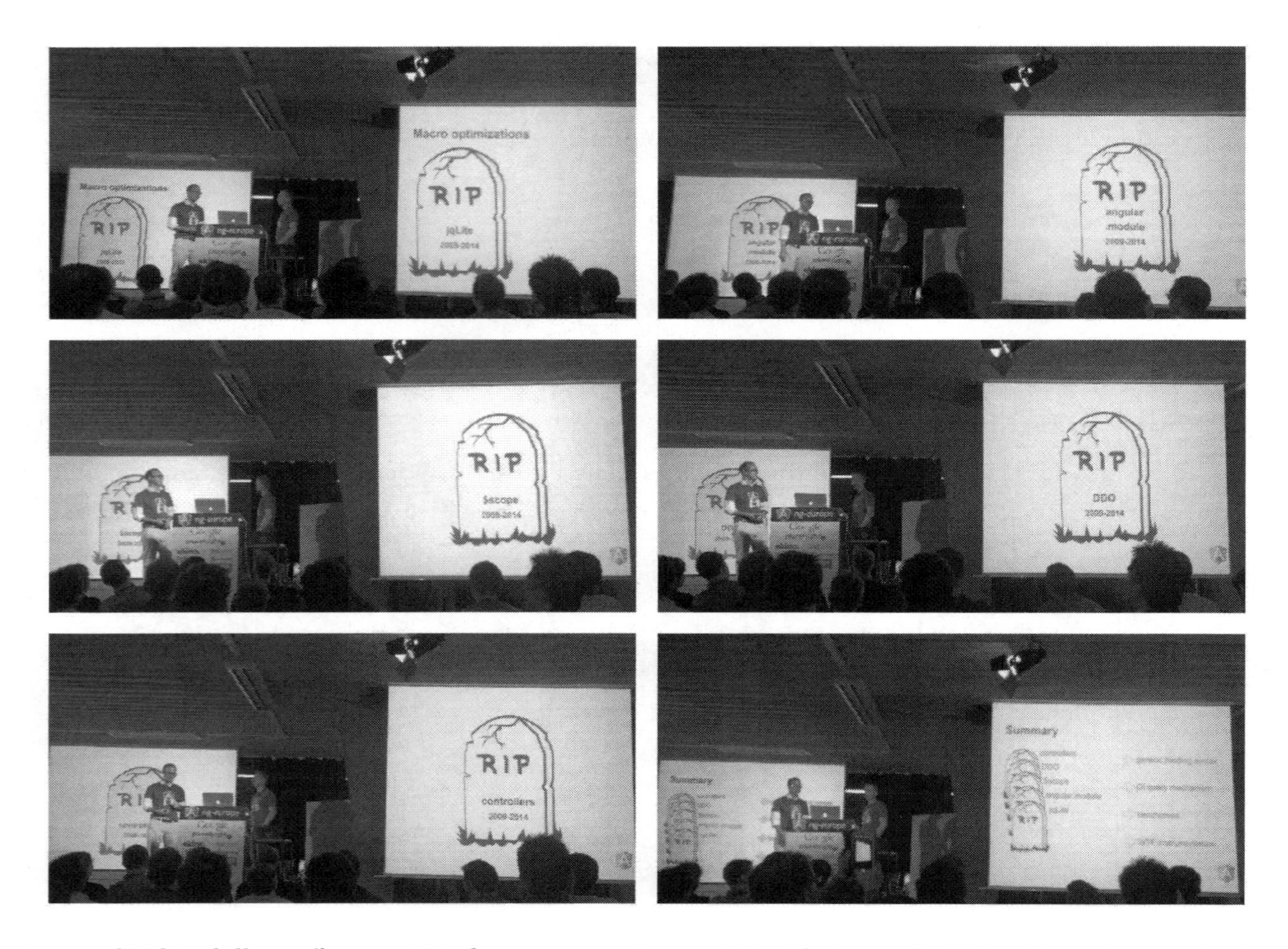


**Igor and Tobias killing off many APIs from 1.x. at ngEurope 2014. Photo Credit: Michael Bromley (used with permission)**

As someone who's built Angular 1 apps and is used to thinking in ng1, we might ask: if we take those things away, what is left? How can you build Angular apps without Controllers and `$scope`?

Well, as much as people like to dramatize how **different** Angular 2 is, it turns out, a lot of the same ideas are still with us and, in fact, Angular 2 provides just as much functionality but with **a much simpler model**.

At a high-level Angular 2 core is made up of:

- Components (think "directives") and
- Services

Of course there's tons of infrastructure required to make those things work. For instance, you need Dependency Injection to manage your Services. And you need a strong change detection library to efficiently propagate data changes to your app. And you need an efficient rendering layer to handle rendering the DOM at the right time.

## Requirements for Interoperability

So given these two different systems, what features do we need for easy interoperability?

- **Use Angular 2 Components in Angular 1**: The first thing that comes to mind is that we need to be able to write new ng2 components, but use them within our ng1 app.
- **Use Angular 1 Components in Angular 2**: It's likely that we won't replace a whole branch of our component-tree with all ng2 components. We want to be able to re-use any ng1 components we have *within* a ng2 component.
- **Service Sharing**: If we have, say, a `UserService` we want to share that service between both ng1 and ng2. Services are normally plain Javascript objects so, more generally, what we need is an interoperable **dependency injection** system.
- **Change Detection**: If we make changes in one side, we want those changes to propagate to the other.

Angular 2 provides solutions for all of these situations and we'll cover them in this chapter.

In this chapter we're going to do the following:

- Describe the ng1 app we'll be converting
- Explain how to setup your hybrid app by using ng2's `UpgradeAdapter`
- Explain step-by-step how to share components (directives) and services between ng1 and ng2 by converting the ng1 app to a hybrid app

## The Angular 1 App

To set the stage, let's go over the Angular 1 version of our app.

This chapter assumes some knowledge of Angular 1 and ui-router[122]. If you're not comfortable with Angular 1 yet, check out ng-book 1[123].

We won't be diving too deeply into explaining each Angular 1 concept. Instead, we're going to review the structure of the app to prepare for our upgrade to a ng2/hybrid app.

To run the ng1 app, `cd` into `conversion/ng1` in the code samples, install the dependencies, and run the app.

---

[122] https://github.com/angular-ui/ui-router

[123] http://ng-book.com

```
1  cd code/conversion/ng1  # change directories
2  npm install             # install dependencies
3  npm run go              # run the app
```

If your browser doesn't open automatically, open the url: http://localhost:8080[124].

In this app, you can see that our user is collecting puppets. We can hover over an item and click the heart to "fav" a pin.

**Red heart indicates a faved pin**

We can also go to the `/add` page and add a new pin. Try submitting the default form.

Handling image uploads is more complex than we want to handle in this demo. For now, just paste the full URL to an image if you want to try a different image.

[124]http://localhost:8080

## The ng1-app HTML

The index.html in our ng1 app uses a common structure:

**code/conversion/ng1/index.html**

```
<!DOCTYPE html>
<html ng-app='interestApp'>
<head>
  <meta charset="utf-8">
  <title>Interest</title>
  <link rel="stylesheet" href="css/bootstrap.min.css">
  <link rel="stylesheet" href="css/sf.css">
  <link rel="stylesheet" href="css/interest.css">
</head>
<body class="container-fullwidth">

  <div class="page-header">
    <div class="container">
      <h1>Interest <small>what you're interested in</small></h1>

      <div class="navLinks">
        <a ui-sref='home' id="navLinkHome">Home</a>
        <a ui-sref='add' id="navLinkAdd">Add</a>
      </div>
    </div>
  </div>

  <div id="content">
    <div ui-view=''></div>
  </div>

  <script src="js/vendor/lodash.js"></script>
  <script src="js/vendor/angular.js"></script>
  <script src="js/vendor/angular-ui-router.js"></script>
  <script src="js/app.js"></script>
</body>
</html>
```

- Notice that we're using ng-app in the html tag to specify that this app uses the module interestApp.
- We load our javascript with script tags at the bottom of the body.
- The template contains a page-header which stores our navigation

- We're using `ui-router` which means we:
  - Use `ui-sref` for our links (`Home` and `Add`) and
  - We use `ui-view` where we want the router to populate our content.

## Code Overview

We'll look at each section in code, but first, let's briefly describe the moving parts.

In our app, we have two routes:

- `/` uses the `HomeController`
- `/add` uses the `AddController`

We use a `PinsService` to hold an array of all of the current pins. `HomeController` renders the list of pins and `AddController` adds a new element to that list.

Our root-level route uses our `HomeController` to render pins. We have a `pin` directive that renders each pin.

The `PinsService` stores the data in our app, so let's look at the `PinsService` first.

### ng1: `PinsService`

**code/conversion/ng1/js/app.js**

```
1  angular.module('interestApp', ['ui.router'])
2  .service('PinsService', function($http, $q) {
3    this._pins = null;
4
5    this.pins = function() {
6      var self = this;
7      if(self._pins == null) {
8        // initialize with sample data
9        return $http.get("/js/data/sample-data.json").then(
10         function(response) {
11           self._pins = response.data;
12           return self._pins;
13         })
14     } else {
15       return $q.when(self._pins);
16     }
17   }
18
```

```
19    this.addPin = function(newPin) {
20      // adding would normally be an API request so lets mock async
21      return $q.when(
22        this._pins.unshift(newPin)
23      );
24    }
25  })
```

The `PinsService` is a `.service` that stores an array of pins in the property `_.pins`.

The method `.pins` returns a promise that resolves to the list of pins. If `_.pins` is `null` (i.e. the first time), then we will load sample data from `/js/data/sample-data.json`.

**code/conversion/ng1/js/data/sample-data.json**

```
1   [
2     {
3       "title": "sock puppets",
4       "description": "from:\nThe FunCraft Book of Puppets\n1976\nISBN: 0-590-11936\
5   -2",
6       "user_name": "tofutti break",
7       "avatar_src": "images/avatars/42826303@N00.jpg",
8       "src": "images/pins/106033588_167d811702_o.jpg",
9       "url": "https://www.flickr.com/photos/tofuttibreak/106033588/",
10      "faved": false,
11      "id": "106033588"
12    },
13    {
14      "title": "Puppet play.",
15      "description": "My wife's handmade.",
16      "user_name": "MIKI Yoshihito (´□ω□)",
17      "avatar_src": "images/avatars/7940758@N07.jpg",
18      "src": "images/pins/4422575066_7d5c4c41e7_o.jpg",
19      "url": "https://www.flickr.com/photos/mujitra/4422575066/",
20      "faved": false,
21      "id": "4422575066"
22    },
23    {
24      "title": "easy to make puppets - oliver owl (detail)",
25      "description": "from easy to make puppets by joyce luckin (1975)",
26      "user_name": "gilliflower",
27      "avatar_src": "images/avatars/26265986@N00.jpg",
28      "src": "images/pins/6819859061_25d05ef2e1_o.jpg",
```

```
29      "url": "https://www.flickr.com/photos/gilliflower/6819859061/",
30      "faved": false,
31      "id": "6819859061"
32    },
```

*Snippet from Sample Data*

The method `.addPin` simply adds the new pin to the array of pins. We use `$q.when` here to return a promise, which is likely what would happen if we were doing a real async call to a server.

## ng1: Configuring Routes

We're going to configure our routes with `ui-router`.

If you're unfamiliar with `ui-router` you can read the docs here[125].

As we mentioned, we're going to have two routes:

**code/conversion/ng1/js/app.js**

```
26  .config(function($stateProvider, $urlRouterProvider) {
27    $stateProvider
28      .state('home', {
29        templateUrl: '/templates/home.html',
30        controller: 'HomeController as ctrl',
31        url: '/',
32        resolve: {
33          'pins': function(PinsService) {
34            return PinsService.pins();
35          }
36        }
37      })
38      .state('add', {
39        templateUrl: '/templates/add.html',
40        controller: 'AddController as ctrl',
41        url: '/add',
42        resolve: {
43          'pins': function(PinsService) {
44            return PinsService.pins();
45          }
```

[125] https://github.com/angular-ui/ui-router/wiki

```
46          }
47        })
48
49        $urlRouterProvider.when('', '/') ;
50    })
```

The first route / maps to the `HomeController`. It has a template, which we'll look at in a minute. Notice that we also are using the `resolve` functionality of `ui-router`. This says that before we load this route for the user, we want to call `PinsService.pins()` and inject the result (the list of pins) into the controller (`HomeController`).

The `/add` route as similarly, except that it has a different template and a different controller.

Let's first look at our `HomeController`.

## ng1: `HomeController`

Our `HomeController` is straightforward. We save `pins`, which is injected because of our `resolve`, to `$scope.pins`.

**code/conversion/ng1/js/app.js**

```
60  .controller('HomeController', function(pins) {
61    this.pins = pins;
62  })
```

## ng1: / `HomeController` template

Our home template is small: we use an `ng-repeat` to repeat over the pins in `$scope.pins`. Then we render each pin with the `pin` directive.

**code/conversion/ng1/templates/home.html**

```
1  <div class="container">
2    <div class="row">
3      <pin item="pin" ng-repeat="pin in ctrl.pins">
4      </pin>
5    </div>
6  </div>
```

Let's dive deeper and look at this `pin` directive.

## ng1: `pin` Directive

The `pin` directive is restricted to matching an element (`E`) and has a `template`.

We can input our `pin` via the `item` attribute, as we did in the `home.html` template.

Our `link` function, defines a function on the scope called `toggleFav` which toggles the pin's `faved` property.

**code/conversion/ng1/js/app.js**

```
 92 })
 93 .directive('pin', function() {
 94   return {
 95     restrict: 'E',
 96     templateUrl: '/templates/pin.html',
 97     scope: {
 98       'pin': "=item"
 99     },
100     link: function(scope, elem, attrs) {
101       scope.toggleFav = function() {
102         scope.pin.faved = !scope.pin.faved;
103       }
104     }
105   }
106 })
```

This directive shouldn't be taken as an example of directive best-practices in 2016. For instance, if I was writing this component anew (in ng1) I would probably use the new `.component` directive in Angular 1.5. At the very least, I'd probably use `controllerAs` instead of `link` here.

But this section is less about how to write ng1 code, as much as how to work with the ng1 code you already have.

## ng1: `pin` Directive template

The template `templates/pin.html` renders an individual pin on our page.

**code/conversion/ng1/templates/pin.html**

```
1  <div class="col-sm-6 col-md-4">
2    <div class="thumbnail">
3      <div class="content">
4        <img ng-src="{{pin.src}}" class="img-responsive">
5        <div class="caption">
6          <h3>{{pin.title}}</h3>
7          <p>{{pin.description | truncate:100}}</p>
8        </div>
9        <div class="attribution">
10         <img ng-src="{{pin.avatar_src}}" class="img-circle">
11         <h4>{{pin.user_name}}</h4>
12       </div>
13     </div>
14     <div class="overlay">
15       <div class="controls">
16         <div class="heart">
17           <a ng-click="toggleFav()">
18             <img src="/images/icons/Heart-Empty.png" ng-if="!pin.faved"></img>
19             <img src="/images/icons/Heart-Red.png"   ng-if="pin.faved"></img>
20           </a>
21         </div>
22       </div>
23     </div>
24   </div>
25 </div>
```

The directives we use here are ng1 built-ins:

- We use `ng-src` to render the `img`.
- Next we show the `pin.title` and `pin.description`.
- We use `ng-if` to show either the red or empty heart

The most interesting thing here is the `ng-click` that will call `toggleFav`. `toggleFav` changes the `pin.faved` property and thus the red or empty heart will be shown accordingly.

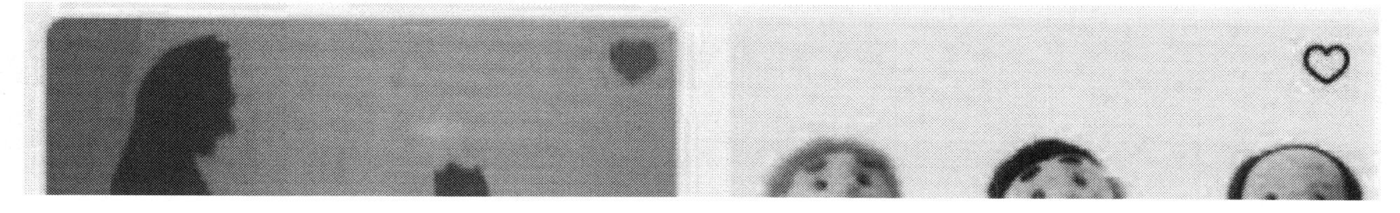

**Red vs. Black Heart**

Now let's turn our attention to the AddController.

## ng1: AddController

Our AddController has a bit more code than the HomeController. We open by defining the controller and specifying the services it will inject:

**code/conversion/ng1/js/app.js**

```
63  .controller('AddController', function($state, PinsService, $timeout) {
64    var ctrl = this;
65    ctrl.saving = false;
```

We're using controllerAs syntax in our router and template, which means we set properties on this instead of on $scope. Scoping this in ES5 Javascript can be tricky, so we assign var ctrl = this; which helps disambiguate when we're referencing the controller in nested functions.

**code/conversion/ng1/js/app.js**

```
67    var makeNewPin = function() {
68      return {
69        "title": "Steampunk Cat",
70        "description": "A cat wearing goggles",
71        "user_name": "me",
72        "avatar_src": "images/avatars/me.jpg",
73        "src": "/images/pins/cat.jpg",
74        "url": "http://cats.com",
75        "faved": false,
76        "id": Math.floor(Math.random() * 10000).toString()
77      }
78    }
79
80    ctrl.newPin = makeNewPin();
```

We create a function makeNewPin that contains the default structure and data for a pin.

We also initialize this controller by setting ctrl.newPin to the value of calling this function.

The last thing we need to do is define the function to submit a new pin:

**code/conversion/ng1/js/app.js**

```
82    ctrl.submitPin = function() {
83      ctrl.saving = true;
84      $timeout(function() {
85        PinsService.addPin(ctrl.newPin).then(function() {
86          ctrl.newPin = makeNewPin();
87          ctrl.saving = false;
88          $state.go('home');
89        });
90      }, 2000);
91    }
92  })
```

Essentially, this article is calling out to `PinService.addPin` and creating a new pin. But there's a few other things going on here.

In a real application, this would almost certainly call back to a server. We're mimicking that effect by using `$timeout`. (That is, you could remove the `$timeout` function and this would still work. It's just here to deliberately slow down the app to give us a chance to see the "Saving" indicator.)

We want to give some indication to the user that their pin is saving, so we set the `ctrl.saving = true`.

We call `PinsService.addPin` giving it our `ctrl.newPin`. `addPin` returns a promise, so in our promise function we

1. revert `ctrl.newPin` to the original value
2. we set `ctrl.saving` to `false`, because we're done saving the pin
3. we use the `$state` service to redirect the user to the homepage where we can see our new pin

Here's the whole code of the `AddController`:

**code/conversion/ng1/js/app.js**

```
63  .controller('AddController', function($state, PinsService, $timeout) {
64    var ctrl = this;
65    ctrl.saving = false;
66
67    var makeNewPin = function() {
68      return {
69        "title": "Steampunk Cat",
70        "description": "A cat wearing goggles",
71        "user_name": "me",
```

```
72        "avatar_src": "images/avatars/me.jpg",
73        "src": "/images/pins/cat.jpg",
74        "url": "http://cats.com",
75        "faved": false,
76        "id": Math.floor(Math.random() * 10000).toString()
77      }
78    }
79
80    ctrl.newPin = makeNewPin();
81
82    ctrl.submitPin = function() {
83      ctrl.saving = true;
84      $timeout(function() {
85        PinsService.addPin(ctrl.newPin).then(function() {
86          ctrl.newPin = makeNewPin();
87          ctrl.saving = false;
88          $state.go('home');
89        });
90      }, 2000);
91    }
92  })
```

## ng1: `AddController` template

Our `/add` route renders the `add.html` template.

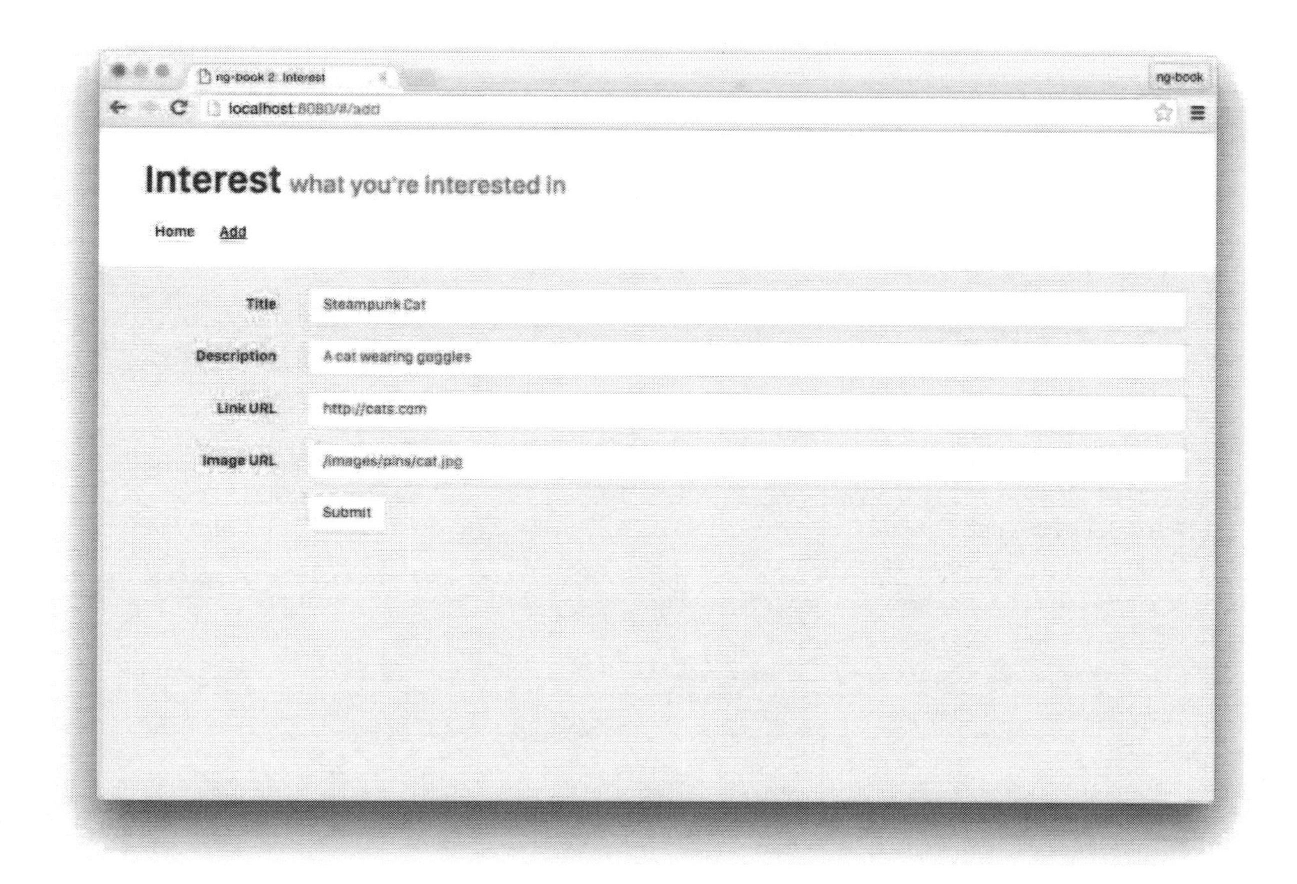


**Adding a New Pin Form**

The template uses `ng-model` to bind the `input` tags to the properties of the `newPin` on the controller.

The interesting things here are that:

- We use `ng-click` on the submit button to call `ctrl.submitPin` and
- We show a "Saving…" message if `ctrl.saving` is truthy

**code/conversion/ng1/templates/add.html**

```
1  <div class="container">
2    <div class="row">
3
4      <form class="form-horizontal">
5
6        <div class="form-group">
7          <label for="title"
8                 class="col-sm-2 control-label">Title</label>
```

```
 9          <div class="col-sm-10">
10            <input type="text"
11                   class="form-control"
12                   id="title"
13                   placeholder="Title"
14                   ng-model="ctrl.newPin.title">
15          </div>
16        </div>
17
18        <div class="form-group">
19          <label for="description"
20                 class="col-sm-2 control-label">Description</label>
21          <div class="col-sm-10">
22            <input type="text"
23                   class="form-control"
24                   id="description"
25                   placeholder="Description"
26                   ng-model="ctrl.newPin.description">
27          </div>
28        </div>
29
30        <div class="form-group">
31          <label for="url"
32                 class="col-sm-2 control-label">Link URL</label>
33          <div class="col-sm-10">
34            <input type="text"
35                   class="form-control"
36                   id="url"
37                   placeholder="Link URL"
38                   ng-model="ctrl.newPin.url">
39          </div>
40        </div>
41
42        <div class="form-group">
43          <label for="url"
44                 class="col-sm-2 control-label">Image URL</label>
45          <div class="col-sm-10">
46            <input type="text"
47                   class="form-control"
48                   id="url"
49                   placeholder="Image URL"
50                   ng-model="ctrl.newPin.src">
```

```
51        </div>
52      </div>
53
54      <div class="form-group">
55        <div class="col-sm-offset-2 col-sm-10">
56          <button type="submit"
57                  class="btn btn-default"
58                  ng-click="ctrl.submitPin()">Submit</button>
59        </div>
60      </div>
61      <div ng-if="ctrl.saving">
62        Saving...
63      </div>
64    </form>
65
66  </div>
67 </div>
```

### ng1: Summary

There we have it. This app has just the right amount of complexity that we can start porting it to Angular 2.

## Building A Hybrid

Now we're ready to start putting some Angular 2 in our Angular 1 app.

Before we start using Angular 2 in our browser, we're going to need to make some modifications to our project structure.

You can find the code for this example in `code/conversion/hybrid`.

### Hybrid Project Structure

The first step to creating a hybrid app is to make sure you have both ng1 and ng2 loaded as dependencies. Everyone's situation is going to be slightly different.

In this example we've **vendored** the Angular 1 libraries (in `js/vendor`) and we're loading the Angular 2 libraries from `npm`.

In your project, you might want to vendor them both, use bower[126], etc. However, using `npm` is very convenient for Angular 2, and so we suggest using `npm` to install Angular 2.

## Dependencies with `package.json`

You install dependencies with `npm` using the `package.json` file. Here's our `package.json` for the hybrid example:

**code/conversion/hybrid/package.json**

```
1  {
2    "name": "ng-hybrid-pinterest",
3    "version": "0.0.1",
4    "description": "toy pinterest clone in ng1/ng2 hybrid",
5    "contributors": [
6      "Nate Murray <nate@fullstack.io>",
7      "Felipe Coury <felipe@ng-book.com>"
8    ],
9    "main": "index.js",
10   "private": true,
11   "scripts": {
12     "clean": "rm -f ts/*.js ts/*.js.map ts/components/*.js ts/components/*.js.ma\
13 p ts/services/*.js ts/services.js.map",
14     "tsc": "./node_modules/.bin/tsc",
15     "tsc:w": "./node_modules/.bin/tsc -w",
16     "serve": "./node_modules/.bin/live-server --host=localhost --port=8080 .",
17     "go": "concurrent \"npm run tsc:w\" \"npm run serve\" "
18   },
19   "dependencies": {
20     "@angular/common": "2.2.0-rc.0",
21     "@angular/compiler": "2.2.0-rc.0",
22     "@angular/core": "2.2.0-rc.0",
23     "@angular/forms": "2.2.0-rc.0",
24     "@angular/http": "2.2.0-rc.0",
25     "@angular/platform-browser": "2.2.0-rc.0",
26     "@angular/platform-browser-dynamic": "2.2.0-rc.0",
27     "@angular/router": "3.2.0-rc.0",
28     "@angular/upgrade": "2.0.0-rc.6",
29     "core-js": "2.4.1",
30     "es6-shim": "0.35.0",
31     "reflect-metadata": "0.1.3",
32     "rxjs": "5.0.0-beta.12",
```

[126] http://bower.io/

```
33       "systemjs": "0.19.6",
34       "ts-helpers": "1.1.1",
35       "tslint": "3.7.0-dev.2",
36       "typescript": "1.9.0-dev.20160409",
37       "typings": "0.8.1",
38       "zone.js": "0.6.21"
39     },
40     "devDependencies": {
41       "concurrently": "1.0.0",
42       "karma": "0.12.22",
43       "karma-chrome-launcher": "0.1.4",
44       "karma-jasmine": "0.1.5",
45       "live-server": "0.9.0",
46       "typescript": "1.7.3"
47     }
48   }
```

If you're unfamiliar with what one of these packages does, it's a good idea to find out. `rxjs`, for example, is the library that provides our observables. `systemjs` provides the module loader that we're going to use in this chapter.

Once you've added the Angular 2 dependencies, run the command `npm install` to install them.

## Compiling our code

You'll notice that in the `package.json` `"scripts"` key we have another key that specifies `"tsc"`. This means we can run the comment `npm run tsc` and it will call out to the TypeScript compiler and compile our code.

We're going to be using TypeScript in this example alongside our Javascript Angular 1 code.

To do this, we're going to put all of our TypeScript code in the folder `ts/` and our Javascript code in the folder `js/`.

We configure the TypeScript compiler by using the `tsconfig.json` file. The important thing to know right now about that file is that in the `filesGlob` key we're specifying a glob of: `"./ts/**/*.ts"` which means "when we run the TypeScript compiler, we want to compile all files ending in `.ts` in the `ts/` directory".

In this project **our browser will only load Javascript**. We're going to use the TypeScript compiler (`tsc`) to compile our code to Javascript and then we will load our ng1 and ng2 *JavaScript* in our browser.

### Loading index.html dependencies

Now that we have our dependencies and our compiler setup, we need to load these Javascript files into our browser. We do that by adding `script` tags:

**code/conversion/ng1/hybrid/index.html**

```
23   <div id="content">
24     <div ui-view=''></div>
25   </div>
26 
27   <!-- Libraries -->
28   <script src="node_modules/core-js/client/shim.min.js"></script>
29   <script src="node_modules/zone.js/dist/zone.js"></script>
30   <script src="node_modules/reflect-metadata/Reflect.js"></script>
31   <script src="node_modules/systemjs/dist/system.src.js"></script>
32 
33   <script src="js/vendor/angular.js"></script>
34   <script src="js/vendor/angular-ui-router.js"></script>
```

The files we loaded from `node_modules/` are Angular 2 and its dependencies. Similarly, the files we loaded from `js/vendor/` are Angular 1 and its dependencies.

But you'll notice here we didn't load any of *our* code in these tags. To load our code we're going to use System.js.

### Configuring System.js

We're going to use System.js as the module loader for this example.

We could use `Webpack` (as we do in other examples in this book) or a variety of other loaders (requirejs etc.). However System.js is a wonderful and flexible loader that is often used with Angular 2. This chapter will provide a nice example of how you can use Angular 2 with System.js

To configure System.js we do the following in a `<script>` tag in our `index.html`:

```
1 <script src="resources/systemjs.config.js"></script>
2 System.import('ts/app.js')
3       .then(null, console.error.bind(console));
```

`System.import('ts/app.js')` says that the entry point of our app will be the file `ts/app.js`. When we write hybrid ng2 apps **the Angular 2 code becomes the entry point**. This makes sense because it's Angular 2 that's providing the backwards compatibility with Angular 1. We'll talk more about how to bootstrap the app in a minute.

Another thing to notice here is that we're loading a `.js` file from the `ts/` directory. Why? Because our TypeScript compiler will have compiled this file down to Javascript by the time this page loads.

We have configured System.js in `resources/systemjs.config.js`. That file contains a mostly-standard configuration, but since we have to be able to load our ng1 app in our ng2 code we've added a special key `interestAppNg1` that points to our ng1 app. This option lets us do the following in our TypeScript code:

```
1  import 'interestAppNg1'; // "bare import" for side-effects
```

The module loader will see the string `'interestAppNg1'` and load our Angular 1 app at `./js/app.js`.

The `packages` key specifies that files in the `ts` "packages" will have the extension `.js` and use the System.js `register` module format.

There are a bunch of module formats your TypeScript compiler can output. The System.js `format` needs to match the module format you're compiling to. So in this case,t he `register` module format will work with our TypeScript because we specified `compilerOptions.module` as `"system"` in our `tsconfig.json`

Configuring System.js is fairly advanced and there are a lot of potential options here.

This isn't a book on module loaders and, in-fact, it would probably take a whole book to explore in-depth how to configure System.js and other Javascript module loaders.

For now, we're not going to talk much more about module loading, but you can read up more on System.js here[127]

*Would you like to read a book on Javascript module loaders? We're considering writing one. If you'd like to be notified when it's ready, put in your email here*[128]

[127]https://github.com/systemjs/systemjs/blob/master/docs/config-api.md

[128]http://eepurl.com/bMOaEX

## Bootstrapping our Hybrid App

Now that we have our project structure in place, let's bootstrap the app.

If you recall, with Angular 1 you can bootstrap the app in 1 of two ways:

1. You can use the `ng-app` directive, such as `ng-app='interestApp'`, in your HTML or
2. You can use `angular.bootstrap` in Javascript

In hybrid apps we use a **new bootstrap** method that comes from an `UpgradeAdapter`.

Since we'll be bootstrapping the app in code, **make sure you remove the `ng-app` from your `index.html`**.

Here's what a minimal bootstrapping of our code would look like:

```
1  // code/conversion/hybrid/ts/app.ts
2  import {
3    NgModule,
4    forwardRef
5  } from '@angular/core';
6  import { CommonModule } from '@angular/common';
7  import { BrowserModule } from '@angular/platform-browser';
8
9  import { UpgradeAdapter } from '@angular/upgrade';
10 declare var angular: any;
11 import 'interestAppNg1'; // "bare import" for side-effects
12
13 /*
14  * Create our upgradeAdapter
15  */
16 const upgradeAdapter: UpgradeAdapter = new UpgradeAdapter(
17   forwardRef(() => MyAppModule)); // <-- notice forward reference
18
19 // ...
20 // upgrade and downgrade components in here
21 // ...
22
23 /*
24  * Create our app's entry NgModule
25  */
26 @NgModule({
27   declarations: [ MyNg2Component, ... ],
28   imports: [
```

```
29      CommonModule,
30      BrowserModule
31    ],
32    providers: [ MyNg2Services, ... ]
33  })
34  class MyAppModule { }
35
36  /*
37   * Bootstrap the App
38   */
39  upgradeAdapter.bootstrap(document.body, ['interestApp']);
```

We start by importing the `UpgradeAdapter` and then we create an instance of it: `upgradeAdapter`.

However, the constructor of `UpgradeAdapter` requires an `NgModule` that we'll be using for our Angular 2 up - but we haven't defined it yet! To get around this we use the `forwardRef` function which allows us to take a 'forward reference' to our `NgModule` which we declare below.

When we define our `NgModule` `MyAppModule` (or specifically in this app it will be `InterestAppModule`), we define it like we would any other Angular 2 `NgModule`: we put in our declarations, imports, providers, etc.

Lastly, we tell the `upgradeAdapter` to `bootstrap` our app on the element `document.body` and we specify the module name of our **angular 1 app**.

This will bootstrap our Angular 1 app within our Angular 2 app! Now we can start replacing pieces with Angular 2.

## What We'll Upgrade

Let's discuss what we're going to port to ng2 in this example and what will stay in ng1.

### The Homepage

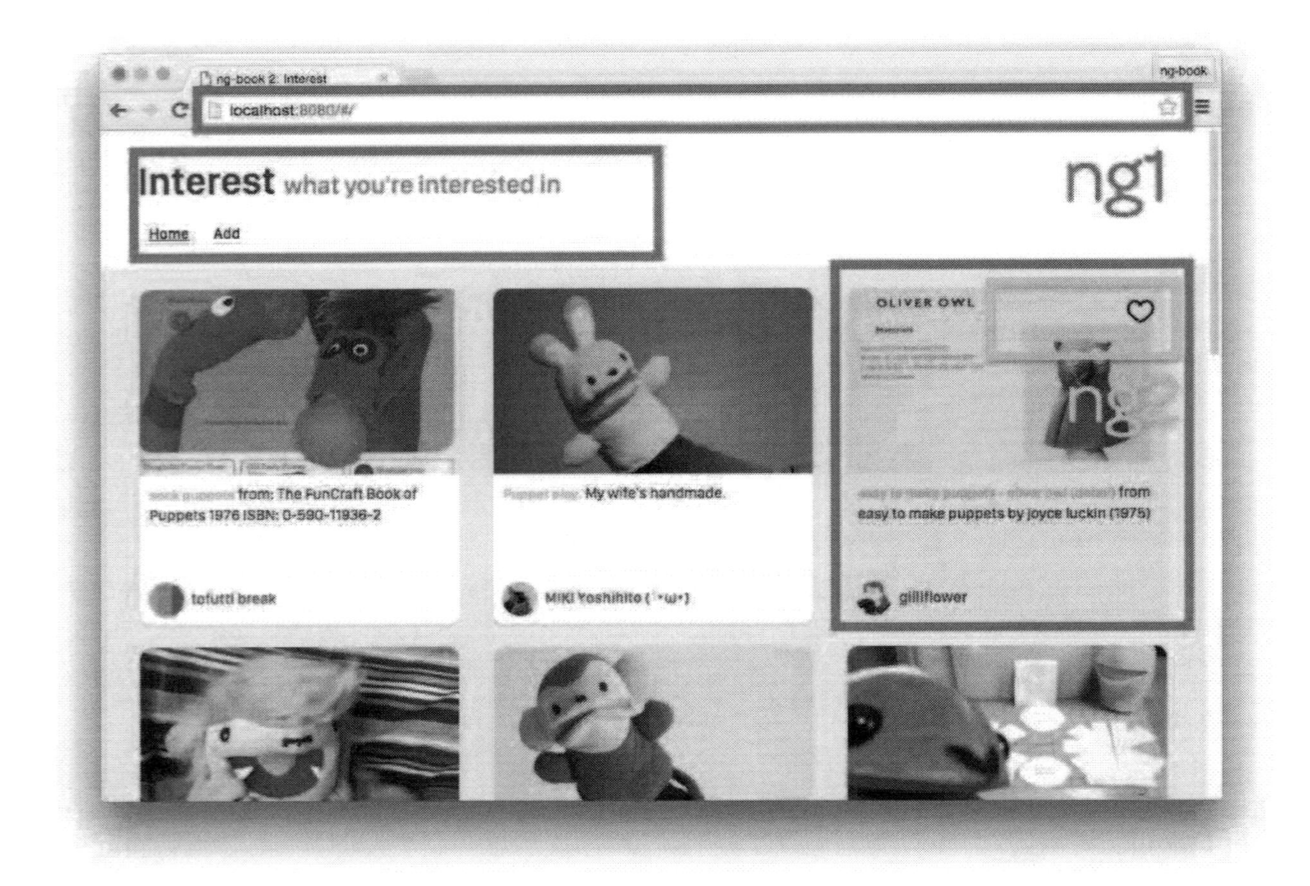


**Homepage ng1 and ng2 Components**

The first thing to notice is that we're going to continue to manage routing with ng1. Of course, Angular 2 has its own routing, which you can read about in our routing chapter. But if you're building a hybrid app, you probably have lots of routes configured with Angular 1 and so in this example we'll continue to use `ui-router` for the routing.

On the homepage, we're going to nest a ng2 component within an ng1 directive. In this case, we're going to convert the "pin controls" to a ng2 component. That is, our ng1 `pin` directive, will call out to the ng2 `pin-controls` component and `pin-controls` will render the fav heart.

It's a small example that shows a powerful idea: how to seamlessly exchange data between ng versions.

### The About Page

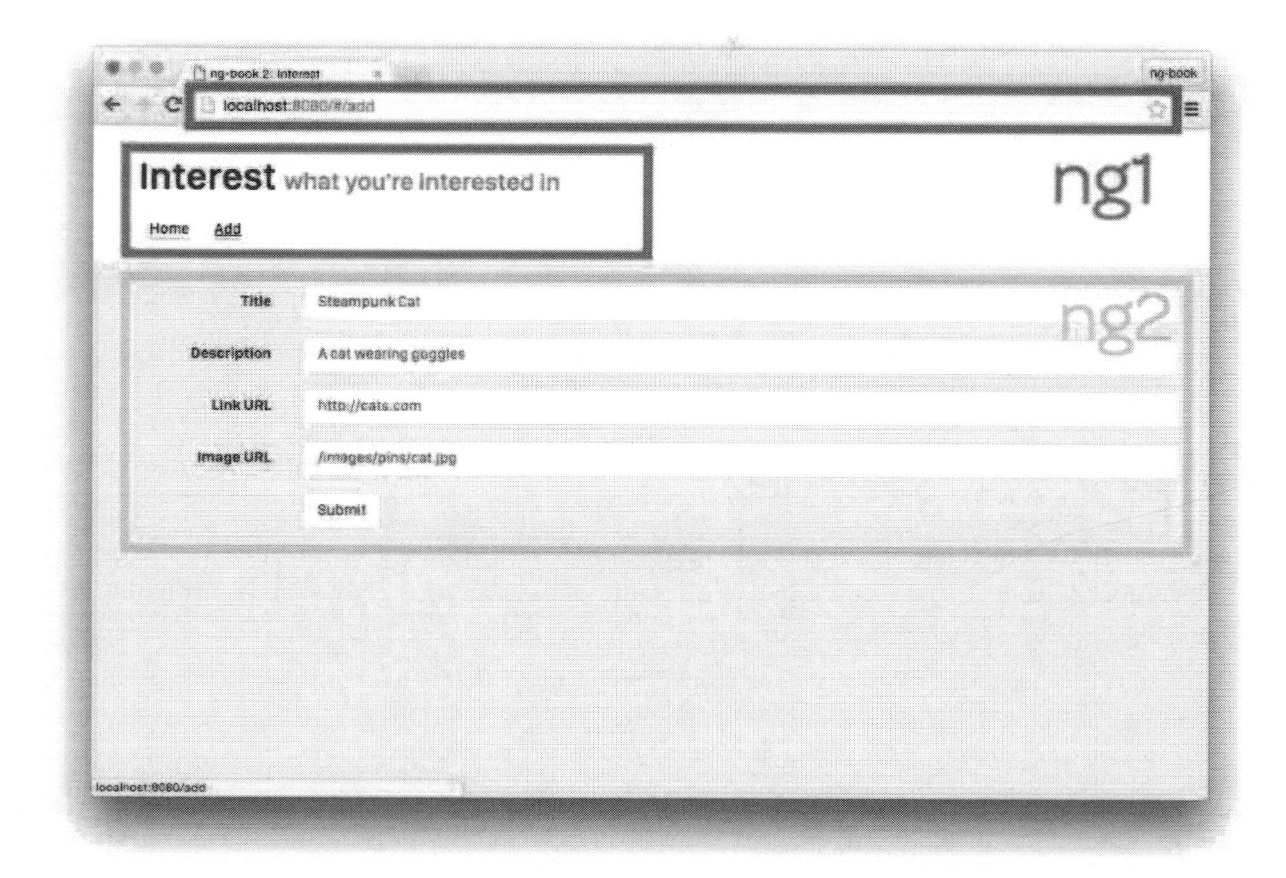


**About Page ng1 and ng2 Components**

We're going to use ng1 for the router and header on the about page as well. However on the about page, we're going to replace the whole form with a ng2 component: `AddPinComponent`.

If you recall, the form will add a new pin to the `PinsService`, and so in this example we're going to need to somehow make the (ng1) `PinsService` accessible to the (ng2) `AddPinComponent`.

Also, remember that when a new pin is added, the app should be redirected to the homepage. However, to change routes we need to use the `ui-router` `$state` service (ng1) in the `AddPinComponent` (ng2). So we also need to make sure the `$state` service can be used in `AddPinComponent` as well.

### Services

So far we've talked about two ng1 services that will be *upgraded* to ng2:

- `PinsService` and
- `$state`

We also want to explore "downgrading" a ng2 service to be used by ng1. For this, later on in the chapter, we'll create an `AnalyticsService` in TypeScript/ng2 that we share with ng1.

### Taking Inventory

So to recap we're going to "cross-expose" the following:

- Downgrade the ng2 `PinControlsComponent` to ng1 (for the fav buttons)
- Downgrade the ng2 `AddPinComponent` to ng1 (for the add pin page)
- Downgrade the ng2 `AnalyticsService` to ng1 (for recording events)
- Upgrade the ng1 `PinsService` to ng2 (for adding new pins)
- Upgrade the ng1 `$state` service to ng2 (for controlling routes)

## A Minor Detour: Typing Files

One of the great things about TypeScript is the compile-time typing. However, if you're building a hybrid app, I suspect that you've got a lot of untyped Javascript code that you're going to be integrating into this project.

When you try to use your Javascript code from TypeScript you may get compiler errors because the compiler doesn't know the structure of your Javascript objects. You could try casting everything to `<any>` but that is ugly and error prone.

The better solution is to, instead, provide your TypeScript compiler with custom *type annotations*. Then the compiler will be able to enforce the types of your Javascript code.

For instance, remember how in our `ng1` app we created a pin object in `makeNewPin`?

**code/conversion/ng1/js/app.js**

```
67    var makeNewPin = function() {
68      return {
69        "title": "Steampunk Cat",
70        "description": "A cat wearing goggles",
71        "user_name": "me",
72        "avatar_src": "images/avatars/me.jpg",
73        "src": "/images/pins/cat.jpg",
74        "url": "http://cats.com",
75        "faved": false,
76        "id": Math.floor(Math.random() * 10000).toString()
77      }
78    }
79
80    ctrl.newPin = makeNewPin();
```

It would be nice if we could tell the compiler about the structure of these objects and not resort to using `any` everywhere.

Furthermore, we're going to be using the `ui-router` `$state` service in Angular 2 / TypeScript, and we need to tell the compiler what functions are available there, too.

So while providing TypeScript custom type definitions is a TypeScript (and not an Angular-specific) chore, it's a chore we need to do nonetheless. And it's something that many people haven't done yet because TypeScript is, at time of publishing, relatively new.

So in this section I want to walk through how you deal with custom typings in TypeScript.

If you're already familiar with how to create and use TypeScript type definition files, you can safely skim this section.

## Typing Files

In TypeScript we can describe the structure of our code by writing *typing definition files*. Typing definition files generally end in the extension `.d.ts`.

Generally, when you write TypeScript code, you don't need to write a `.d.ts` because your TypeScript code itself contains types. We write `.d.ts` files when we have some external Javascript code that we want to add typing to after the fact.

For instance, in describing our pin object, we could write an `interface` for it like so:

**code/conversion/hybrid/js/app.d.ts**

```
3  export interface Pin {
4    title: string;
5    description: string;
6    user_name: string;
7    avatar_src: string;
8    src: string;
9    url: string;
10   faved: boolean;
11   id: string;
12 }
```

Notice that we're not declaring a class, and we're not creating an instance. Instead, we're defining the shape (types) of an interface.

In order to use `.d.ts` files, you need to tell the TypeScript compiler where they are. The easiest way to do this is by modifying the `tsconfig.json` file. For instance, if we had a file `js/app.d.ts` we could add it like this:

```
// tsconfig.json
"compilerOptions": { ... },
"files": [
  "ts/app.ts",
  "js/app.d.ts"
],
// more...
```

Look closely at the paths of the files in this case. We're loading our TypeScript `ts/app.ts`. And we're loading `app.d.ts` from `js/`. This is because the `js/app.d.ts` is the typing file for `js/app.js` (the ng1 Javascript file, not the ng2 TypeScript).

We'll write `app.d.ts` in a little bit. First, let's explore a tool that exists to help us with third-party TypeScript definition files: `typings`.

### Third-party libraries with `typings`

`typings` is a tool for managing TypeScript type definition files for libraries that may not have them otherwise.

We're going to use `angular-ui-router` with our app, so let's install the `typings` typings for `angular-ui-router`. Here's how to get it setup.

You need to have `typings` installed, which you can do with `npm install -g typings`.

Next we configure a `typings.json` file, which you can create with `typings init` (or use the one provided).

Then we install the package we need by running: `typings install angular-ui-router --save`.

Notice that `typings` created a `typings` directory that contains a file `browser.d.ts`. This `browser.d.ts` is the entry point for the rest of the typings that are **managed by `typings`**. That is, if you write your own typings files, they're not going to be here, but any of the typings files you install via `typings` will be loaded via the `reference` tag in that file.

Don't modify the `typings/browser.d.ts` file directly! `typings` manages this file for you and if you change it your changes may be overwritten.

Now that we have the typings file `typings/browser.d.ts`, how do we use it? We have to tell our compiler about it, and we do that via the `tsconfig.json` file.

```
// tsconfig.json
"compilerOptions": { ... },
"files": [
  "typings/browser.d.ts",
  "ts/app.ts",
  "js/app.d.ts"
],
// more...
```

Notice that we added `typings/browser.d.ts` to the `files` array. This tells our compiler that we want to include our `typings` typings at compile time.

What if we were loading a different library, such as `underscore` and we needed to load it from System.js as well?

The idea is that you have to 1. make the typings available to the compiler at compile time and 2. make the code available at runtime

One way is like this:

1. `typings install underscore` - installs the typings file
2. `npm install underscore` - installs the javascript file in `node_modules`
3. In your `index.html` where you call `System.config`, add a new entry to the `paths` key like: `underscore: './node_modules/underscore/underscore.js'`
4. Then you can import underscore in your TypeScript using: `import * as _ from 'underscore';`
5. Use underscore like so: `let foo = _.map([1,2,3], (x) => x + 1);`

We've already done a `typings install` for you for this application so you don't need to install the dependencies yourself.

In fact, if you do run `typings install` you may find that you get the error:

```
1 node_modules/angular2/typings/angular-protractor/angular-protractor.d.ts(1679,13\
2 ): error TS2403: Subsequent variable declarations must have the same type.  Vari\
3 able '$' must be of type 'JQueryStatic', but here has type 'cssSelectorHelper'.
```

This is due to a bug between the `jquery` and the `angular` typings both trying to assign a type to the dollar sign `$`. At time of publishing, the hacky workaround is to open `typings/jquery/jquery.d.ts` and comment out this line:

```
1 // declare var $: JQueryStatic; // - ng-book told me to comment this
```

Of course, this will cause problems if you're trying to use jQuery-specific typings via `$` in TypeScript (but we aren't for this example).

## Custom Typing Files

Being able to use third-party typing files is great, but there are going to be situations where typing files don't already exist: especially in the case of our own code.

Generally, when we write custom typing files we co-locate the file alongside its respective Javascript code. So let's create the file `js/app.d.ts`:

**code/conversion/hybrid/js/app.d.ts**

```
1  declare module interestAppNg1 {
2
3    export interface Pin {
4      title: string;
5      description: string;
6      user_name: string;
7      avatar_src: string;
8      src: string;
9      url: string;
10     faved: boolean;
11     id: string;
12   }
13
14   export interface PinsService {
15     pins(): Promise<Pin[]>;
```

```
16      addPin(pin: Pin): Promise<any>;
17    }
18
19  }
20
21  declare module 'interestAppNg1' {
22    export = interestAppNg1;
23  }
```

When we use the `declare` keyword, that is called making an "ambient declaration" and the idea is that we're defining a variable that didn't originate from a TypeScript file. In this case, we're defining two interfaces:

1. `Pin`
2. `PinsService`

The `Pin` interface describes the keys and value-types of a pin object.

The `PinsService` interface describes the types of our two methods on our `PinsService`.

- `pins()` returns a `Promise` of an array of `Pins`
- `addPin()` takes a `Pin` as an argument and returns a `Promise`

**Learn More about Writing Type Definition Files**

If you'd like to learn more about writing `.d.ts` files, checkout these helpful links:

- TypeScript Handbook: Working with other Javascript Libraries[129]
- TypeScript Handbook: Writing definition files[130]
- Quick tip: Typescript declare keyword[131]

You might have noticed that we don't declare the token `interestAppNg1` anywhere in our ng1 Javascript code. `interestAppNg1` is just an identifier we use on the TypeScript side to specify this javascript code.

Now that we have this file setup, we can import these types like so:

[129]http://www.typescriptlang.org/Handbook#modules-working-with-other-javascript-libraries

[130]https://github.com/Microsoft/TypeScript-Handbook/blob/master/pages/Writing%20Definition%20Files.md

[131]http://blogs.microsoft.co.il/gilf/2013/07/22/quick-tip-typescript-declare-keyword/

```
1 import { Pin, PinsService } from 'interestAppNg1';
```

## Writing ng2 `PinControlsComponent`

Now that we have the typings figured out, let's turn our attention back to the hybrid app.

The first thing we're going to do is write the ng2 `PinControlsComponent`. This will be an ng2 component nested within an ng1 directive. The `PinControlsComponent` displays the fav hearts and toggles fav'ing a pin.

Let's start by importing our `Pin` type, along with a few other constants that we'll need:

**code/conversion/hybrid/ts/components/PinControlsComponent.ts**

```
1  /*
2   * PinControls: a component that holds the controls for a particular pin
3   */
4  import {
5    Component,
6    Input,
7    Output,
8    EventEmitter
9  } from '@angular/core';
10 import { NgIf } from '@angular/common';
11 import { Pin } from 'interestAppNg1';
```

Next, let's write the `@Component` annotation:

**code/conversion/hybrid/ts/components/PinControlsComponent.ts**

```
13 @Component({
14   selector: 'pin-controls',
15   template: `
16 <div class="controls">
17   <div class="heart">
18     <a (click)="toggleFav()">
19       <img src="/images/icons/Heart-Empty.png" *ngIf="!pin.faved" />
20       <img src="/images/icons/Heart-Red.png"   *ngIf="pin.faved" />
21     </a>
22   </div>
23 </div>
24   `
25 })
```

Notice here that we'll match the element `pin-controls`.

Our template looks very similar to the ng1 version except we're using the ng2 template syntax for `(click)` and `*ngIf`.

Now the component definition class:

**code/conversion/hybrid/ts/components/PinControlsComponent.ts**

```
26  export class PinControlsComponent {
27    @Input() pin: Pin;
28    @Output() faved: EventEmitter<Pin> = new EventEmitter<Pin>();
29
30    toggleFav(): void {
31      this.faved.next(this.pin);
32    }
33  }
```

Notice that instead of specifying `inputs` and `outputs` in the `@Component` annotation, in this case we're annotating the properties on the class directly with the `@Input` and `@Output` annotations. This is a convenient way to us to provide typings to these properties.

This component will take an input of `pin`, which is the `Pin` object we're controlling.

This component specifies an output of `faved`. This is a little bit different than how we did it in the ng1 app. If you look at `toggleFav` all we're doing is emitting (on the `EventEmitter`) the current pin.

The idea here is that we've already implemented how to change the `faved` state in ng1 and we may not want to re-implement that functionality ng2 (you may want to, it just depends on your team conventions).

## Using ng2 `PinControlsComponent`

Now that we have an ng2 `pin-controls` component, we can use it in a template. Here's what our `pin.html` template looks like now:

**code/conversion/hybrid/templates/pin.html**

```
1  <div class="col-sm-6 col-md-4">
2    <div class="thumbnail">
3      <div class="content">
4        <img ng-src="{{pin.src}}" class="img-responsive">
5        <div class="caption">
6          <h3>{{pin.title}}</h3>
7          <p>{{pin.description | truncate:100}}</p>
8        </div>
```

```
 9        <div class="attribution">
10          <img ng-src="{{pin.avatar_src}}" class="img-circle">
11          <h4>{{pin.user_name}}</h4>
12        </div>
13      </div>
14      <div class="overlay">
15        <pin-controls [pin]="pin"
16                      (faved)="toggleFav($event)"></pin-controls>
17      </div>
18    </div>
19  </div>
```

This template is for an ng1 directive, and we can use ng1 directives such as `ng-src`. However, notice the line where we use our ng2 `pin-controls` component:

```
1  <pin-controls [pin]="pin"
2                (faved)="toggleFav($event)"></pin-controls>
```

What's interesting here is that we're using the ng2 input bracket syntax `[pin]` and the ng2 output parenthesis syntax `(faved)`.

In a hybrid app **when you use ng2 directives in ng1, you still use the ng2 syntax**.

With our input `[pin]` we're passing the `pin` which comes from the `scope` of the ng1 directive.

With our output `(faved)` we're calling the `toggleFav` function on the `scope` of the ng1 directive. Notice what we did here: we didn't modify the `pin.faved` state within the ng2 directive (although, we could have). Instead, we asked the ng2 `PinControlsComponent` to simply emit the pin when `toggleFav` is called there. (If this is confusing, take a second look at `toggleFav` of `PinControlsComponent`.)

Again, the reason we do this is because we're showing how you can keep your existing functionality (`scope.toggleFav`) in ng1, but start porting over components to ng2. In this case, the ng1 `pin` directive listens for the `faved` event on the ng2 `PinControlsComponent`.

If you refresh your page now, you'll notice that it doesn't work. That's because there's one more thing we need to do: downgrade `PinControlsComponent` to ng1.

## Downgrading ng2 `PinControlsComponent` to ng1

The final step to using our components across ng2/ng1 borders is to use our `UpgradeAdapter` to downgrade our components (or upgrade, as we'll see in a bit).

We perform this downgrade in our `app.ts` file (where we called `upgradeAdapter.bootstrap`).

First we need to import the necessary `angular` libraries:

**code/conversion/hybrid/ts/app.ts**

```
 9  import {
10    NgModule,
11    forwardRef
12  } from '@angular/core';
13  import { CommonModule } from '@angular/common';
14  import {
15    FormsModule,
16  } from '@angular/forms';
17  import { BrowserModule } from "@angular/platform-browser";
18  import { UpgradeAdapter } from '@angular/upgrade';
19  declare var angular: any;
20  import 'interestAppNg1'; // "bare import" for side-effects
```

Then we create a `.directive` in (almost) the normal ng1 way:

**code/conversion/hybrid/ts/app.ts**

```
34  angular.module('interestApp')
35    .directive('pinControls',
36               upgradeAdapter.downgradeNg2Component(PinControlsComponent))
```

Above, remember that when we `import 'interestAppNg1'` this will load up our ng1 app, which calls `angular.module('interestApp', [])`. That is, our ng1 app has already registered the `interestApp` module with `angular`.

Now we want to look up that module by calling `angular.module('interestApp')` and then add directives to it, just like we do in ng1 normally.

**`angular.module` getter and setter syntax**

If you recall, when we pass an array as the second argument to `angular.module`, we are *creating* a module. That is, `angular.module('foo', [])` will *create* the module `foo`. Informally, we call this the "setter" syntax.

Similarly, if we omit the array we are *getting* a module (that is assumed to already exist). That is, `angular.module('foo')` will *get* the module `foo`. We call this the "getter" syntax.

In this example, if you forget this distinction and call `angular.module('interestApp', [])` in `app.ts` (ng2) then you will accidentally overwrite your existing `interestApp` module and your app won't work. Careful!

We're calling `.directive` and creating a directive called `'pinControls'`. This is standard ng1 practice. For the second argument, the directive definition object (DDO), we don't create the DDO manually. Instead, we call `upgradeAdapter.downgradeNg2Component`.

`downgradeNg2Component` will convert our `PinControlsComponent` into an ng1-compatible directive. Pretty neat.

Now if you try refreshing, you'll notice that our faving works just like before, only now we're using ng2 embedded in ng1!

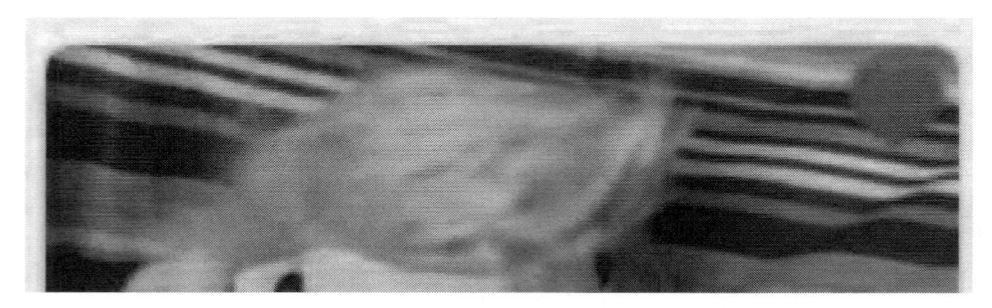

**Faving works like a charm**

## Adding Pins with ng2

The next thing we want to do is upgrade the add pins page with an ng2 component.

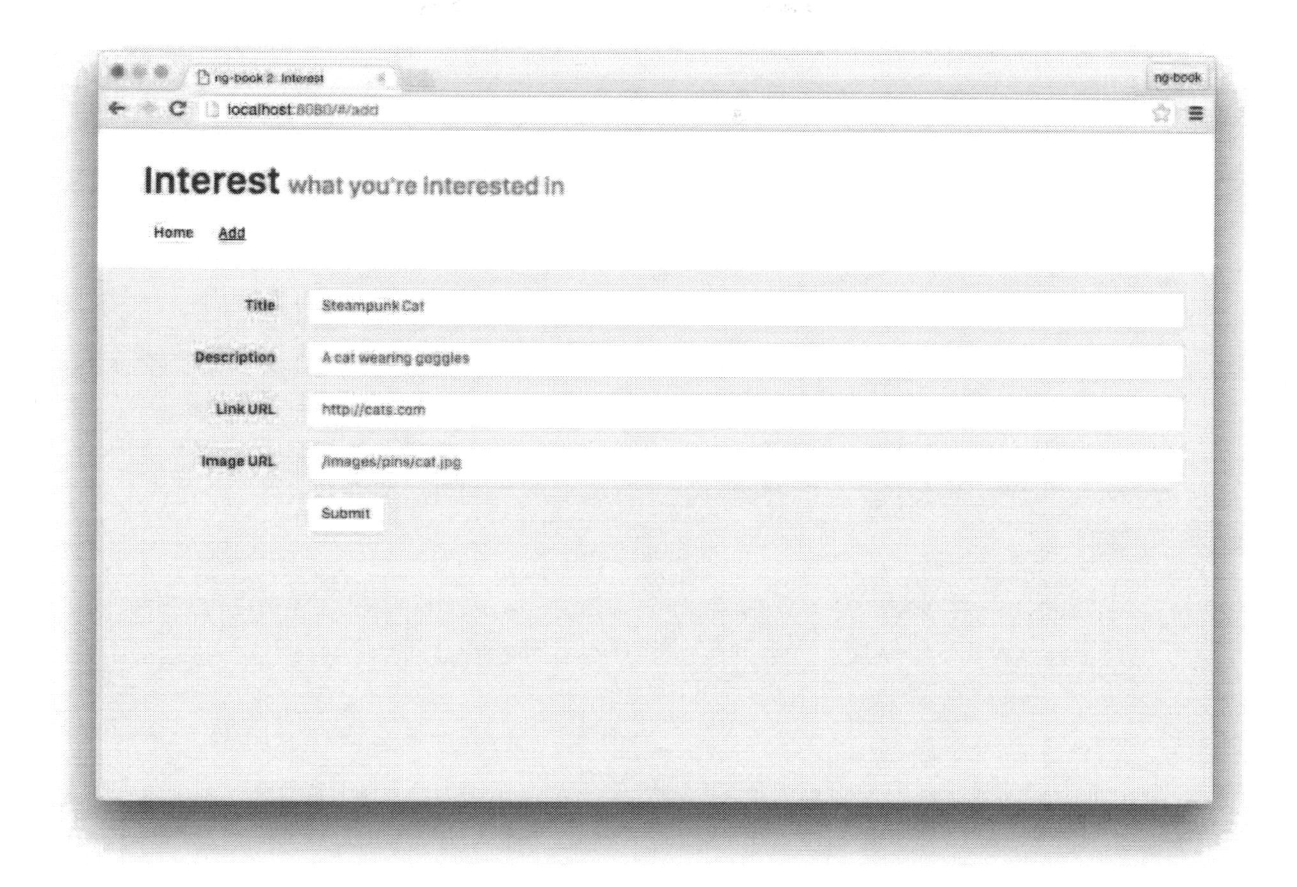


**Adding a New Pin Form**

If you recall, this page does three things:

1. Present a form to the user for describing the pin
2. Use the `PinsService` to add the new pin to the list of pins
3. Redirect the user to the homepage

Let's think through how we're going to do these things from ng2.

Angular 2 provides a robust forms library. So there's no complication here. We're going to write a straight ng2 form.

However the `PinsService` comes from ng1. Often we have many existing services in ng1 and we don't have time to upgrade them all. So for this example, we're going to keep `PinsService` as an ng1 object, and *inject it into ng2*.

Similarly, we're using `ui-router` in ng1 for our routing. To change pages in `ui-router` we have to use the `$state` service, which is an ng1 service.

So what we're going to do is **upgrade** the `PinsService` and the `$state` service from ng1 to ng2. And this couldn't be any easier.

## Upgrading ng1 `PinsService` and `$state` to ng2

To upgrade ng1 services we call `upgradeAdapter.upgradeNg1Provider`:

**code/conversion/hybrid/ts/app.ts**

```
44 /*
45  * Expose our ng1 content to ng2
46  */
47 upgradeAdapter.upgradeNg1Provider('PinsService');
48 upgradeAdapter.upgradeNg1Provider('$state');
```

And that's it. Now we can `@Inject` our ng1 services into ng2 components like so:

```
1 class AddPinComponent {
2   constructor(@Inject('PinsService') public pinsService: PinsService,
3               @Inject('$state') public uiState: IStateService) {
4   }
5   // ...
6   // now you can use this.pinsService
7   // or this.uiState
8   // ...
9 }
```

In this constructor, there's a few things to look at:

The `@Inject` annotation, says that we want the next variable to be assigned the value of what the injection will resolve to. In the first case, that would be our ng1 `PinsService`.

In TypeScript, in a `constructor` when you use the `public` keyword, it is a shorthand for assigning that variable to `this`. That is, here when we say `public pinsService` what we're saying is, 1. declare a property `pinsService` on instances of this class and 2. assign the constructor argument `pinsService` to `this.pinsService`.

The result is that we can access `this.pinsService` throughout our class.

Lastly we define the type of both services we're injecting: `PinsService` and `IStateService`.

`PinsService` comes from the `app.d.ts` we defined previously:

**code/conversion/hybrid/js/app.d.ts**

```
14    export interface PinsService {
15      pins(): Promise<Pin[]>;
16      addPin(pin: Pin): Promise<any>;
17    }
```

And `IStateService` comes from the typings for `ui-router`, which we installed with `typings`.

By telling TypeScript the types of these services we can enjoy type-checking as we write our code.

Let's write the rest of our `AddPinComponent`.

## Writing ng2 `AddPinComponent`

We start by importing the types we need:

**code/conversion/hybrid/ts/components/AddPinComponent.ts**

```
1   /*
2    * AddPinComponent: a component that controls the "add pin" page
3    */
4   import {
5     Component,
6     Inject
7   } from '@angular/core';
8   import { Pin, PinsService } from 'interestAppNg1';
9   import { IStateService } from 'angular-ui-router';
```

Again, notice that we're importing our custom types `Pin` and `PinsService`. And we're also importing `IStateService` from `angular-ui-router`.

#### `AddPinComponent @Component`

Our `@Component` annotation is straightforward:

**code/conversion/hybrid/ts/components/AddPinComponent.ts**

```
11 @Component({
12   selector: 'add-pin',
13   templateUrl: '/templates/add-ng2.html'
14 })
```

#### `AddPinComponent` template

We're loading our template using a `templateUrl`. In that template, we setup our form much like the ng1 form, only we're using ng2 form directives.

We're not going to describe `ngModel` / `ngSubmit` deeply here. If you'd like to know more about how Angular 2 forms work, checkout the forms chapter, where we describe forms in depth.

**code/conversion/hybrid/templates/add-ng2.html**

```
1  <div class="container">
2    <div class="row">
3
4      <form (ngSubmit)="onSubmit()"
5            class="form-horizontal">
6
7        <div class="form-group">
8          <label for="title"
9                 class="col-sm-2 control-label">Title</label>
10         <div class="col-sm-10">
11           <input type="text"
12                  class="form-control"
13                  id="title"
14                  name="title"
15                  placeholder="Title"
16                  [(ngModel)]="newPin.title">
17         </div>
```

We're using two directives here: `ngSubmit` and `ngModel`.

We use `(ngSubmit)` on the form to call the `onSubmit` function when the form is submitted. (We'll define `onSubmit` on the `AddPinComponent` controller below.)

We use `[(ngModel)]` to bind the value of the `title` input tag to the value of `newPin.title` on the controller.

Here's the full listing of the template:

**code/conversion/hybrid/templates/add-ng2.html**

```
1  <div class="container">
2    <div class="row">
3
4      <form (ngSubmit)="onSubmit()"
5            class="form-horizontal">
6
7        <div class="form-group">
8          <label for="title"
9                 class="col-sm-2 control-label">Title</label>
10         <div class="col-sm-10">
11           <input type="text"
12                  class="form-control"
13                  id="title"
14                  name="title"
15                  placeholder="Title"
16                  [(ngModel)]="newPin.title">
17         </div>
18       </div>
19
20       <div class="form-group">
21         <label for="description"
22                class="col-sm-2 control-label">Description</label>
23         <div class="col-sm-10">
24           <input type="text"
25                  class="form-control"
26                  id="description"
27                  name="description"
28                  placeholder="Description"
29                  [(ngModel)]="newPin.description">
30         </div>
31       </div>
32
33       <div class="form-group">
34         <label for="url"
35                class="col-sm-2 control-label">Link URL</label>
36         <div class="col-sm-10">
37           <input type="text"
38                  class="form-control"
39                  id="url"
40                  name="url"
41                  placeholder="Link URL"
```

```
42                  [(ngModel)]="newPin.url">
43          </div>
44        </div>
45
46        <div class="form-group">
47          <label for="url"
48                 class="col-sm-2 control-label">Image URL</label>
49          <div class="col-sm-10">
50            <input type="text"
51                   class="form-control"
52                   id="url"
53                   name="url"
54                   placeholder="Image URL"
55                   [(ngModel)]="newPin.src">
56          </div>
57        </div>
58
59        <div class="form-group">
60          <div class="col-sm-offset-2 col-sm-10">
61            <button type="submit"
62                    class="btn btn-default"
63                    >Submit</button>
64          </div>
65        </div>
66        <div *ngIf="saving">
67          Saving...
68        </div>
69      </form>
```

### `AddPinComponent` Controller

Now we can define `AddPinComponent`. We start by setting up two instance variables:

**code/conversion/hybrid/ts/components/AddPinComponent.ts**

```
15 export class AddPinComponent {
16   saving: boolean = false;
17   newPin: Pin;
```

We use `saving` to indicate to the user that the save is in progress and we use `newPin` to store the `Pin` we're working with.

**code/conversion/hybrid/ts/components/AddPinComponent.ts**

```
19    constructor(@Inject('PinsService') private pinsService: PinsService,
20                @Inject('$state') private uiState: IStateService) {
21      this.newPin = this.makeNewPin();
22    }
```

In our constructor we Inject the services, as we discussed above. We also set this.newPin to the value of makeNewPin, which we'll define now:

**code/conversion/hybrid/ts/components/AddPinComponent.ts**

```
24    makeNewPin(): Pin {
25      return {
26        title: 'Steampunk Cat',
27        description: 'A cat wearing goggles',
28        user_name: 'me',
29        avatar_src: 'images/avatars/me.jpg',
30        src: '/images/pins/cat.jpg',
31        url: 'http://cats.com',
32        faved: false,
33        id: Math.floor(Math.random() * 10000).toString()
34      };
35    }
```

This looks a lot like how we defined it in ng1, only now we have the benefit of it being typed.

When the form is submitted, we call onSubmit. Let's define that:

**code/conversion/hybrid/ts/components/AddPinComponent.ts**

```
37    onSubmit(): void {
38      this.saving = true;
39      console.log('submitted', this.newPin);
40      setTimeout(() => {
41        this.pinsService.addPin(this.newPin).then(() => {
42          this.newPin = this.makeNewPin();
43          this.saving = false;
44          this.uiState.go('home');
45        });
46      }, 2000);
47    }
```

Again, we're using a timeout to *simulate* the effect of what would happen if we had to call out to a server to save this pin. Here, we're using setTimeout. Compare that to how we defined this function in ng1:

**code/conversion/ng1/js/app.js**

```
82    ctrl.submitPin = function() {
83      ctrl.saving = true;
84      $timeout(function() {
85        PinsService.addPin(ctrl.newPin).then(function() {
86          ctrl.newPin = makeNewPin();
87          ctrl.saving = false;
88          $state.go('home');
89        });
90      }, 2000);
91    }
```

Notice that in ng1 we had to use the `$timeout` service. Why is that? Because ng1 is based around the digest loop. If you use `setTimeout` in ng1, then when the callback function is called, it's "outside" of angular and so your changes aren't propagated unless something kicks off a digest loop (e.g. using `$scope.apply`).

However in ng2, we can use `setTimeout` directly because change detection in ng2 uses Zones and is therefore, more or less automatic. We don't need to worry about the digest loop in the same way, which is really nice.

In `onSubmit` we're calling out to the `PinsService` by:

```
1 this.pinsService.addPin(this.newPin).then(() => {
2 // ...
```

Again, the `PinsService` is accessible via `this.pinsService` because of how we defined the `constructor`. The compiler doesn't complain because we said that `addPin` takes a `Pin` as the first argument in our `app.d.ts`:

**code/conversion/hybrid/js/app.d.ts**

```
14    export interface PinsService {
15      pins(): Promise<Pin[]>;
16      addPin(pin: Pin): Promise<any>;
17    }
```

And we defined `this.newPin` to be a Pin.

After `addPin` resolves, we reset the pin using `makeNewPin` and set `this.saving = false`.

To go back to the homepage, we use the `ui-router` `$state` service, which we stored as `this.uiState`. So we can change states by calling `this.uiState.go('home')`.

### Using AddPinComponent

Now let's use the AddPinComponent.

#### Downgrade ng2 AddPinComponent

To use AddPinComponent we need to downgrade it:

**code/conversion/hybrid/ts/app.ts**

```
34 angular.module('interestApp')
35   .directive('pinControls',
36              upgradeAdapter.downgradeNg2Component(PinControlsComponent))
37   .directive('addPin',
38              upgradeAdapter.downgradeNg2Component(AddPinComponent));
```

This will create the addPin directive in ng1, which will match the tag <add-pin>.

#### Routing to add-pin

In order to use our new AddPinComponent page, we need to place it somewhere within our ng1 app. What we're going to do is take the add state in our router and just set the <add-pin> directive to be the template:

**code/conversion/hybrid/js/app.js**

```
39     .state('add', {
40       template: "<add-pin></add-pin>",
41       url: '/add',
42       resolve: {
43         'pins': function(PinsService) {
44           return PinsService.pins();
45         }
46       }
47     })
```

## Exposing an ng2 service to ng1

So far we've downgraded ng2 components to be used in ng2, and upgraded ng1 services to be used in ng2. But as our application start converting over to ng2, we'll probably start writing services in Typescript/ng2 that we'll want to expose to our ng1 code.

Let's create a simple service in ng2: an "analytics" service that will record events.

The idea is that we have an AnalyticsService in our app that we use to recordEvents. In reality, we're just going to console.log the event and store it in an array. But it gives us a chance to focus on what's important: describing how we share a ng2 service with ng1.

## Writing the `AnalyticsService`

Let's take a look at the `AnalyticsService` implementation:

**code/conversion/hybrid/ts/services/AnalyticsService.ts**

```
1  import { Injectable } from '@angular/core';
2
3  /**
4   * Analytics Service records metrics about what the user is doing
5   */
6  @Injectable()
7  export class AnalyticsService {
8    events: string[] = [];
9
10   public recordEvent(event: string): void {
11     console.log(`Event: ${event}`);
12     this.events.push(event);
13   }
14 }
15
16 export var analyticsServiceInjectables: Array<any> = [
17   { provide: AnalyticsService, useClass: AnalyticsService }
18 ];
```

There are two things to note here: 1. `recordEvent` and 2. being `Injectable`

`recordEvent` is straightforward: we take an `event: string`, log it, and store it in `events`. In your application you would probably send the event to an external service like Google Analytics or Mixpanel.

To make this service injectable, we do two things: 1. Annotate the class with `@Injectable` and 2. `bind` the token `AnalyticsService` to this class.

Now Angular will manage a singleton of this service and we will be able to inject it where we need it.

## Downgrade ng2 `AnalyticsService` to ng1

Before we can use the `AnalyticsService` in ng1, we need to downgrade it.

The process of downgrading an ng2 service to ng1 is similar to the process of downgrading a directive, but there is one extra step: we need to make sure `AnayticsService` is in the list of `providers` for our `NgModule`:

**code/conversion/hybrid/ts/app.ts**

```
50 @NgModule({
51   declarations: [
52     PinControlsComponent,
53     AddPinComponent
54   ],
55   imports: [
56     CommonModule,
57     BrowserModule,
58     FormsModule
59   ],
60   providers: [
61     AnalyticsService,
62   ]
63 })
64 class InterestAppModule { }
```

Then we can use `downgradeNg2Provider`:

**code/conversion/hybrid/ts/app.ts**

```
40 angular.module('interestApp')
41   .factory('AnalyticsService',
42            upgradeAdapter.downgradeNg2Provider(AnalyticsService));
```

We call `angular.module('interestApp')` to get our ng1 module and then call `.factory` like we would in ng1. To downgrade the service, we call

`upgradeAdapter.downgradeNg2Provider(AnalyticsService)`, which wraps our `AnalyticsService` in a function that adapts it to an ng1 factory.

## Using `AnalyticsService` in ng1

Now we can inject our ng2 `AnalyticsService` into ng1. Let's say we want to record whenever the `HomeController` is visited. We could record this event like so:

**code/conversion/hybrid/js/app.js**

```
60  .controller('HomeController', function(pins, AnalyticsService) {
61    AnalyticsService.recordEvent('HomeControllerVisited');
62    this.pins = pins;
63  })
```

Here we inject `AnalyticsService` as if it was a normal ng1 service we call `recordEvent`. Fantastic!

We can use this service anywhere we would use injection in ng1. For instance, we can also inject the `AnalyticsService` into our ng1 `pin` directive:

**code/conversion/hybrid/js/app.js**

```
64  .directive('pin', function(AnalyticsService) {
65    return {
66      restrict: 'E',
67      templateUrl: '/templates/pin.html',
68      scope: {
69        'pin': "=item"
70      },
71      link: function(scope, elem, attrs) {
72        scope.toggleFav = function() {
73          AnalyticsService.recordEvent('PinFaved');
74          scope.pin.faved = !scope.pin.faved;
75        }
76      }
77    }
78  })
```

# Summary

Now you have all the tools you need to start upgrading your ng1 app to a hybrid ng1/ng2 app. The interoperability between ng1 and ng2 works very well and we owe a lot to the Angular team for making this so easy.

Being able to exchange directives and services between ng1 and ng2 make it super easy to start upgrading your apps. We can't always upgrade our apps to ng2 overnight, but the `UpgradeAdapter` lets us start using ng2 - without having to throw our old code away.

# References

If you're looking to learn more about hybrid Angular apps, here are a few resources:

- The Official Angular Upgrade Guide[132]
- The Angular2 Upgrade Spec Test[133]
- The Angular2 Source for `DowngradeNg2ComponentAdapter`[134]

[132]https://angular.io/docs/ts/latest/guide/upgrade.html
[133]https://github.com/angular/angular/blob/master/modules/angular2/test/upgrade/upgrade_spec.ts
[134]https://github.com/angular/angular/blob/master/modules/angular2/src/upgrade/downgrade_ng2_adapter.ts

Made in the USA
San Bernardino, CA
02 December 2016